Foundations of Chemical Kinetics

A hands-on approach

Online at: https://doi.org/10.1088/978-0-7503-5321-2

Foundations of Chemical Kinetics

A hands-on approach

Marc R Roussel
Department of Chemistry and Biochemistry, University of Lethbridge, Lethbridge, Canada

IOP Publishing, Bristol, UK

ISBN 978-0-7503-5321-2 (ebook)
ISBN 978-0-7503-5319-9 (print)
ISBN 978-0-7503-5322-9 (myPrint)
ISBN 978-0-7503-5320-5 (mobi)

DOI 10.1088/978-0-7503-5321-2

Supplementary material is available for this book from https://doi.org/10.1088/978-0-7503-5321-2.

Version: 20230801

IOP ebooks

British Library Cataloguing-in-Publication Data: A catalogue record for this book is available from the British Library.

Published by IOP Publishing, wholly owned by The Institute of Physics, London

IOP Publishing, No.2 The Distillery, Glassfields, Avon Street, Bristol, BS2 0GR, UK

US Office: IOP Publishing, Inc., 190 North Independence Mall West, Suite 601, Philadelphia, PA 19106, USA

This book is dedicated to my kinetics teachers at the University of Toronto:

George Burns†

Simon J Fraser

Raymond Kapral

Stuart Whittington

Contents

Preface

I usually think of chemical kinetics as one of the three pillars of physical chemistry, along with thermodynamics and quantum mechanics, with statistical mechanics linking up much of the structure (Roussel, 2012, p 2). Students usually get some practical chemical kinetics in their introductory physical chemistry course, and maybe a glimpse at transition-state theory, one of the great theories of kinetics, but it is left to advanced courses to discuss the theoretical basis of kinetics. This is a book written for those advanced courses. It is assumed that students coming to this course will have had an introduction to mass-action kinetics and will be familiar with its basic ideas and equations. Chapter 1 briefly reviews some of these concepts, but it is just that: a review.

Some great books are already available for advanced chemical kinetics courses: Pilling and Seakins' *Reaction Kinetics* (Oxford, 1995) and Steinfeld, Francisco, and Hase's *Chemical Kinetics and Dynamics*, 2nd edition (Prentice Hall, 1999) both spring to mind. You will notice the age of these books. This was part of the impetus for writing my book. A lot has changed since the 1990s, notably the widespread availability of cheap and powerful desktop and laptop computers.

I'm the kind of person who likes to teach something new once in a while. When I was asked to put on a course for senior undergraduates and graduate students in 2012, rather than repeat something I had taught before, I decided to go back to my roots and to offer a course in advanced chemical kinetics. I chose the Steinfeld, Francisco, and Hase book, which is both well written and comprehensive. But I found myself a bit dissatisfied with the experience. Books of this vintage contain a lot of detail about the theories of chemical kinetics but relatively little in the way of tangible exercises that students can engage with, for the simple reason that sufficiently powerful computers and user-friendly software were not routinely available at the time.

When the opportunity came for me to teach my advanced kinetics course again in 2021, I thought I should write what I had come to think of as the missing book: one that, whenever possible, would use modern computer tools to show students how the theories worked in practice. I had a pretty good set of lecture notes from the previous offering of the course. All I needed to do was to write those up in the form of book chapters, add the necessary practical material, and deliver the chapters to the students on a weekly basis. I would just need to keep a bit ahead of the classes with my writing as the term went on, and everything would be OK. I'm not ever doing that again. But it did result in the book you have in your hands now.

To write a book is to make choices—about topics and, in practically oriented books like this one, about tools. In terms of topics, I tried to cover what I think are the major theoretical ideas that a chemist can lean on for insight when thinking about chemical reactions. I teach at a small university. One of the challenges we have is that we tend to offer core courses once a year, and we can't offer advanced courses on a regular rotation. So, for instance, given that I also teach the statistical mechanics course, a few years could separate a statistical mechanics course and the advanced kinetics course. There is no way I can set statistical mechanics as a prerequisite for the kinetics course. To maximize the audience for my course and, not coincidentally, for this book, I chose to build in a lot of background that some students may already have but that many, at least at the University of Lethbridge, will not. There are accordingly a couple of chapters on basic statistical mechanics (chapters 2 and 6) and a focused introduction to computational chemistry in chapter 4. If your department is able to set prerequisites in statistical mechanics or in computational chemistry, you will be able to drop some or all of these chapters from your syllabus, and this will leave more time for the kinetics material. But the material is there in a sequence that you will hopefully find logical so that you can incorporate it into your course as and when needed. And students, even if your instructor skips some of these sections, the material is there for you should you need a refresher.

For the tools, I knew I needed an *ab initio* package, and I chose Gaussian with the GaussView interface because this suite is widely available in chemistry departments worldwide. I do recognize that this is commercial software, and I hope that not too many instructors and their students will be excluded by this choice. It is very likely, if you are using a different software package, that you will have the same functionality that Gaussian provides, and hopefully your instructors will be able to bridge you over from my Gaussian/GaussView instructions to whatever software you are using. Having said that, there is no doubt that this book will be easier to use if you happen to be studying on a Gaussian campus.

I also think that students should have a bit of exposure to computer programming during their studies. Obviously, I can't teach a whole computer programming course while covering kinetics, but I can show you how to build a small program using a few basic instructions. For this part of the work, I chose Matlab. Matlab is also commercial software, but a free clone called Octave is available, so here you should be able to work through the examples and exercises even if your campus doesn't have a Matlab license. Matlab can be used interactively for some things, and it's an interpreted language, so you can rapidly try things, then fix a program that doesn't work as intended and try again.

To save you a bit of typing, the Matlab codes presented in the text are available from https://doi.org/10.1088/978-0-7503-5321-2. Each Matlab program is named based on the section where it appears in the text and, when a program was developed incrementally, the name is modified according to whether it was the first, second or third version.

As I noted above, writing a book is all about choices. I chose to focus on smaller systems for which the computations are quick and easy. I stayed away from problems that would require high-performance computing, because that would limit the audience of the book even more than the choice of software. This book will not teach you how to do a cutting-edge calculation. It will teach you the ideas of chemical kinetics theory and show you how some of the calculations are done. I will point out from time to time how things could be done better. But if you are to become an expert in some area of chemical kinetics, you will likely need to study the topic with a specialist. My objective in writing this book was to expose a broader audience of chemists to the theory of chemical kinetics so that they can appreciate the factors that determine, at a very elementary level, the rates of chemical reactions. I like hands-on calculations because I think you get a better sense of a theory by applying it than by reading about it. It will be up to you to determine how successful this approach has been.

Whether you are a course instructor or a student, if you have any comments about this book, by all means get in touch. It's always interesting to hear what people think, and who knows? Maybe there will be a second edition someday.

Marc R Roussel (roussel@uleth.ca)
University of Lethbridge
April 1, 2023

Reference

Roussel M R 2012 *A Life Scientist's Guide to Physical Chemistry* (Cambridge: Cambridge University Press)

Acknowledgments

I want to thank Gaussian, and particularly James Hess, for permission to use Gaussian and GaussView screenshots in this book. It is often much easier to show than to tell.

I especially want to thank the students of Chemistry 4000 who inspired this book. You were a great group to teach, and I hope you enjoyed the journey through chemical kinetics.

About the author

Marc R Roussel

Marc Roussel earned a bachelor's degree in chemical physics from Queen's University in 1988. He then went on to graduate work in the Chemical Physics Theory Group at the University of Toronto under the supervision of Simon J Fraser, earning an MSc in 1990 and a PhD in 1994. His graduate work focused on using the theory of invariant manifolds to understand steady-state kinetics, with a particular emphasis on enzyme kinetics. For this work he was awarded the D J LeRoy Award in 1993, 'for excellence in graduate work in physical chemistry at the University of Toronto.'

Marc held an NSERC postdoctoral fellowship with Professor Michael C Mackey in the Physiology Department at McGill University in 1994–95, where he worked on the application of delay-differential equations in chemical modeling. In 1995, Marc was hired as a tenure-track Assistant Professor by the Department of Chemistry[1] at the University of Lethbridge (UofL). He was promoted to Associate Professor in 2000 and to Professor in 2005. He teaches a broad range of courses in introductory and physical chemistry, as well as more specialized courses in nonlinear dynamics and mathematical biology. He has served as biochemistry coordinator (1997–2004) and as department chair (2013–6). He has authored two textbooks, *A Life Scientist's Guide to Physical Chemistry* (Cambridge University Press, 2012) and *Nonlinear Dynamics: A Hands-On Introductory Survey* (IOP Concise Physics Series, Morgan & Claypool, 2019).

Marc has a broad research program whose threads are connected by the kinetics of biochemical systems. His research uses methods from both nonlinear dynamics and stochastic kinetics. His recent research interests include the detailed modeling of events in gene expression (transcription, splicing, and translation), the modeling of gene expression networks, developmental modeling, graph-theoretic analysis of reaction networks, and the development of model reduction methods.

From 2009 to 2011, Marc was the first holder of the UofL University Scholar (Science) designation for excellence in research. Marc is one of the founding members of the Alberta RNA Research and Training Institute (2011). Since 2022, Marc has been the Associate Dean in the UofL's School of Graduate Studies.

[1] Renamed the Department of Chemistry and Biochemistry in 1999.

Symbols

ε	Energy (J)
ε_a	Activation energy (J)
γ_i	Activity coefficient
η	Viscosity (Pa s)
κ	Stochastic rate constant (s^{-1})
μ	Reduced mass (kg)
μ_m	Reduced molar mass (kg mol^{-1})
μ_i	Chemical potential of species i (J mol^{-1})
μ_3	Third central moment of a distribution
ν	Frequency (Hz)
ω	Angular frequency (s^{-1})
Ω	Solid angle
σ	Collision cross-section (m^2)
σ	Standard deviation
σ	Symmetry number
σ^2	Variance
θ, ϕ	Spherical polar coordinate angles
υ	Mobility (m^2 V^{-1} s^{-1})
a_i	Activity of a chemical species
A	Pre-exponential factor
A, B, C	Rotational constants (J)
$\mathcal{A}$	Space of allowed energies
c_i	Concentration
c°	Standard concentration
d	distance (m)
e	Elementary charge ($1.602\ 176\ 634 \times 10^{-18}$ C)
$\mathbf{E}$	Electric field (V m^{-1})
E_a	Activation energy (J mol^{-1})
f	Frictional coefficient (N s m^{-1})
g_i	Degeneracy of an energy level
$g(\varepsilon)$	Density of states
h	Planck's constant ($6.626\ 070\ 015 \times 10^{-34}$ J Hz^{-1})
$\hbar$	$h/2\pi$ ($1.054\ 571\ 818 \times 10^{-34}$ J s)
I_R	Differential cross-section (m^2)
I_{xx}, I_{yy}, I_{zz}	Principal moments of inertia (kg m^2)
I_A, I_B, I_C	Principal moments of inertia ($I_A \leqslant I_B \leqslant I_C$; kg m^2)
J	Rotational quantum number
$\mathbf{J}$	Flux (mol $m^{-2}s^{-1}$)
k	Rate constant
K	Equilibrium constant
K	z-axis angular momentum quantum number
k_B	Boltzmann's constant ($1.380\ 649 \times 10^{-23}$ J K^{-1})
L	Avogadro's number ($6.022\ 140\ 76 \times 10^{23}$ mol^{-1})
L_x, L_y, L_z	Dimensions of container (m)
m	Particle mass (kg)
M	Molar mass (kg mol^{-1})
n_x, n_y, n_z	Translational quantum numbers

N	Number of molecules
$\mathcal{N}$	Composition space
$\mathcal{N}$	Coordination number
$\mathcal{N}$	Reactions not in reactive set
$\mathcal{N}(1)$	Normally distributed random number with unit variance
p	Pressure
p	Probability density
p°	Standard pressure (1 bar)
P	Probability
$\mathbf{P}$	Probability distribution
q	Molecular partition function
q	Volumic partition function
$\mathcal{q}$	Generalized coordinate
$\hat{q}$	'Gaussian-style' partition function
r	Radius (m)
r_{AB}	Radius of collision cross-section (m)
R	Bond length (m)
R	Ideal gas constant (8.314 462 618 J $K^{-1}mol^{-1}$)
$\mathbf{R}$	Nuclear coordinate vector
$\mathcal{R}$	Reactive set
$\mathcal{R}$	Region of space over which a probability density is defined
$\mathcal{R}$	Set of reactions that can occur in a given system
S	Cumulative probability distribution
$\mathcal{S}$	Set of allowed states
t	Time (s)
T	Temperature (K)
u	Speed ($m\ s^{-1}$)
u_x, u_y, u_z	Velocity components ($m\ s^{-1}$)
$\langle u_{rel} \rangle$	Average relative speed ($m\ s^{-1}$)
U	Potential energy (J)
v	Rate of reaction
v	Vibrational quantum number
V	Volume
$\hat{V}$	Volume per molecule
w_{rs}	Transition rate from state r to state s
z_i	Charge of an ion in elementary units
Z_{AB}	Rate of collisions between A and B molecules ($m^{-3}s^{-1}$)

Part I

Gas-phase kinetics

Chapter 1

A review of basic concepts in chemical kinetics

This course is intended to follow an introduction to chemical kinetics, typically studied in a general physical chemistry course. In this chapter, we will review some topics to which you should have been exposed in your previous course. If any of these topics are unfamiliar to you, I recommend that you familiarize yourself with them by reading the relevant sections of any available physical chemistry textbook. If you don't know where to start, some textbook suggestions are provided at the end of this chapter.

1.1 Some basic definitions

Chemical kinetics is about how fast things change during a chemical reaction. In many laboratory experiments, we hold the temperature constant and mainly focus on the rates of change of the concentrations. However, non-isothermal kinetics is also an important topic where we consider the feedback between the temperature change caused by the reaction and the changes in the rates caused by the temperature change.

Before we proceed to talk about rates in detail, it is useful to distinguish between elementary and complex reactions:

An elementary reaction occurs in one step, exactly as written. As we will see later in this book, once the reactants are committed to forming products, product formation is extremely fast, so that, from the point of view of many experiments, product formation is essentially instantaneous. In other words, an elementary reaction can often be thought of as a dichotomy: either we have reactants, or we have products. Having said that, the state that divides reactants from products, known as the transition state, plays a central role in the theory.

A complex reaction occurs in multiple elementary steps.

doi:10.1088/978-0-7503-5321-2ch1

A reaction mechanism is a list of all the elementary reactions in a complex reaction.

Reaction intermediates, or just **intermediates** for short, are chemical species that are formed and removed in the course of converting reactants to products. For example, in the complex reaction

$$A \rightleftharpoons B \rightleftharpoons C,$$

B is an intermediate in the conversion of the reactant A into the product C.

In many cases, we can talk about a single 'rate of reaction' because of the stoichiometric relationships among the concentrations. This is automatically true for elementary reactions because, if $A \rightarrow 2B$ is an elementary reaction, then the appearance of two molecules of B always coincides with the disappearance of one A. Accordingly, B is produced twice as fast as A is used up, i.e.

$$\frac{d[B]}{dt} = -2\frac{d[A]}{dt}. \tag{1.1}$$

The negative sign is present because [A] decreases as [B] increases, so the two derivatives have opposite signs.

On the other hand, if $A \rightarrow 2B$ describes the overall stoichiometry of a complex reaction, then the simple relationship (1.1) only holds if the intermediates do not accumulate to an appreciable extent. Fortunately, this is often the case.

Equation (1.1) shows that the rates of change of the concentrations of reactants and products differ by factors that depend on the stoichiometry of the reaction. To facilitate discussions of rates in chemical kinetics, the following convention has been adopted: **The rate of reaction is the rate of formation of a product with unit stoichiometric coefficient.** The rate of reaction is normally represented by the symbol v. According to this convention, if we have a reaction

$$aA + bB \rightarrow cC + eE,$$

then

$$v = -\frac{1}{a}\frac{d[A]}{dt} = -\frac{1}{b}\frac{d[B]}{dt} = \frac{1}{c}\frac{d[C]}{dt} = \frac{1}{e}\frac{d[E]}{dt}.$$

Note the negative signs associated with the rates of change of the reactant concentrations. Unless the reaction is running backwards from the sense in which it was written, this definition makes the rate of reaction a positive quantity.

Studying the dependence of the rate of reaction on the concentrations of chemical species is a very common activity for chemists because of the connection between reaction mechanisms and rates. A **rate law** is a relationship between a rate of reaction and the concentrations of reactants, products, catalysts (if any), and inhibitors (again, if any). Some rate laws take a particularly simple monomial form: $v = k[X]^x[Y]^y\ldots$, where X, Y, $\cdots$ can be any reactant, product, catalyst, or inhibitor. The exponents x, y, ... are known as (partial) **orders of reaction** with respect to the corresponding chemical species, while k is called a **rate constant**.

We sometimes also talk of the overall order of reaction, which is just the sum of the partial orders, in this example, $x + y + \cdots$ An elementary reaction always has a simple rate law due to the

Law of mass action: The rate of an elementary reaction is proportional to the product of the concentrations of the *reactants*. The proportionality constant is called an elementary rate constant.

For example, if $A + 2B \rightarrow A + B_2$ were an elementary reaction, then the rate of reaction would be

$$v = k[A][B]^2.$$

There are several points to note from this simple example:

1. The stoichiometric coefficients of an elementary reaction are the partial orders of reaction.
2. By definition, rate constants are always positive.
3. This is an example of a reaction in which A acts as a catalyst. In this case, it is likely that the role of A is to carry away the energy released on bond formation, which is often an important process in preventing the re-dissociation of a newly formed molecule.
4. In most cases, k is independent of the concentrations of reactants, products, and catalysts. However, k generally depends on many details of the reaction conditions: temperature, ionic strength, crowding by macromolecules, etc. These quantities define the chemical environment in which a reaction occurs. Since the variables defining the chemical environment depend on the chemical composition of a reacting mixture, the rate constant can sometimes depend on the concentrations of the reactants or products. For example, if B is a charged species, then the reaction changes the ionic strength of the solution and thus the rate constant. Unless appropriate experimental precautions are taken, this leads to an apparent non-simple rate law.
5. While we tend to just write 'k' when there is no possible ambiguity, we always have to keep in mind that each elementary reaction has its own rate constant. When necessary, we generally distinguish rate constants using subscripts, e.g. k_1, k_Q, etc.

Let's talk a bit more about writing down rate equations. Suppose that $2A \rightarrow B$ is an elementary reaction with rate constant k. From the definition of the rate of reaction, we have

$$v = \frac{d[B]}{dt} = -\frac{1}{2}\frac{d[A]}{dt}.$$

From the law of mass action, we also have

$$v = k[A]^2.$$

Putting the two together, we have

$$\frac{d[B]}{dt} = k[A]^2$$

and $$\frac{d[A]}{dt} = -2k[A]^2.$$

Note where the stoichiometric coefficient ends up in the rate equation for [A]. A simple way to think about this is that the rate equation for [A] needs to take into account that each occurrence of the reaction removes two molecules of A. Particularly when writing down rate equations for complex mechanisms, it is useful to be able to automatically write down the stoichiometric coefficients in the correct places without going all the way back to the definition of the rate of reaction.

1.2 The steady-state and equilibrium approximations

Elementary reactions are the bread and butter of this book because we will be trying to understand, on a molecular level, what causes different reactions to have different rates. This requires that we look deeply into the individual reaction events, i.e. the elementary reactions. Nevertheless, the techniques used to treat complex reactions will often be used. Let's consider a simple two-step reaction:

$$A + B \underset{k_{-1}}{\overset{k_1}{\rightleftharpoons}} C \overset{k_2}{\rightarrow} P.$$

The overall reaction is

$$A + B \rightarrow P.$$

Thus[1],

$$v = \frac{d[P]}{dt} = k_2[C].$$

The trouble is that we often can't measure the very small concentration of the intermediate, C, so this is not a terribly useful equation. Moreover, we're typically interested in how the rate of reaction depends on the concentrations of things we put into the reaction vessel (the reactants, any catalysts added) or that accumulate to a significant extent during the reaction (i.e. the product(s)). To address this problem, we typically use one of two approximations to eliminate the concentration(s) of the intermediate(s).

The first of these is the steady-state approximation. If we think that the product-forming step might be 'fast' (in some sense)[2], then after a little while, the

[1] The equality $v = k_2[C]$ would have held regardless of how many molecules of product there were. See if you can convince yourself of this.

[2] The best way to think about what we mean by a 'fast step' derives from the theory of singular perturbations. The basics of singular perturbation theory are presented in chapter 7 of reference [1].

intermediate will be removed as fast as it can be made. Accordingly, its concentration becomes approximately steady, i.e.

$$\frac{d[C]}{dt} \approx 0.$$

Applying the law of mass action and taking into account the additive effect of all of the reactions that change the concentration of C, we obtain the **steady-state approximation**:

$$\frac{d[C]}{dt} = k_1[A][B] - k_{-1}[C] - k_2[C] \approx 0.$$

$$\therefore [C] \approx \frac{k_1[A][B]}{k_{-1} + k_2}.$$

$$\therefore v = k_2[C] \approx \frac{k_1 k_2[A][B]}{k_{-1} + k_2}.$$

Alternatively, suppose that we believe that the elementary reaction $C \rightarrow A + B$ (with rate constant k_{-1}) is fast and in fact much faster than the product-forming step. We would then expect an equilibrium to be established between A + B and C, with the product-forming step acting as a small disturbance (or perturbation) to this equilibrium. If the first step tends rapidly to equilibrium, then

$$k_1[A][B] \approx k_{-1}[C].$$

This is called the **equilibrium approximation**. Again, we want to use this approximation to eliminate [C] from the rate law:

$$[C] \approx k_1[A][B]/k_{-1}.$$

$$\therefore v = k_2[C] \approx k_1 k_2[A][B]/k_{-1}.$$

1.3 The temperature dependence of rate constants

Over large temperature ranges, the temperature dependence of an *elementary* rate constant is given by the Arrhenius equation:

$$k = A \exp\left(-\frac{E_a}{RT}\right) = A \exp\left(-\frac{\epsilon_a}{k_B T}\right).$$

In this equation, A is (unimaginatively) called the pre-exponential factor, E_a is the molar activation energy, R is the ideal gas constant, and T is the temperature. The second form of the equation uses the activation energy per molecule, ϵ_a, with Boltzmann's constant, k_B. Each elementary reaction has its own positive constants A and E_a (or ϵ_a).

As we will see later in this book, over *very* large temperature ranges, we expect the Arrhenius relationship to break down. We will be interested both in the microscopic origin of the Arrhenius equation and in deviations from this equation.

Further reading

This chapter is only intended to remind you of some basic facts about chemical kinetics, and should contain ideas that are already familiar to you. If you need to refresh yourself on some aspect of chemical kinetics, I recommend any of the following:

- Engel T and Reid P 2013 *Thermodynamics, Statistical Thermodynamics, & Kinetics* 3rd edn (Boston, MA: Pearson) ch 18 and 19
- Laidler K J, Meiser J H and Sanctuary B C 2003 *Physical Chemistry* (Boston, MA: Houghton Mifflin) ch 9 and 10
- Roussel M R 2012 *A Life Scientist's Guide to Physical Chemistry* (Cambridge: Cambridge University Press) part three, especially ch 11, 12, 14, and 17 https://doi.org/10.1017/CBO9781139017480
- Vallance C 2017 *An Introduction to Chemical Kinetics (IOP Concise Physics)* (San Rafael, CA: Morgan & Claypool) https://doi.org/10.1088/978-1-6817-4664-7

Note that almost any other general textbook in physical chemistry will contain chapters that adequately address these introductory topics.

Exercises

1.1 The gas-phase reaction $NO_3 + NO \rightarrow 2NO_2$ is known to be elementary.
   (a) According to the law of mass action, what should the rate of reaction be? What is the order of the reaction with respect to $[NO_3]$? What is the overall order of the reaction?
   (b) What is the predicted relationship between the rate of change of the concentration of NO_2 and the reactant concentrations?
   (c) In gas-phase reactions, concentrations are often given in molecules per cubic centimeter (molecules cm^{-3}). Suppose that we use those units for the concentration and measure time in seconds. What are the units of the rate of reaction? What are the units of the rate constant?
   (d) Another option is to use pressures instead of concentrations.
      (i) Using the ideal gas law, show that pressures and concentrations are proportional to each other (neglecting nonideal behavior) at fixed temperature.
      (ii) Give the units of the rate and rate constant if the reactant and product pressures are measured in bar.
   (e) Draw Lewis structures of all reactants and products. Explain why it is plausible that this reaction is elementary.

1.2
   (a) An often-heard rule of thumb suggests that the rate of reaction typically doubles for every ten-degree increase in the temperature. What value of the activation energy does this rule imply?

(b) By what factor would the rate of reaction increase for a ten-degree increase in temperature if the activation energy was half as large as that calculated in part (a)?
(c) What if the activation energy was twice as large as in part (a)?

Reference

[1] Roussel M R 2019 *Nonlinear Dynamics: A Hands-On Introductory Survey (IOP Concise Physics)* (Morgan & Claypool: San Rafael, CA)

IOP Publishing

Foundations of Chemical Kinetics

A hands-on approach

Marc R Roussel

Chapter 2

The Boltzmann distribution

2.1 Statistical ideas in molecular science

A large part of the theory of chemical reactions is statistical in nature. This shouldn't be surprising: an A + B reaction, for instance, requires a collision between a molecule of A and one of B. There may also be some orientational requirements, since most molecules are not spherical. And we will see later that many other factors enter into determining whether or not a reaction occurs, notably that the molecules have sufficient energy localized within the molecules in an appropriate way. All of these factors are affected by the collisions undergone by molecules as they fly through space in a gas or jostle their way through a solution in the process of diffusion. At each collision, energy and momentum are exchanged. These exchanges depend on the details of the collision (relative orientations, speeds, angular momenta, etc.) so that the path of a particular molecule becomes essentially unpredictable (i.e. 'random') after a few collisions, even if we had very precise data about its position and velocity at a particular moment in time[1]. Note that this has nothing to do with quantum mechanics: If molecules were hard spheres (like billiard balls) subject to purely Newtonian mechanics, we would still lose our ability to predict the path of a molecule after a few collisions.

Fortunately, typical kinetics experiments involve large numbers of molecules, so that statistical arguments can be used to predict the behavior of the system observed on a macroscopic scale. This chapter will arm you with some basic tools of statistical mechanics that we will need through the first few chapters of our study. Additional developments will be presented in chapter 6.

[1] For the historians of science among you, this is the essence of Boltzmann's molecular chaos hypothesis. This idea has been put on a solid footing through the development of the theory of nonlinear dynamics, starting with the work of Poincaré in the late 19th century. In the latter half of the 20th century, a technical definition of chaos was developed in which sensitive dependence on initial conditions features prominently. We will return to the molecular chaos hypothesis in section 10.1.

doi:10.1088/978-0-7503-5321-2ch2

2.2 The Boltzmann distribution

We start with a discussion of the statistics of molecular energetics. As mentioned at the beginning of this chapter, frequent collisions between molecules result in an exchange of energy, both between molecules and between different modes of energy storage (translational, vibrational, rotational, or electronic). Under typical conditions, an equilibrium is rapidly established between the different modes of energy storage such that the average amount of energy stored in each mode summed over all of the molecules in the system becomes approximately constant, even though the energy stored in any given molecule changes with each collision. Equilibrium statistical mechanics builds on this observation to develop a statistical theory of matter and its thermal properties.

One of the key concepts in statistical mechanics is that of an **ensemble**. An ensemble is an imaginary set of copies of a system all held under identical conditions. This set is usually taken to be infinite in size. For illustration, here are a few examples of ensembles:

- We could have an ensemble consisting of, say, copies of a 1 L cylinder containing 0.004 mol of oxygen at a temperature of 25 °C. This would be called an NVT ensemble (fixed number of molecules, volume, and temperature) or, as it is known in the business, a **canonical ensemble**. Note that we would not normally be in possession of such an ensemble and certainly not if we insist on an ensemble being an infinite set. Our lab might contain just one physical realization, i.e. a single member of the ensemble. For the purpose of developing a statistical theory however, it is useful to imagine an ensemble of cylinders that all share the same N, V and T.
- Sometimes, a useful tactic for treating gas-phase phenomena is to think of an ensemble of molecules held at fixed temperature. This is sensible for gases in the ideal-behavior regime, where intermolecular forces are negligible except during collisions, and where, therefore, the energy of the system can be written as a sum of individual molecular energies[2]. Provided we specify the volume of the container in which the molecules are held, this **molecular ensemble** is really just a canonical ensemble with $N = 1$. From a practical perspective, a container holding of the order of 10^{23} molecules of gas now comes very close to the ideal of an infinite molecular ensemble.
- If you ever get involved with molecular dynamics simulations, you are likely to run into the NpT ensemble which, as its name suggests, consists of a large number of replicas of a system with fixed values of the number of molecules, pressure, and temperature.

You can imagine many other ways to define an ensemble, each corresponding to holding different system properties constant.

[2] There are other situations in which a molecular ensemble makes sense, but this is by far the simplest one to think about.

In the spirit of not getting too deeply into the subject, let's skip straight to a key result of statistical mechanics, due to Boltzmann. The probability that any given molecule in a molecular ensemble has energy ε_i is given by the **Boltzmann distribution**:

$$P(\varepsilon_i) = g_i \exp\left(-\frac{\varepsilon_i}{k_B T}\right) \Big/ q, \tag{2.1}$$

where k_B is Boltzmann's constant, which you can think of as the ideal-gas constant expressed on a per-molecule basis[3], g_i is the degeneracy of energy ε_i, and q is a normalization factor known as the **partition function** chosen so that the probabilities over all possible energy levels of the molecule sum to one:

$$\sum_i P(\varepsilon_i) = \sum_i g_i \exp\left(-\frac{\varepsilon_i}{k_B T}\right) \Big/ q = 1;$$

$$\therefore \frac{1}{q} \sum_i g_i \exp\left(-\frac{\varepsilon_i}{k_B T}\right) = 1;$$

$$\therefore q = \sum_i g_i \exp\left(-\frac{\varepsilon_i}{k_B T}\right). \tag{2.2}$$

The partition function turns out to be a central quantity in statistical mechanics, but again, we don't need the full machinery of this theory here. It is nevertheless useful to understand what the partition function measures. For this purpose, suppose that, at some temperature of interest, $k_B T \gg \varepsilon_i$ for $i \leqslant n$ and $k_B T \ll \varepsilon_i$ for $i > n$. In other words, assume that there is a large gap in energy between ε_n and ε_{n+1}, and that $k_B T$ falls somewhere in that gap. Then, for states whose energies are well below $k_B T$, $\exp(-\varepsilon_i/k_B T) \approx 1$, while for states well above $k_B T$, $\exp(-\varepsilon_i/k_B T) \approx 0$. Thus,

$$q \approx \sum_{i=1}^{n} g_i.$$

In words: the partition function counts the number of states with energy well below $k_B T$. In general, we can say that the partition function gives a rough count of the number of states accessible at temperature T.

Example 2.1. *Systems with just two energy levels are often useful simplified models of complex systems. For example, we can use them to gain a basic understanding of protein folding and thermal denaturation with a low-energy folded state and a higher-energy unfolded state [1]. Suppose that we have a two-level system with* $\varepsilon_1 = 0$, $g_1 = 1$ *and* $\varepsilon_2 = 1 \times 10^{-19}$ J, $g_2 = 3$. *The partition function for such a system is*

[3] If you read enough papers or books in statistical physics, you will sooner or later see the exponential in the Boltzmann distribution written $e^{\beta\varepsilon_i}$, where $\beta = (k_B T)^{-1}$. There are lots of reasons why people do this, but I tend to avoid it since it doesn't save a lot of writing and somewhat obscures the dependence on T.

$$q = 1 + 3\exp\left(-\frac{\varepsilon_2}{k_B T}\right) = 1 + 3\exp(-7243\ \mathrm{K}/T).$$

This partition function is plotted in figure 2.1. Note that q is about one when T is sufficiently below $\varepsilon_2/k_B = 7243$ K and rises towards the limiting value of four at high temperatures. The values of q at intermediate values of T reflect the bias towards occupation of the ground state, which decreases as T increases.

The interpretation of q given above implicitly assumes that we set the ground-state energy to zero. Since the zero of energy is an arbitrary quantity, we could add any arbitrary constant to all of the energies. The effect of this is to multiply each term in q by $e^{-\delta\varepsilon/k_B T}$, where $\delta\varepsilon$ is the arbitrary energy offset. This factor would cancel out in calculating probabilities using equation (2.1), so it has no effect on the probabilities. It does, however, affect the numerical value of the partition function and thus the simple interpretation as a count of accessible states. Still, it's useful to think of q as quantifying (give or take a scale factor) the number of accessible energy levels at temperature T.

In equation (2.2), the sum is over energy levels. It is sometimes convenient to sum over quantum states (i.e. over the different sets of quantum numbers). The following is a completely equivalent equation for the partition function:

$$q = \sum_{s \in \mathrm{states}} \exp\left(-\frac{\varepsilon_s}{k_B T}\right).$$

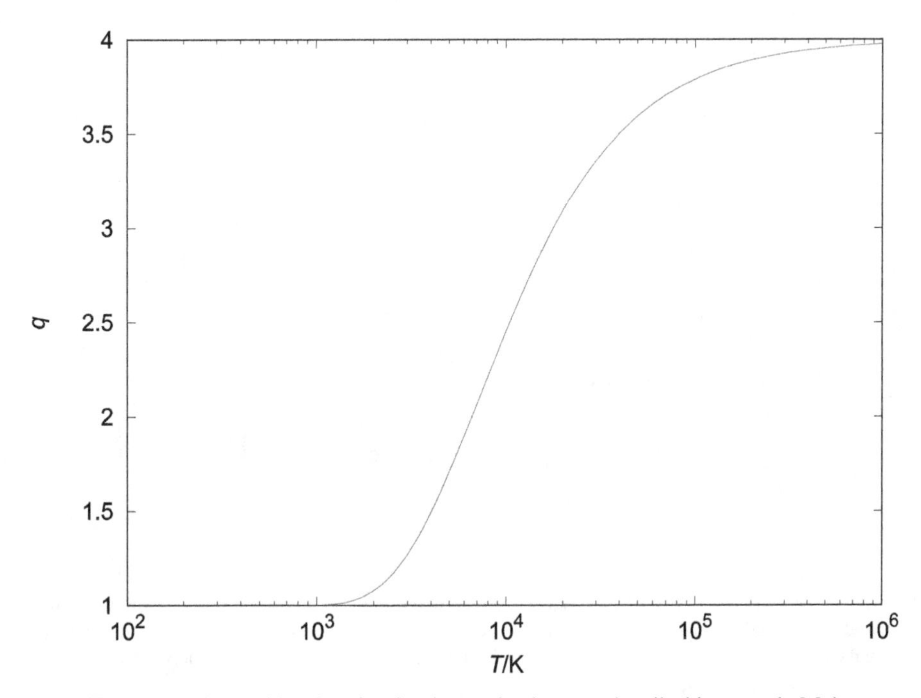

Figure 2.1. The partition function for the two-level system described in example 2.2.1.

It should be clear from the context whether we are summing over energy levels or states.

Now suppose that the molecular energy can be decomposed into a sum of independent contributions, i.e. $\varepsilon = \varepsilon^{(1)} + \varepsilon^{(2)} + \cdots$ Using the 'sum over quantum states' form of the partition function, we then have

$$\begin{aligned} q &= \sum_i \sum_j \sum_k \cdots \exp\left(-\frac{\varepsilon_i^{(1)} + \varepsilon_j^{(2)} + \varepsilon_k^{(3)} + \cdots}{k_B T}\right) \\ &= \sum_i \sum_j \sum_k \cdots \exp\left(-\frac{\varepsilon_i^{(1)}}{k_B T}\right) \exp\left(-\frac{\varepsilon_j^{(2)}}{k_B T}\right) \exp\left(-\frac{\varepsilon_k^{(3)}}{k_B T}\right) \cdots \\ &= \sum_i \exp\left(-\frac{\varepsilon_i^{(1)}}{k_B T}\right) \sum_j \exp\left(-\frac{\varepsilon_j^{(2)}}{k_B T}\right) \sum_k \exp\left(-\frac{\varepsilon_k^{(3)}}{k_B T}\right) \cdots \\ &= q^{(1)} q^{(2)} q^{(3)} \ldots \end{aligned}$$

We see that the partition function factors in these cases, and we say that the partition function is **separable**.

2.3 The Boltzmann distribution in classical mechanics

One of the main differences between classical and quantum mechanics is that energy is not quantized in the classical theory. Accordingly, in classical mechanics, it doesn't make sense to ask for the probability that a molecule has *exactly* energy ε. The answer to this question is zero[4]. The correct question to ask is: what is the probability that a molecule has an energy between two specified limits, i.e. what is $P(\varepsilon_{low} \leqslant \varepsilon \leqslant \varepsilon_{high})$?

Answering questions like this requires some new mathematical machinery: We define the **density of states** $g(\varepsilon)$ such that $g(\varepsilon)\,d\varepsilon$ is the number of states between energies ε and $\varepsilon + d\varepsilon$.

For simplicity, let's think about an ideal gas made up of molecules whose energies have a continuous component. The translational kinetic energy, for example, behaves classically in typical situations. Let ε_o be an observation of this continuous energy component for a randomly chosen molecule. The probability that the observed value falls within the interval $(\varepsilon, \varepsilon + d\varepsilon)$ is given by the continuous (classical) Boltzmann distribution:

$$P(\varepsilon \leqslant \varepsilon_o \leqslant \varepsilon + d\varepsilon) = p(\varepsilon)\,d\varepsilon = g(\varepsilon)\exp\left(-\frac{\varepsilon}{k_B T}\right) d\varepsilon / q, \tag{2.3}$$

[4] This has to do with the fact that the classical energy is a real number, and that the probability that a number on the real line exactly equals some specified value is zero. If you collect enough decimal places (as an intellectual exercise, not in the real world), there will always be some digit that differs between a measurement and a specified number.

where $p(\varepsilon)$ is the probability density, i.e. the probability per unit energy. If we want to know the probability that the observed value falls within a finite interval $(\varepsilon_{\text{low}}, \varepsilon_{\text{high}})$, then we just have to integrate this equation:

$$P(\varepsilon_{\text{low}} \leqslant \varepsilon_{\text{o}} \leqslant \varepsilon_{\text{high}}) = \frac{1}{q}\int_{\varepsilon_{\text{low}}}^{\varepsilon_{\text{high}}} g(\varepsilon)\exp\left(-\frac{\varepsilon}{k_{\text{B}}T}\right)\text{d}\varepsilon.$$

Although we focused on a molecule in an ideal gas, this equation would apply equally to realizations of any canonical ensemble with a separable continuous (i.e. non-quantized) energy component.

The probability depends once again on a partition function. By requiring the probabilities to sum to one, we can again determine the partition function. The only difference is that, in an energy continuum, 'summing' means 'integrating,' so we integrate the probability over the entire space of allowed energies, denoted by $\mathcal{A}$:

$$q = \int_{\mathcal{A}} g(\varepsilon)\exp\left(-\frac{\varepsilon}{k_{\text{B}}T}\right)\text{d}\varepsilon. \tag{2.4}$$

2.4 Vibrational energy levels and the density of states

The density of states often comes up in semiclassical treatments of quantum systems. Let's start with a simple example, namely a calculation of the density of states associated with a vibrational mode of a molecule. Recall that the energy levels of a vibrational mode are given, to a first approximation, by the energy levels of a harmonic oscillator, which are

$$\varepsilon_v = \hbar\omega_0\left(v + \frac{1}{2}\right),$$

where $\hbar = h/2\pi$, $v = 0, 1, 2, \ldots$ is a quantum number, and ω_0 is the fundamental frequency of the oscillator, here expressed as an angular frequency with units of s^{-1}. Note that when the molecule is in its ground vibrational state ($v = 0$), there is a zero-point energy of $\hbar\omega_0/2$. Figure 2.2 shows the harmonic-oscillator energy levels with the simple parabolic potential energy curve which is the basis of this model.

Molecules with more than two atoms have several vibrational modes, each with its own frequency. For the moment, however, let's focus on one particular nondegenerate vibrational mode and work out the corresponding density of states.

The density of states is the number of energy states per unit energy change. For the harmonic oscillator, the energy changes by $\hbar\omega_0$ whenever the quantum number, which fully defines the state, changes by one unit. Thus,

$$g(\varepsilon_v) = \frac{1}{\hbar\omega_0}.$$

The constant density of states reflects the fact that the harmonic-oscillator energy levels are equally spaced. Needless to say, we will have to do a bit more work when this isn't the case.

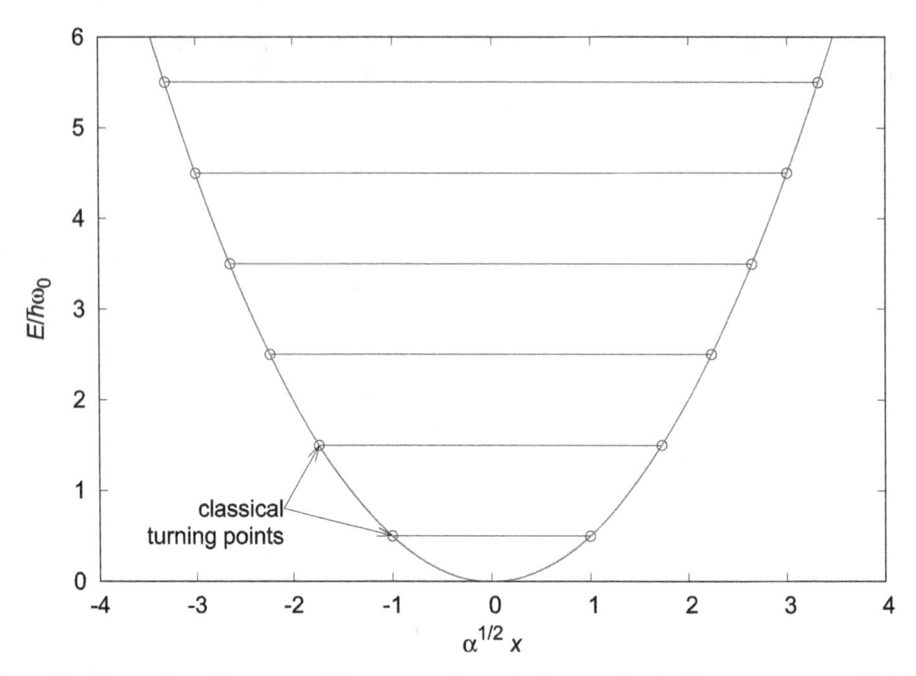

Figure 2.2. Harmonic-oscillator potential energy and quantized energy levels. The parameter α, which has units of inverse length squared, is given by $\sqrt{\hbar\omega_0/k}$. The two coordinates of this graph are thus dimensionless. The classical turning points are those points where the potential energy equals the total energy of the oscillator. Classically, these would be the points where the oscillator would come to rest before reversing its motion, either going from compression to expansion, or vice versa.

Note that we used a quantum mechanical equation for the energy to obtain the density of states, which technically only makes sense if the states form a continuum. The density of states we just obtained is therefore only useful when the energy is sufficiently high that the differences in energy between the vibrational levels are negligibly small. This is the sense in which the foregoing calculation is semiclassical.

2.5 The derivation of the Arrhenius equation

We can derive the Arrhenius equation using what we have learned so far. Figure 2.3 gives a useful mental picture of a chemical reaction. This kind of potential energy profile is particularly appropriate to an isomerization. The reaction coordinate x would, in general, represent some measure of the progress of the reaction. For example, in the case of an intramolecular proton transfer, x might be the distance between the proton and one of the two basic sites. Each of the two potential energy wells represents a set of states of the system that we can associate with an isomer. We arbitrarily label one of them as the reactant, and the other as the product. (Which is which depends on the experiment.) Each of the two isomers has its own set of vibrational energy levels, which are drawn as horizontal lines. To get from 'reactant' to 'product,' the system needs to pass over the potential energy barrier, which has height ε_a. We neglect the zero-point energy, since we are going to attack

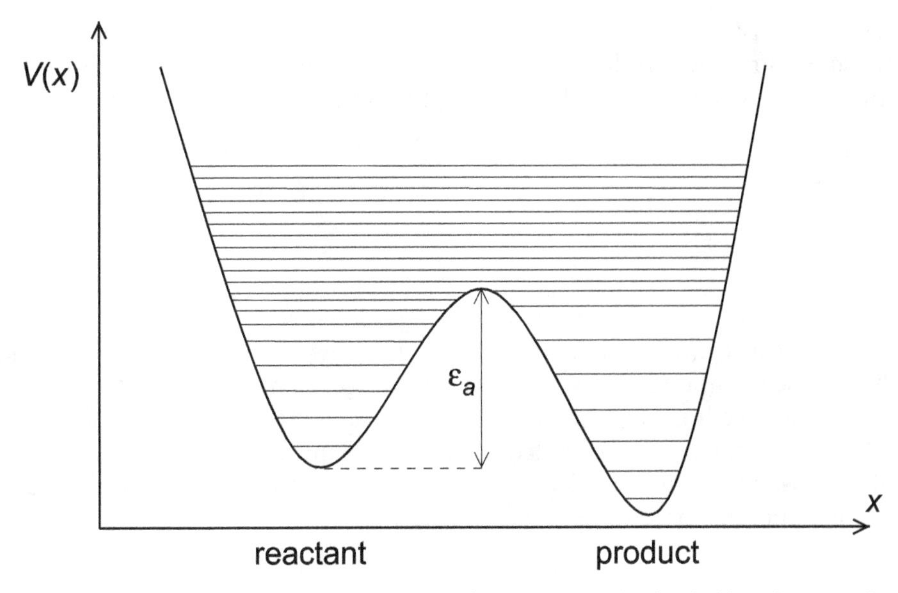

Figure 2.3. A sketch of the potential energy and vibrational energy levels of a double-well system. We can think of the two wells as representing a reactant and a product of a reaction. A classical activation barrier ε_a separates the two wells along the reaction coordinate x.

the barrier-crossing problem as a classical problem for now. The point right at the top of the barrier is called the **transition state**. In a double-well system, there are also vibrational states above the barrier. The vibrational states above the barrier are much closer together than those below the barrier. This is a general phenomenon connected to the correspondence principle: roughly speaking, as a system is able to explore a larger region of space, quantum mechanics eventually tends to a classical limit, where we would have a continuum of possible energies.

We now turn to the question of how often we would expect barrier crossings from reactant to product in an ensemble of molecules at temperature T. According to the law of mass action, this rate takes the form k[R], where [R] is the reactant concentration. If we divide this rate by [R], we get the specific rate of crossing, i.e. the number of left-to-right barrier crossings per molecule per unit time, which is just the rate constant, k. We can decompose k into three contributions, namely the probability that a molecule has an energy above ε_a (relative to the bottom of the reactant well) times the rate at which molecules pass over the barrier from left to right, de-energize, and thus drop into the product well[5]. We group the latter two contributions and write

$$k = P(\varepsilon > \varepsilon_a) \times (\text{rate of crossing \& de–energization}).$$

Let us call the latter contribution k_{cd}, so we have

$$k = k_{cd} P(\varepsilon > \varepsilon_a).$$

[5] If you're worried about the molecule dropping back down into the reactant well, hold that thought. We'll get to it when we discuss transition-state theory.

Arrhenius theory focuses on the factor $P(\varepsilon > \varepsilon_a)$, which we can calculate from the Boltzmann distribution. If the barrier is sufficiently high, such that the steps between the states, of size $\hbar\omega_0$, are small compared to ε_a, then we can treat the vibrational states roughly as a continuum. We then have

$$P(\varepsilon > \varepsilon_a) = \frac{\int_{\varepsilon_a}^{\infty} g(\varepsilon)\exp(-\varepsilon/k_BT)\,d\varepsilon}{\int_0^{\infty} g(\varepsilon)\exp(-\varepsilon/k_BT)\,d\varepsilon}.$$

Referring now to our earlier discussion of figure 2.3, we note that the density of states necessarily changes for energies above vs below the barrier. As illustrated in the figure, the vibrational levels get closer together even before we get to the top of the barrier. Neglecting the small range of energies close to the top of the barrier, the density of states can be approximated as being constant. There is a different but also roughly constant density of states above the barrier. We therefore introduce the following approximation to the density of states:

$$g(\varepsilon) = \begin{cases} g_b & \text{for} \quad \varepsilon < \varepsilon_a \quad (b = \text{'below'}), \\ g_a & \text{for} \quad \varepsilon > \varepsilon_a \quad (a = \text{'above'}). \end{cases}$$

This leads to the following expression for the probability that the energy of the molecule lies above that of the transition state:

$$P(\varepsilon > \varepsilon_a) = \frac{g_a\int_{\varepsilon_a}^{\infty} \exp(-\varepsilon/k_BT)\,d\varepsilon}{g_b\int_0^{\varepsilon_a} \exp(-\varepsilon/k_BT)\,d\varepsilon + g_a\int_{\varepsilon_a}^{\infty} \exp(-\varepsilon/k_BT)\,d\varepsilon}.$$

For a high barrier, very few states above the barrier will be populated compared to the number of states in the reactant well. Thus,

$$P(\varepsilon > \varepsilon_a) \approx \frac{g_a\int_{\varepsilon_a}^{\infty} \exp(-\varepsilon/k_BT)\,d\varepsilon}{g_b\int_0^{\varepsilon_a} \exp(-\varepsilon/k_BT)\,d\varepsilon}.$$

For the same reason, we make only a small error by extending the range of integration in the denominator to infinity:

$$\begin{aligned} P(\varepsilon > \varepsilon_a) &\approx \frac{g_a\int_{\varepsilon_a}^{\infty} \exp(-\varepsilon/k_BT)\,d\varepsilon}{g_b\int_0^{\infty} \exp(-\varepsilon/k_BT)\,d\varepsilon} \\ &= \frac{-g_a k_BT\exp(-\varepsilon/k_BT)\big|_{\varepsilon_a}^{\infty}}{-g_b k_BT\exp(-\varepsilon/k_BT)\big|_0^{\infty}} \\ &= \frac{g_a}{g_b}\exp(-\varepsilon_a/k_BT). \end{aligned}$$

The rate constant becomes

$$k = k_{\mathrm{cd}} \frac{g_{\mathrm{a}}}{g_{\mathrm{b}}} \exp(-\varepsilon_{\mathrm{a}}/k_{\mathrm{B}}T) = A \exp(-\varepsilon_{\mathrm{a}}/k_{\mathrm{B}}T),$$

which is the familiar Arrhenius equation. We thus see that the Arrhenius equation is a consequence of the Boltzmann distribution. We haven't really said anything about how to compute the pre-exponential factor, but at least we now know where the exponential term comes from.

The derivation presented here assumed that reaction requires the molecule to climb up through its vibrational levels. Of course, it is possible for a reaction to depend on other forms of energy storage. For bimolecular reactions, the kinetic energy might be important. In other cases, rapid rotation might facilitate a reaction. And of course the electronic state of the molecule might be important. Regardless of the case we consider, we always end up with a rate constant that contains a Boltzmann–Arrhenius dependence on the temperature, essentially for the reason seen above: if the molecule has to cross an energy barrier, the Boltzmann distribution controls the number of molecules that have sufficient energy to do so at any given time.

2.6 The Maxwell–Boltzmann distribution

In this section, we derive the Maxwell–Boltzmann distribution of molecular speeds. The Boltzmann distribution itself applies to any kind of energy, including the translational kinetic energies of molecules. As you know, the kinetic energy is given by

$$K = \frac{1}{2}mu^2 = \frac{m}{2}\left(u_x^2 + u_y^2 + u_z^2\right).$$

(We use u rather than v for the speed and velocity components to avoid confusion with the reaction rate later on.) As this is a sum of terms, we can apply the Boltzmann distribution to each component independently. We have, for each component of the velocity,

$$p(u_i)\,\mathrm{d}u_i = \frac{1}{q}\exp\left(\frac{-mu_i^2}{2k_{\mathrm{B}}T}\right)\mathrm{d}u_i. \tag{2.5}$$

We get equation (2.5) from the discrete probability distribution (2.1) and not, as you might expect, from the continuous distribution (2.3). The problem is that we don't know what to use, *a priori*, for the density of states in equation (2.3). What we do instead is imagine an infinitesimally small cube of dimensions $\mathrm{d}u_x\,\mathrm{d}u_y\,\mathrm{d}u_z$ around a particular point in velocity space (u_x, u_y, u_z). For sufficiently small $\mathrm{d}u_x$, $\mathrm{d}u_y$ and $\mathrm{d}u_z$, this infinitesimal volume contains a single 'state,' and we can then write equation (2.5) as a direct continuous analog of (2.3).

The partition function is easy to write down[6]:

$$q = \int_{-\infty}^{\infty} \exp\left(\frac{-mu_i^2}{2k_BT}\right) du_i = \sqrt{2\pi k_BT/m}.$$

The probability distribution of the molecular velocity components is therefore

$$p(u_i)\, du_i = \sqrt{\frac{m}{2\pi k_BT}} \exp\left(\frac{-mu_i^2}{2k_BT}\right) du_i. \tag{2.6}$$

The full distribution in the three-dimensional velocity space is

$$p(u_x, u_y, u_z)\, du_x\, du_y\, du_z = \left(\frac{m}{2\pi k_BT}\right)^{3/2} \exp\left(\frac{-m(u_x^2 + u_y^2 + u_z^2)}{2k_BT}\right) du_x\, du_y\, du_z. \tag{2.7}$$

To emphasize: this is the probability per unit volume in velocity space of finding the velocity within a cube of dimensions $du_x\, du_y\, du_z$ surrounding a given velocity vector (u_x, u_y, u_z).

For most purposes, that's not quite what we want. We usually don't care much about the direction a molecule is traveling in. What we usually care about is the speed at which the molecule is traveling. The speed, u, is given by

$$u^2 = u_x^2 + u_y^2 + u_z^2.$$

For a given u, this is the equation of a sphere in the velocity space. In equation (2.7), $du_x\, du_y\, du_z$ is a volume element. If we want to know the probability that the speed is within du of u, we need to add up the volumes of all of the volume elements inside a shell of radius u and thickness du (figure 2.4). The area of a shell of radius u is $4\pi u^2$. The volume of the shell is the area times the thickness, so $4\pi u^2 du$. The result of integrating equation (2.7) over such a shell is therefore

$$p(u)\, du = 4\pi u^2 \left(\frac{m}{2\pi k_BT}\right)^{3/2} \exp\left(\frac{-mu^2}{2k_BT}\right) du. \tag{2.8}$$

This equation is called the **Maxwell–Boltzmann distribution**.

Equation (2.8) is written in terms of the mass of a single molecule, m. It is often more convenient to use the molar mass. Since $m = M/L$, where L is Avogadro's number, and $k_B = R/L$, we have the equivalent equation

$$p(u)\, du = 4\pi u^2 \left(\frac{M}{2\pi RT}\right)^{3/2} \exp\left(\frac{-Mu^2}{2RT}\right) du.$$

[6] The integral $\int e^{-x^2}dx$ is not one you can evaluate analytically using the techniques you would have learned in your calculus course. If you have taken some statistics, you may recognize that this integral is related to properties of the normal distribution, and perhaps you even learned that its value is related to the 'error function' erf(x). For our purposes, we only need to know that $\int_{-\infty}^{\infty} e^{-x^2}dx = \sqrt{\pi}$ in order to evaluate this integral.

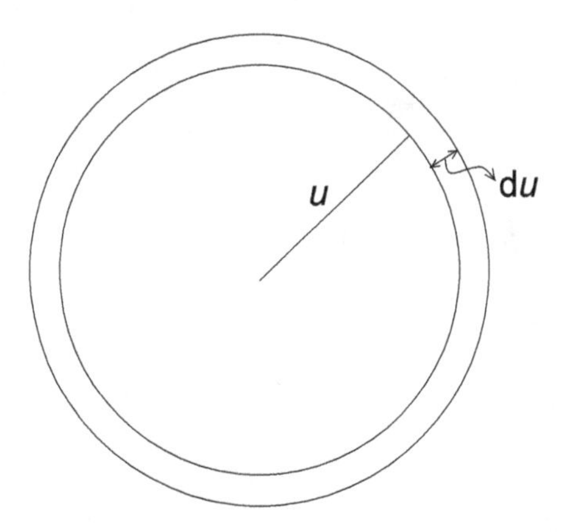

Figure 2.4. A cut through a spherical shell of radius u and thickness du.

Note that you have to be a bit careful about the units. All of the quantities in this equation are in SI units, so the molar mass has to be in units of kg mol^{-1}.

The Maxwell–Boltzmann distribution is plotted for a few different temperatures in figure 2.5. As you might have expected, molecules move faster, on average, when the temperature increases. The width of the distribution also increases.

Once we know the probability density of a variable, we can calculate the averages of quantities depending on that variable. Suppose that $p(x)$ is the probability density for some variable x, and let $f(x)$ be a quantity that depends on x. The average value of f is then given by

$$\langle f \rangle = \int_{\mathcal{R}} f(x)\, p(x)\, \mathrm{d}x,$$

where $\mathcal{R}$ is the region over which the probability density is defined.

For example, we can calculate the average speed of a molecule in a gas as follows:

$$\begin{aligned}\langle u \rangle &= \int_0^\infty u\, p(u)\, \mathrm{d}u \\ &= \int_0^\infty 4\pi u^3 \left(\frac{m}{2\pi k_\mathrm{B} T}\right)^{3/2} \exp\left(\frac{-mu^2}{2k_\mathrm{B}T}\right) \mathrm{d}u \\ &= \sqrt{\frac{8k_\mathrm{B}T}{\pi m}} \equiv \sqrt{\frac{8RT}{\pi M}}.\end{aligned}$$

In many problems in kinetics, we actually need the relative speeds of two molecules rather than the speed of any one molecule. Classical mechanics tells us that, in two-body problems, the relative motion obeys essentially identical equations to those of a single particle, except that the mass is replaced by the reduced mass μ, which is defined as follows:

$$\frac{1}{\mu} = \frac{1}{m_1} + \frac{1}{m_2}, \tag{2.9}$$

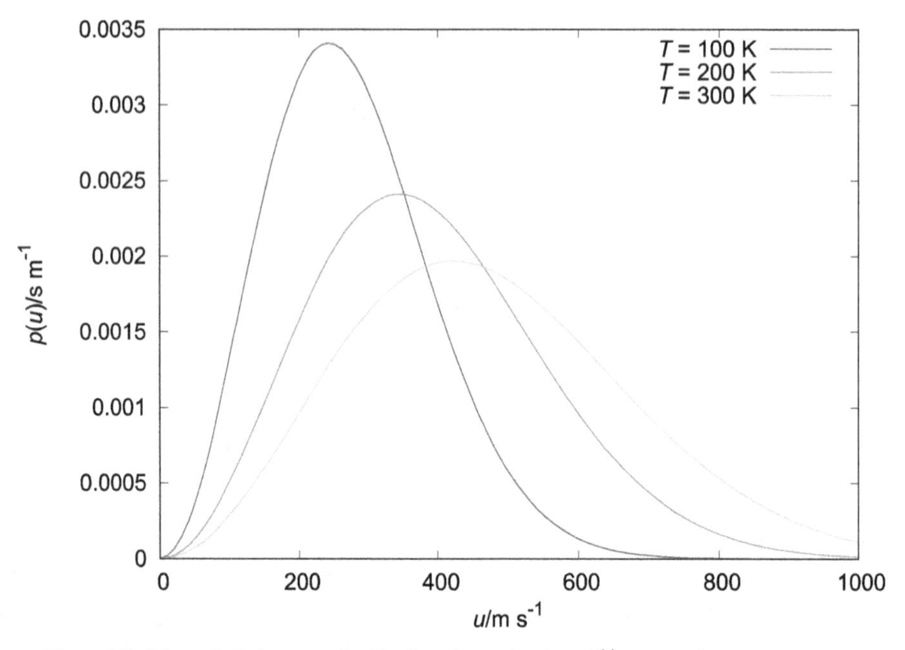

Figure 2.5. Maxwell–Boltzmann distributions for molecules of $^{14}N_2$ at various temperatures.

where m_1 and m_2 are the masses of the two particles. The same thing is true, incidentally, in non-relativistic quantum mechanics. Using this simple idea, we obtain the Maxwell–Boltzmann distribution for the relative speed:

$$\begin{aligned} p(u_{\rm rel})\,\mathrm{d}u_{\rm rel} &= 4\pi u_{\rm rel}^2\left(\frac{\mu}{2\pi k_{\rm B}T}\right)^{3/2}\exp\left(\frac{-\mu u_{\rm rel}^2}{2k_{\rm B}T}\right)\mathrm{d}u_{\rm rel} \\ &= 4\pi u_{\rm rel}^2\left(\frac{\mu_{\rm m}}{2\pi RT}\right)^{3/2}\exp\left(\frac{-\mu_{\rm m}u_{\rm rel}^2}{2RT}\right)\mathrm{d}u_{\rm rel}, \end{aligned} \tag{2.10}$$

where $\mu_{\rm m}$ is the reduced molar mass. Using the same trick, we find that the average relative speed of two molecules is

$$\langle u_{\rm rel}\rangle = \sqrt{\frac{8k_{\rm B}T}{\pi\mu}} = \sqrt{\frac{8RT}{\pi\mu_{\rm m}}}. \tag{2.11}$$

Exercise

2.1

(a) What are the units of $\hbar$?
(b) Show that the units of $\omega = E/\hbar$ are appropriate units for a frequency.

2.2 The nuclear magnetic resonance (NMR) effect in spin-$\frac{1}{2}$ nuclei (such as the ^{1}H nucleus) is a simple example of a two-level system. The energies of the two spin states are given by

$$\varepsilon_{\pm\frac{1}{2}} = \mp\frac{1}{2}\hbar\gamma B_z,$$

where γ is the magnetogyric ratio of the nucleus and B_z is the field strength of the NMR magnet. For clarity: a nucleus whose spin is aligned with the field ($m_s = +\frac{1}{2}$) has the lower (negative) energy. For a proton, $\gamma = 2.675\,22 \times 10^{8}\ \mathrm{T^{-1}s^{-1}}$. The University of Lethbridge's 700 MHz NMR has a field strength of 16.4 T. At 25 °C, by how much does the probability that the proton is to be found in the lower energy spin state exceed $\frac{1}{2}$? In other words, calculate $P(\varepsilon_{+}) - \frac{1}{2}$.

Note: The sample magnetization detected in NMR is due to the difference in population between the two spin states. You will find this difference surprisingly small, so small that it is difficult to evaluate it directly using a calculator. Use the following Taylor expansion to approximate the exponentials:

$$e^{x} \approx 1 + x$$

for small values of x.

2.3 In section 2.6, we worked out the average speed. The standard deviation of the speed is

$$\sigma_u = \sqrt{\langle u^2\rangle - \langle u\rangle^2}.$$

Note the difference between $\langle u^2\rangle$, the average of (u^2), and $\langle u\rangle^2$, which is the square of the average of u. These are not the same thing.

(a) Work out an equation for the standard deviation of the molecular speed.

Useful integral:

$$\int_0^\infty x^4 e^{-x^2} dx = 3\sqrt{\pi}/8.$$

(b) The coefficient of variation (CV) is the ratio of the standard deviation to the average. In this case,

$$\mathrm{CV} = \sigma_u/\langle u\rangle.$$

Work out an equation for the CV.

Reference

[1] Dill K A, Bromberg S, Yue K, Fiebig K M, Yee D P, Thomas P D and Chan H S 1995 Principles of protein folding–a perspective from simple exact models *Protein Sci.* **4** 561–602

IOP Publishing

Foundations of Chemical Kinetics
A hands-on approach
Marc R Roussel

Chapter 3

Gas-phase collision theory

The basic idea of collision theory is very simple: in a bimolecular reaction in the gas phase, the two reactants have to collide in order to react. Maybe if we calculate the rate of collision, it will give us the pre-exponential factor? Surprisingly, advanced versions of collision theory can do much more, even giving us some insight into the origin of the activation energy.

3.1 Simple collision theory

For a bimolecular reaction, we can think of the Arrhenius equation as having a simple interpretation, in accord with the discussion of section 2.5: the Boltzmann–Arrhenius factor is related to the number of molecules that have sufficient energy to surpass the activation barrier. The pre-exponential factor can be interpreted as the collision rate that *could* lead to reaction *if* the molecules had sufficient energy. The factors that determine whether or not a reaction can occur might include the relative orientations of the molecules, the way energy is distributed into rotational and vibrational modes, electronic effects, etc. If we ignore all of the latter effects and just calculate a collision rate, we should end up with a maximum value for the pre-exponential factor.

We are going to treat an elementary gas-phase reaction

$$A + B \rightarrow \text{product(s)}.$$

For simplicity, we assume spherical molecules of radii r_A and r_B and define $r_{AB} = r_A + r_B$. Suppose we have n_A and n_B moles of molecules of A and B, respectively, in a container of volume V. The simplest way to calculate the collision rate is to imagine that the B molecules are stationary and to calculate how often a particular molecule of A is likely to collide with B molecules, assuming that the molecules are uniformly distributed in the container. We will later correct for the fact that the molecules of B are moving, and of course we will also take into account the fact that we have many molecules of A.

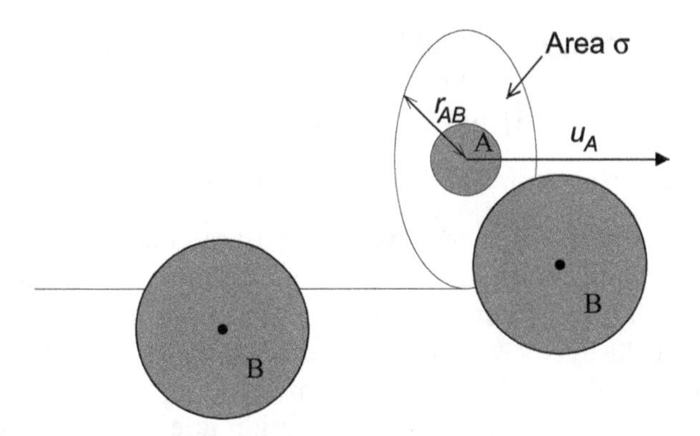

Figure 3.1. A molecule of A moving with speed u_A through a field of stationary B molecules collides with a B when their centers pass within a distance of r_{AB}. The collision cross-section σ is the area of a disc of radius r_{AB}, which we imagine to be traveling with A. Note that the center of the molecule of B on the left was outside the disc and did not collide with A, while the molecule of B on the right is about to collide with A, since the center of this molecule of B will shortly intersect the disc co-moving with A.

Figure 3.1 sketches the situation described above. In the figure, a molecule of A is moving from left to right at a speed u_A. It has just missed one molecule of B (the one on the lower left), and is about to collide with another. We can see that A will collide with B if, treating A and B as intangible particles that can pass right through each other, the center of A will pass within a distance r_{AB} of the center of a B molecule. One way to picture the condition for a collision is that the center of a B molecule has to cross a disc of radius r_{AB} traveling with an A molecule. The area of this disc is called the **collision cross-section**, σ. Thus, what we need to do is to determine the number of occurrences per unit time of the centers of B molecules passing through a disc of area σ traveling at speed u_A.

The speed at which the disc is traveling is a distance traveled per unit time. The speed times the area of the disc is therefore the volume of a cylinder swept out by the disc per unit time: volume/time = σu_A. The number of B molecules per unit volume is $n_B L/V$, where L is Avogadro's number. The number of B molecules whose centers pass through our imaginary disc per unit time is therefore

$$\frac{\text{number of B}}{\text{volume}} \times \frac{\text{volume}}{\text{time}} = \sigma u_A n_B L/V.$$

This is the number of collisions per unit time of a single molecule of A with molecules of B. Since there are $n_A L$ molecules of A, the total number of collisions for all molecules of A per unit time is just $\sigma u_A n_A n_B L^2/V$. But of course, the molecules of B are moving too. There is an easy fix for this. Instead of using the speed of a single A molecule, we use the average relative speed from equation (2.11). This gives the rate of collisions $\sigma\langle u_{\text{rel}}\rangle n_A n_B L^2/V$. This is almost what we want. This rate has units of[1] collisions s^{-1}. In kinetics, in which we normally express rates in units of

[1] 'Collisions' are of course dimensionless, but it's useful to include them in the units to help us with our mental bookkeeping.

concentration per unit time, we would want the rate of collisions to be a number of collisions per unit time per unit volume, so we divide by an additional factor of V. We then note that $n_A/V = [A]$ and $n_B/V = [B]$ to obtain the rate of collisions

$$Z_{AB} = \sigma\langle u_{rel}\rangle L^2[A][B]. \qquad \text{(Units: collisions m}^{-3}\text{s}^{-1}\text{)} \tag{3.1}$$

In order for the units to work out, we have to use concentration units that are compatible with the units of length found in $\sigma\langle u_{rel}\rangle$. This would normally mean using the SI unit of length throughout and thus concentrations in mol m^{-3}, which would give us a rate of collisions in the indicated units of collisions m^{-3}s^{-1}. Many other consistent choices for the units are of course possible.

As noted above, equation (3.1) gives the collision rate as the number of collisions per unit volume per unit time. We often prefer to express rates in units of moles of collisions per unit volume per unit time, which we get by dividing Z_{AB} by Avogadro's number:

$$Z_{AB} = \sigma\langle u_{rel}\rangle L[A][B]. \qquad \text{(Units: mol m}^{-3}\text{s}^{-1}\text{)} \tag{3.2}$$

Again, the difference between equations (3.1) and (3.2) is only one of units. There should be no confusion if you track your units carefully through your calculations.

In simple collision theory, the rate of reaction would be Z_{AB} multiplied by an Arrhenius factor to account for the probability that the molecules have sufficient energy to react:

$$v = \sigma\langle u_{rel}\rangle L e^{-E_a/RT}[A][B].$$

For an elementary bimolecular reaction, the law of mass action tells us the rate should be $k[A][B]$. By comparing the two expressions, we get the collision-theory value of the rate constant, and since $k = Ae^{-E_a/RT}$, we finally obtain the collision-theory pre-exponential factor:

$$A_{ct}^{A+B} = \sigma\langle u_{rel}\rangle L. \tag{3.3}$$

The superscript 'A + B' is necessary for reasons to be discussed in a moment but will be dropped when there is no possible ambiguity. This equation can be used to estimate the maximum possible value of the pre-exponential factor, but it can also be used to determine an effective collision cross-section given a measured pre-exponential factor.

When a reaction has a pre-exponential factor that is similar to the collision-theory value, we say that the reaction is **collision limited**. If a reaction is truly collision limited, then every collision of the reactants with sufficient energy results in the formation of products.

Example 3.1. *The reaction* $O + N_2 \rightarrow N + NO$ *has a very large pre-exponential factor of* 3.0×10^{-10} cm^3 molecule^{-1} s^{-1} *over the temperature range 1400–4000 K. Let's see whether we can predict this value assuming that the reaction is collision limited.*

Because the average relative speed depends on temperature, we need to pick a value of T. My usual rule in these cases is to choose the midpoint of the experimental interval, in this case 2700 K. We can use standard molar masses given that the data provided don't indicate specific isotopes. If we calculate the reduced molar mass, then we just have to remember to use R instead of k_B in the equation for the average relative speed.

$$\begin{aligned}\mu_m^{-1} &= \frac{1}{M_O} + \frac{1}{M_{N_2}} \\ &= \frac{1}{16.000 \text{ g mol}^{-1}} + \frac{1}{28.014 \text{ g mol}^{-1}} = 0.098\,196 \text{ mol g}^{-1}; \\ \therefore \mu_m &= \frac{1}{0.098\,196 \text{ mol g}^{-1}} = 10.184 \text{ g mol}^{-1} \equiv 0.010\,184 \text{ kg mol}^{-1}.\end{aligned}$$

$$\begin{aligned}\langle u_{rel}\rangle &= \sqrt{\frac{8RT}{\pi\mu_m}} \\ &= \sqrt{\frac{8(8.314\,462\,618 \text{ J K}^{-1}\text{mol}^{-1})(2700 \text{ K})}{\pi(0.010\,184 \text{ kg mol}^{-1})}} \\ &= 2369 \text{ m s}^{-1}.\end{aligned}$$

We now need to estimate the collision cross-section. For this, we need the radii of the oxygen atom and nitrogen molecule. There are many possible definitions of the 'radius' of an atom. Perhaps the most widely used, as well as being the most appropriate for gas-phase reactions, is the van der Waals radius. A table of van der Waals radii for the main-group elements is available in a paper by Mantina and coworkers [1]. This table gives a radius of 152 pm for an oxygen atom. The 'radius' of a nitrogen molecule is a trickier quantity. Since molecules in the gas phase rotate rapidly, we can think of the radius of a homonuclear diatomic molecule as being the diameter of one atom (figure 3.2). The covalent radius is defined based on the bond length in a molecule, which makes it a reasonable atomic radius to use for atoms in a

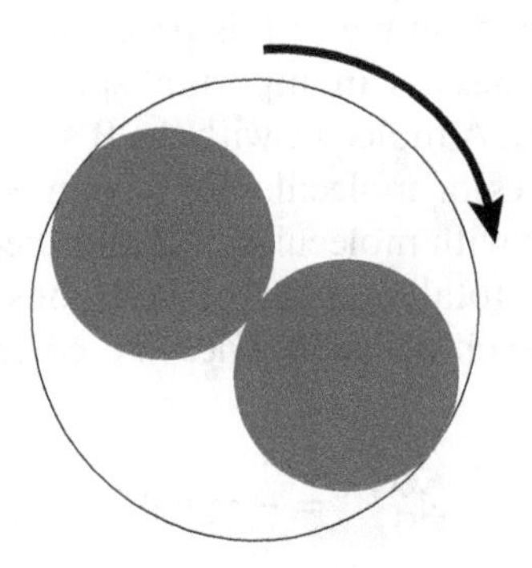

Figure 3.2. A sketch of a rotating homonuclear diatomic molecule. As the molecule rotates, it sweeps out a sphere whose radius is equal to the diameter of one atom. (Note that the molecule rotates around two axes relative to a coordinate system attached to the molecule.)

molecule. The bond length is equal to twice the atomic radius, and thus to the diameter of an atom. The bond length in N_2 *is 109.76 pm. Thus we have* $R_{AB} = 152 + 109.76\text{ pm} = 262\text{ pm}$. *Our estimate of the cross-section is therefore*

$$\sigma = \pi R_{AB}^2 = \pi(262 \times 10^{-12}\text{ m})^2$$
$$= 2.15 \times 10^{-19}\text{ m}^2.$$

I want to emphasize that this is an estimate. *We could have made different assumptions about the radii and obtained different values for the cross-section. In particular, a good argument could have been made for using the van der Waals radius of a nitrogen molecule instead of the bond length. This would have resulted in a cross-section that was about twice as large as calculated here. The problem is that simple collision theory is based on the assumption that molecules are hard spheres, but quantum mechanics tells us this isn't so. Comparisons of collision theory with experimental data are therefore always based on some assumptions about the radii, which we hope are reasonable.*

We are now ready to put all of the pieces together. The experimental value was given to us in units of $\text{cm}^3\text{ molecule}^{-1}\text{ s}^{-1}$. *We can convert the equation for* A_{ct} *to units of* $\text{m}^3\text{ molecule}^{-1}\text{ s}^{-1}$ *by dividing by Avogadro's number. This gives*

$$\begin{aligned} A_{\text{ct}} &= \sigma\langle u_{\text{rel}}\rangle \\ &= (2.15 \times 10^{-19}\text{ m}^2)(2369\text{ m s}^{-1}) \\ &= 5.10 \times 10^{-16}\text{ m}^3\text{ molecule}^{-1}\text{ s}^{-1} \\ &\equiv (5.10 \times 10^{-16}\text{ m}^3\text{ molecule}^{-1}\text{ s}^{-1})(100\text{ cm m}^{-1})^3 \\ &= 5.10 \times 10^{-10}\text{ cm}^3\text{ molecule}^{-1}\text{ s}^{-1}. \end{aligned}$$

This represents a very reasonable agreement with the experimental value given the uncertainty in the appropriate values to use for the radii. We could reasonably conclude from this calculation that the $O + N_2$ *reaction is largely collision limited.*

'Forward' calculations of pre-exponential factors using simple collision theory are rare. More commonly, simple collision theory is used to obtain an effective collision cross-section. We will see an example of this procedure shortly.

First, though, we need to discuss an important special case. When we worked out the number of collisions of an A molecule with the B's, it was implicitly assumed that A and B were different types of molecules. We counted how many times a single molecule of A would collide with molecules of B and then multiplied by the number of A molecules to get the total number of collisions. However, for an A + A reaction, this would count each collision twice. To correct for this, we just have to divide by two:

$$A_{\text{ct}}^{\text{A+A}} = \frac{1}{2}\sigma\langle u_{\text{rel}}\rangle L.$$

Another useful fact is that if the two molecules have identical mass, the reduced mass (equation (2.9)) simplifies to $\mu = m_A/2$.

Example 3.2. *The reaction* $2HI_{(g)} \rightarrow H_{2(g)} + I_{2(g)}$ *has a pre-exponential factor of* 10^{11} L mol^{-1}s^{-1} *for temperatures near 500 K. Using these data, we will calculate a collision cross-section.*

$$\begin{aligned}
\mu_m &= M_{\text{HI}}/2 = (127.912\text{ g mol}^{-1})/2 \\
&= 63.956\text{ g mol}^{-1} \equiv 0.063\,956\text{ kg mol}^{-1}.
\end{aligned}$$

$$\langle u_{\text{rel}}\rangle = \sqrt{\frac{8(8.314\,462\,618\text{ J K}^{-1}\text{mol}^{-1})(500\text{ K})}{\pi(0.063\,956\text{ kg mol}^{-1})}} = 407\text{ m s}^{-1}$$

$$A = \frac{10^{11}\text{ L mol}^{-1}\text{s}^{-1}}{1000\text{ L m}^{-3}} = 10^{8}\text{ m}^3\text{ mol}^{-1}\text{s}^{-1}$$

$$\begin{aligned}
\sigma &= \frac{2A}{\langle u_{\text{rel}}\rangle L} \qquad \text{(A + A reaction)} \\
&= \frac{2(10^{8}\text{ m}^3\text{ mol}^{-1}\text{s}^{-1})}{(407\text{ m s}^{-1})(6.022\,140\,76\times 10^{23}\text{ mol}^{-1})} \\
&= 8\times 10^{-19}\text{ m}^2.
\end{aligned}$$

I find it difficult to interpret areas on a molecular scale, so, to help me think about these quantities, I usually calculate the radius of the cross-section:

$$\begin{aligned}
r_{\text{AB}} &= \sqrt{\sigma/\pi} \\
&= \sqrt{8\times 10^{-19}\text{ m}^2/\pi} \\
&= 5\times 10^{-10}\text{ m} \equiv 5\text{ Å}.
\end{aligned}$$

This is very large. For comparison, the bond length in HI is only 1.61 Å. This large cross-section suggests that some kind of long-distance interaction between molecules of HI results in an elevated rate of reaction. Having done this calculation, one could then think about the possible physical origins of the large reaction cross-section.

An interesting feature of most theories of reaction kinetics is that they predict a temperature-dependent pre-exponential factor. In the case of simple collision theory, the temperature dependence enters through the relative speed, equation (2.11). You may have carried out some experiments in which you found beautifully straight Arrhenius plots. How is that possible if A is temperature dependent? Let us say for the sake of argument that you carried out a set of experiments over the temperature range 300–500 K. Since $\langle u_{\text{rel}}\rangle$ depends on the square root of T, the relative speed changes by a factor of 1.3 over this range. Typical activation energies are in the range of a few tens to a hundred kJ mol^{-1} or so. (Reactions with much larger activation energies are usually too slow to study using standard experimental methods.) Suppose that the activation energy of a reaction is about 50 kJ mol^{-1}. Over the assumed temperature range, the Arrhenius factor changes by a factor of about 3000. The small variation with temperature of the pre-exponential factor is just not detectable against the very large variation in the Boltzmann–Arrhenius factor over moderately large temperature ranges.

3.2 Molecular beam experiments

Simple collision theory can be generalized to consider, among other things, the energy dependence of the collision cross-section and the angular distribution of products, as well as the energy distributions of products. In order to understand the theory, it helps to think about experiments that probe reactions at this level of detail.

First, though, we have to establish some notation relating to spherical coordinates, which are used to describe these experiments. Figure 3.3 shows the definitions of the spherical coordinates r, θ, and ϕ. Any point in three-dimensional space can be designated by these coordinates. In order to have a unique set of coordinates for each point, we limit the angles to the ranges $\theta \in [0, \pi]$ and $\phi \in [0, 2\pi)$.

In the experiments I will describe shortly, the products travel from the point of reaction to a detector. Imagine a sphere surrounding the point of reaction whose radius is equal to the distance from this point to the detector. The detector has a physical size, and we somehow need to quantify what fraction of the sphere it occupies, i.e. what fraction of the straight-line flight paths from the point of reaction intersect the detector. Let's start by thinking about the (hopefully familiar) two-dimensional analog. Consider figure 3.4(a). The angle θ can be defined as the ratio of the length of an arc of a circle to the radius. The angle so defined has units of radians. The full circle corresponds to an angle of 2π. Radians are dimensionless, since they are a ratio of two lengths. In units of physical quantities, we typically leave them out for this reason.

Now turn to figure 3.4(b). We can define a **solid angle**, typically denoted by Ω, as the ratio of the area bounded by a circle drawn on the surface of a sphere to r^2, the square of the radius. This solid angle has units of **steradians**, again a dimensionless measure. The entire sphere has a solid-angle measure of 4π. There is no necessity for the area on the

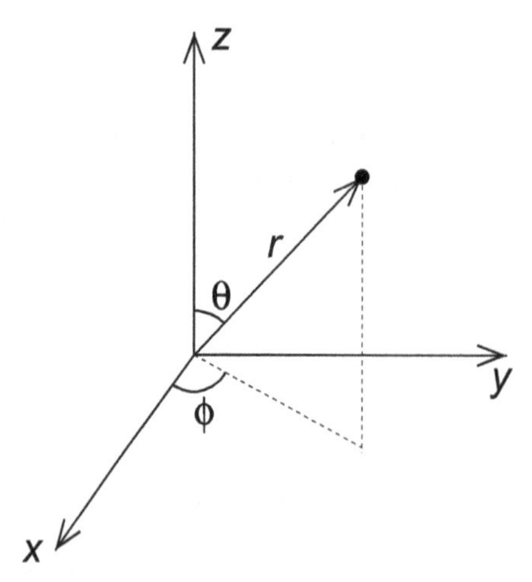

Figure 3.3. The definition of the spherical coordinate system. The coordinates of a point in three-dimensional space (the filled dot) can be defined by the familiar Cartesian (x, y, z) coordinates or by the spherical coordinates (r, θ, ϕ).

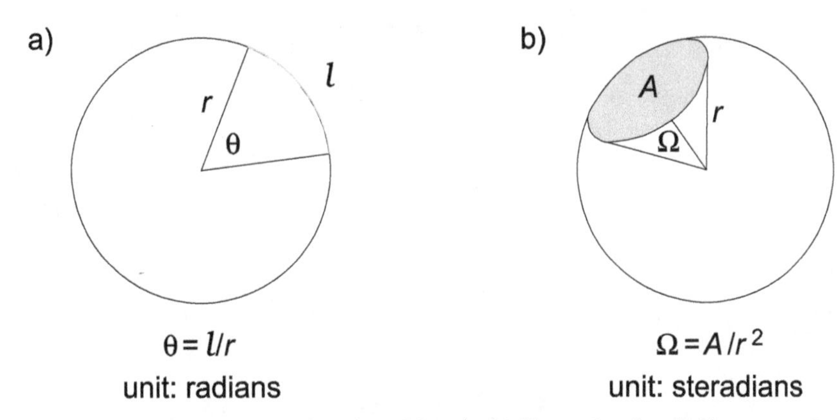

Figure 3.4. An analogy between an angle and a solid angle. (a) The angle subtended by an arc of a circle of length l is the ratio of the length of this arc to the radius of the circle, yielding an angle in radians. (b) The solid angle subtended by a circle drawn on the surface of a sphere of area A is the ratio of this area to the square of the radius, yielding a solid angle in steradians.

sphere to be circumscribed by a circle, although that makes it a bit easier to visualize the meaning of the solid angle. The solid angle of a region of any shape on the surface of a sphere is always given by $\Omega = A/r^2$, where A is the area of the region.

We will sometimes refer to an infinitesimal solid-angle element $d\Omega$. In explicit form, this is

$$d\Omega = \sin\theta \; d\theta d\phi.$$

If we want to study reactions in the kind of detail described above, we have to carefully control where reactions occur and the states of the reactants, especially the kinetic energy but possibly other characteristics as well, such as the vibrational and rotational energies. The fundamental technique in the field of experimental chemical dynamics is the crossed molecular beam experiment. There are many variations on this experiment, of which we will examine only the simplest example. Figure 3.5 shows a sketch of the basic idea. The experiment requires sources of the two reactant molecules, here named A and B. Typically, a collimator, which can be as simple as a pinhole, is used to narrow the molecular beam so that collisions between A and B molecules can only occur in a reaction zone at the intersection of the two beams. After colliding and reacting, the products can emerge at different angles. A movable detector is used to detect the products and, in most experiments, to characterize some of their properties. The figure only shows the in-plane polar angle ϕ, but the detector can usually also be placed at different angles θ. Accordingly, this experiment can provide a fully angle-resolved product distribution.

So how do we make a molecular beam? One simple technique is to boil a sample, heat the gas to a desired temperature, then let the gas escape through a pinhole. This is useful for reactions that are slow at lower temperatures and thus require hot reactants. Another option is to allow a high-pressure gas to escape through a nozzle. Adiabatic expansion of the gas as it exits the nozzle results in substantial cooling. (The same principle is used to make snow on ski hills.) This technique is useful if cold

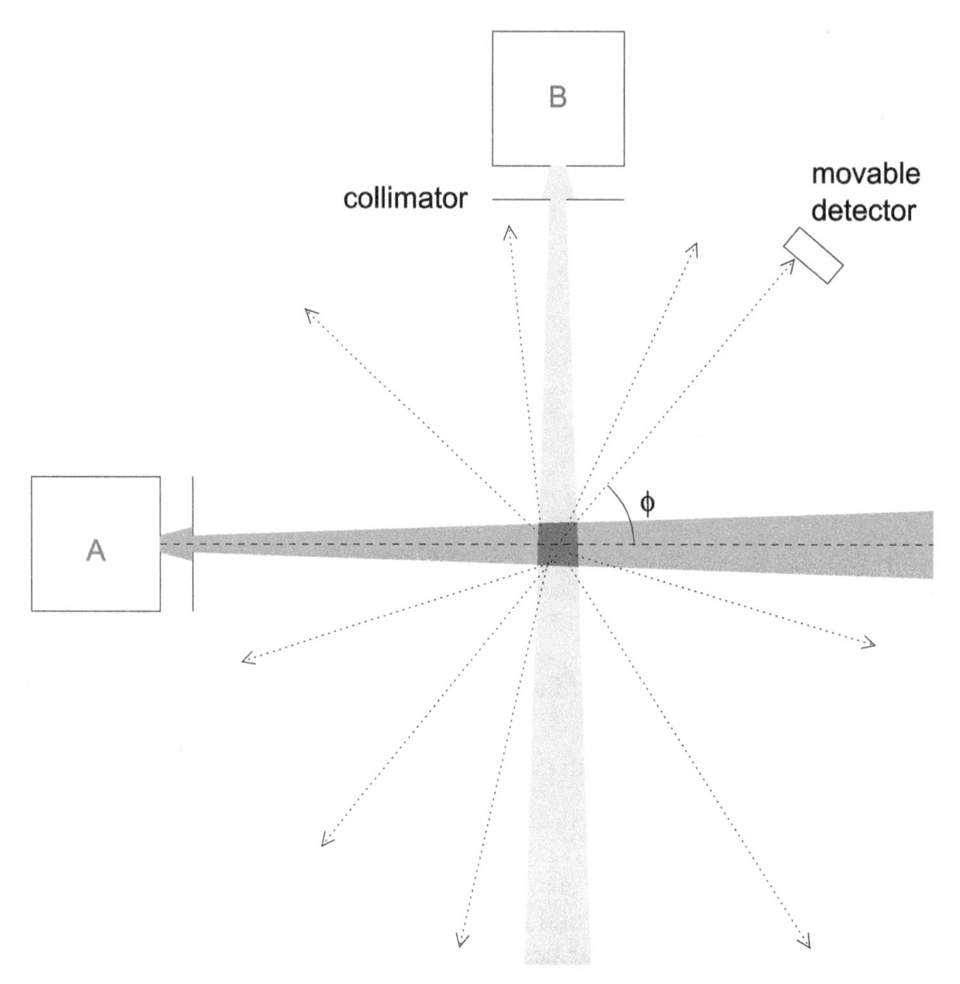

Figure 3.5. A basic molecular beam experiment. Sources of A and B molecules are directed towards a reaction zone after passing through a collimator. The paths of the reaction products are shown as dotted arrows. A movable detector provides data about the reaction products emerging in a particular direction.

reactants are needed. In particular, because of the typical vibrational energy spacings in molecules, molecular beams prepared this way would tend to have almost all of the molecules in their ground vibrational states.

The 'hot' or 'cold' molecular beams produced by these techniques don't give you fine control over molecular kinetic energy. You would still have a Maxwell–Boltzmann distribution of speeds. For this reason, a velocity selector is often added to the apparatus. A simple velocity selector can be constructed from a pair of disks, each of which has a narrow slit that lets molecules pass through when aligned with the incoming molecular beam (figure 3.6). Molecules traveling at speed u travel from one disc to the other in time $t = d/u$. During that time, the disks turn by $t\nu = d\nu/u$ revolutions, where ν is the frequency (in Hz). In order for molecules to pass through the second disc, it has to have turned a whole number of revolutions in this time, so

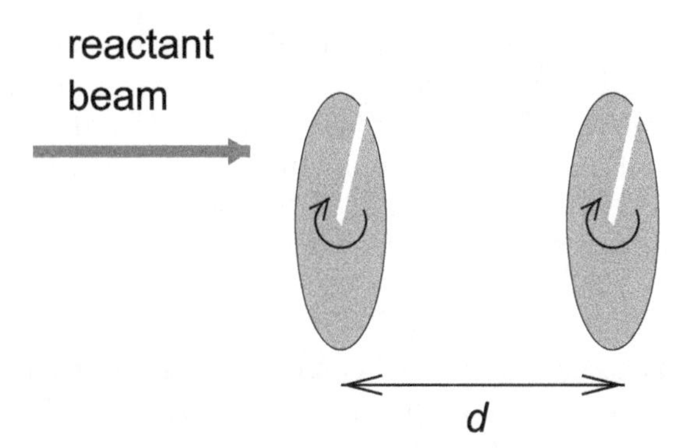

Figure 3.6. A spinning-disc velocity selector. Two disks separated by a distance d spin at a frequency ν. The only molecules that can pass through the second disc, having passed through the first one, are those traveling at a speed such that a whole number of revolutions have occurred as the molecules transit from one disc to the other.

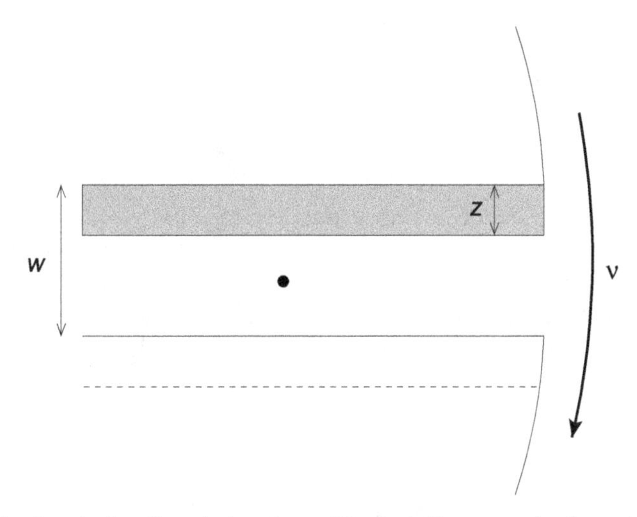

Figure 3.7. The slit of a spinning-disc velocity selector. The dot indicates a molecule passing through the slit. The slit has width w and thickness z. The disc is rotating at frequency ν.

for some integer n, we must have $n = d\nu/u$, which rearranges to $u = d\nu/n$. On the surface, it would seem that there is a flaw with this idea, which is that the rotating disc velocity selector does not select one speed but a harmonic sequence of speeds: $d\nu$, $d\nu/2$, $d\nu/3$, … However, the disks also have a thickness. Any molecules that travel too slowly won't make it all the way through the slit as it rotates. We therefore choose d and ν to select the speed $u = d\nu$. We then design the disks so that anything moving as slowly as $d\nu/2$ will 'crash out' as it moves through the slit. To understand this aspect of a velocity selector, we have to look more closely at the disc and its slit, illustrated in figure 3.7. If a molecule is to make it through the slit, it must travel the distance z (the thickness of the slit) before the disc has rotated far enough for the other side of the slit to sweep it away. The maximum time the molecule has to make it through is the time it takes for one edge of the slit to move a distance w.

This would correspond to a molecule entering the slit just as the bottom edge moves to let it in, and exiting just before the top edge hits it. If the molecular beam enters the slit at a distance r_b from the center of the disc[2], the edges at this distance from the center move at a speed $2\pi\nu r_b$. The maximum time the molecule has to travel through the slit width w is therefore $\Delta t = w/2\pi\nu r_b$. The distance traveled by the particle in this time must be at least z, i.e. $u\Delta t > z$, which rearranges to

$$u > 2\pi\nu r_b z/w. \tag{3.4}$$

Only molecules that exceed this speed will make it through our slit of thickness z. We want to make sure that we exclude any molecules traveling at $d\nu/2$ or slower. Thus, we must have

$$d\nu/2 < 2\pi\nu r_b z/w.$$
$$\therefore z > \frac{dw}{4\pi r_b}.$$

Of course, we also need the selected velocity, $d\nu$, to satisfy the inequality (3.4). Thus, the thickness of the disc must satisfy

$$\frac{dw}{4\pi r_b} < z < \frac{dw}{2\pi r_b}.$$

A single-slit rotating disc is a very crude design for a velocity selector. A quick back-of-the-envelope calculation should convince you that for reasonable molecular speeds (hundreds of meters per second), either the disks have to rotate unreasonably fast or the disks have to be very far apart, which makes it extremely difficult to line them up properly. A smarter idea would be to have multiple slits, and to select the rotation rate and angle between the slits so that the disc only has to rotate a fraction of a turn to line up the next slit. If the disc has n_s equally spaced slits, the velocity selection condition becomes $n/n_s = d\nu/u$, again with n a whole number, or $u = n_s d\nu/n$. To select speed $n_s d\nu$ and exclude speeds corresponding to higher values of n, we would now need disks whose thickness satisfies

$$\frac{n_s dw}{4\pi r_b} < z < \frac{n_s dw}{2\pi r_b}.$$

Note that the larger the value of n_s we choose, the more slowly we can rotate the disks, or the more we can shorten the distance between them. Slower rotation requires thicker disks. Within reason, this is a good thing, since thicker disks can better handle being rotated at high rates than thinner disks.

A multi-slit velocity selector has a further advantage, namely that we get a more intense molecular beam than with a single slit. All the molecules that arrive at the first disc between slits are simply lost, whether they had the correct speed or not. The more slits we have, the more molecules enter the slits and thus the more come out the other end of the device.

[2] In reality, there is a range of r_b values since the beam has a width.

Velocity selection is only one possible aspect of reactant preparation. It is also possible to use lasers to excite the reactants to any desired quantum state prior to reaction. The necessary lasers would intersect the molecular beam relatively close to the reaction zone so that the energy pumped into the molecules is unlikely to be lost through radiation or collisions prior to the reaction.

Having carefully selected the reaction conditions, we still have to detect the products. One attractive method for detecting the products is mass spectrometry, which allows us to identify the products formed in the reaction. Another useful technique is time-of-flight measurements, which allow us to determine the speed at which the products are traveling, and thus their kinetic energies. The simplest version of this technique uses a single spinning disc, much like the ones used in velocity selectors, and a detector. Only molecules that pass through the slit when this slit is aligned with the detector are observed. Knowing the distance from the disc to the detector and the time at which the slit was in the correct orientation allows us to calculate the speed at which the molecule was traveling.

Spectroscopic techniques are also widely used in molecular beam experiments. One key technique is infrared chemiluminescence. Once a reaction has occurred, the products are usually in an excited vibrational state. These excited products emit infrared photons as they shed their excess vibrational energy. These photons can be detected, and the distribution of vibrational states of the products can thus be inferred.

Another useful technique is laser-induced fluorescence. In this technique, a laser tuned to a gap between two energy levels is used to excite the product molecules. Fluorescence occurs by the emission of photons after the loss of some vibrational energy. The fluorescent photons are therefore of slightly lower energy (longer wavelength) than the laser photons. The number of fluorescent photons emitted is directly proportional to the number of laser photons absorbed. These laser photons can only be absorbed if the lower state of the two chosen initially was populated. The intensity of the fluorescence is therefore a readout of the population of the lower state after formation of the products, so again we get information on the distribution of quantum states of the products.

3.3 Reactive scattering

The molecular beam experiment allows us to study the angular distribution of products and their kinetic energies and quantum states. We now need a theory we can use to integrate the information gained from these experiments. This theory is called reactive scattering theory. In the physical sciences, 'scattering' refers to the emergence of particles at an angle to their original direction of travel after a particle–particle interaction. Reactive scattering is therefore the theory of reactions in which the products fly away from the site of reaction in different directions.

Figure 3.8 shows the essential elements of the geometry of a typical molecular beam experiment that could be used to study an A + B → C reaction. We have two molecular beams that intersect in a small region of space, and a small detector subtending a solid angle $d\nu$ that can be positioned at different (θ, ϕ) angles in the

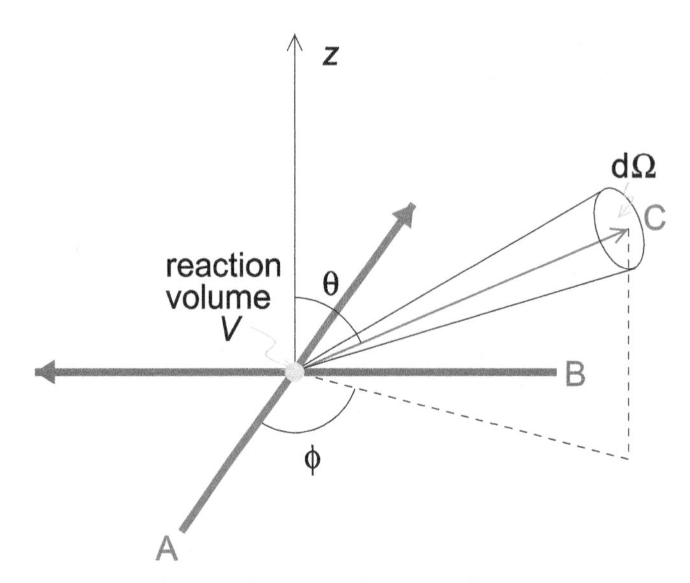

Figure 3.8. The geometry of the molecular beam experiment for an A + B → C reaction. In this case, the two reactants are shown as arriving along the x (A) and y (B) axes. In some experiments, the angle between the two molecular beams can be changed. The product, C, is observed using a detector that subtends a solid angle $d\Omega$.

coordinate system centered on the reaction zone. The data from the experiment, in its simplest form, is a count of molecules of the product C arriving at the detector over some span of time.

Recall equation (3.2), which gives the collision rate from simple collision theory. We want to write down a similar equation that takes into account the dependence of the reaction rate on the relative speed of the reactants and that differentiates the different directions in which the product can travel. The derivation below is heuristic rather than rigorous, but it allows us to focus on the key ideas rather than getting bogged down in mathematics. We define the **differential cross-section** $I_R(\theta, \phi, u_{rel})$ such that $I_R(\theta, \phi, u_{rel})d\Omega$ is the cross-section if the reactants have relative speed u_{rel} counting only events in which the product C emerges from the reaction zone within a range of angles such that it will impact our detector of solid angle $d\Omega$. The reactants have a range of relative speeds governed by the Maxwell–Boltzmann distribution, equation (2.10). By analogy to equation (3.2), the rate of reaction due to reactants with relative speeds within du_{rel} of u_{rel} and with product flying out of the reaction zone toward the detector is

$$dv = I_R(\theta, \phi, u_{rel})d\Omega\, u_{rel}\, p(u_{rel})du_{rel}L[A][B].$$

This equation gives the relationship of the angle-resolved measurements to the theory, telling us the expected rate of event detections due to collisions with relative speed u_{rel} that will be observed at angles (θ, ϕ). From the measurements, the differential cross-section can then be inferred.

Note that the ground has shifted relative to our discussion of simple collision theory: we are now considering not only collisions but also reactions, since our expression for the differential of the rate, dv, includes consideration of the product formed.

To get the rate of reaction from this last equation, we integrate with respect to the angles and u_{rel}:

$$v = L[\mathrm{A}][\mathrm{B}] \int \int \int I_{\mathrm{R}}(\theta, \phi, u_{\mathrm{rel}}) \mathrm{d}\Omega u_{\mathrm{rel}}\, p(u_{\mathrm{rel}}) \mathrm{d}u_{\mathrm{rel}}.$$

Since $v = k[\mathrm{A}][\mathrm{B}]$, we immediately get an equation for the rate constant (and not just the pre-exponential factor):

$$k = L \int \int \int I_{\mathrm{R}}(\theta, \phi, u_{\mathrm{rel}}) \mathrm{d}\Omega u_{\mathrm{rel}}\, p(u_{\mathrm{rel}}) \mathrm{d}u_{\mathrm{rel}}.$$

Substituting the explicit form of $\mathrm{d}\Omega$ into this equation, we get

$$k = L \int_0^{\infty} \int_0^{2\pi} \int_0^{\pi} I_{\mathrm{R}}(\theta, \phi, u_{\mathrm{rel}}) u_{\mathrm{rel}}\, p(u_{\mathrm{rel}}) \sin\theta\ \mathrm{d}\theta\ \mathrm{d}\phi\ \mathrm{d}u_{\mathrm{rel}}.$$

Only $I_{\mathrm{R}}(\theta, \phi, u_{\mathrm{rel}})$ depends on the angles. If we define the **total cross-section** at speed u_{rel} as

$$\sigma_{\mathrm{R}}(u_{\mathrm{rel}}) = \int_0^{2\pi} \int_0^{\pi} I_{\mathrm{R}}(\theta, \phi, u_{\mathrm{rel}}) \sin\theta\ \mathrm{d}\theta\ \mathrm{d}\phi,$$

we can write the rate constant as

$$k = L \int_0^{\infty} \sigma_{\mathrm{R}}(u_{\mathrm{rel}}) u_{\mathrm{rel}}\, p(u_{\mathrm{rel}}) \mathrm{d}u_{\mathrm{rel}}.$$

Substituting in the Maxwell–Boltzmann distribution for the relative speed, we get, after some minor rearrangement,

$$k = 4\pi L \left(\frac{\mu}{2\pi k_{\mathrm{B}} T}\right)^{3/2} \int_0^{\infty} \sigma_{\mathrm{R}}(u_{\mathrm{rel}}) u_{\mathrm{rel}}^2 \exp\left(\frac{-\mu u_{\mathrm{rel}}^2}{2k_{\mathrm{B}}T}\right) u_{\mathrm{rel}}\ \mathrm{d}u_{\mathrm{rel}}.$$

The relative kinetic energy of a collision is given by

$$\varepsilon_k = \frac{1}{2}\mu u_{\mathrm{rel}}^2.$$

$$\therefore\ \mathrm{d}\varepsilon_k = \mu u_{\mathrm{rel}}\ \mathrm{d}u_{\mathrm{rel}}.$$

We can therefore transform our expression for the rate constant into an integral over the relative kinetic energy:

$$k = \frac{L}{k_{\mathrm{B}}T}\sqrt{\frac{8}{\pi \mu k_{\mathrm{B}} T}} \int_0^{\infty} \varepsilon_{\mathrm{k}} \sigma_{\mathrm{R}}(\varepsilon_{\mathrm{k}}) \mathrm{e}^{-\varepsilon_{\mathrm{k}}/k_{\mathrm{B}}T} \mathrm{d}\varepsilon_{\mathrm{k}}. \quad (3.5)$$

The last equation now tells us how the rate constant is related to the energetics of the reaction through the dependence on the total cross-section, which is itself assumed to depend on the energy of the collision. This makes sense: we don't expect the probability of reaction to be the same regardless of the energies of the reactants. The Arrhenius equation already implies that some minimum amount of energy is required in order for a reaction to occur. Equation (3.5) now tells us that the energetic requirement may be more complicated than a simple threshold.

Let's consider some special cases of equation (3.5). Suppose that the energetic requirement for a reaction is a simple cutoff, as suggested by our derivation of the Arrhenius equation in section 2.5: the reaction can occur if the kinetic energy exceeds the activation energy, and not otherwise. In this case, σ_R would be

$$\sigma_R(\varepsilon_k) = \begin{cases} 0 & \text{for} \quad \varepsilon_k < \varepsilon_a, \\ \sigma & \text{for} \quad \varepsilon_k \geqslant \varepsilon_a. \end{cases}$$

In words, this says that there are no reactive events if the energy is below ε_a, and that the reactive cross-section doesn't depend on the energy otherwise. Thus,

$$\begin{aligned} k &= \sigma \frac{L}{k_B T} \sqrt{\frac{8}{\pi \mu k_B T}} \int_{\varepsilon_a}^{\infty} \varepsilon_k e^{-\varepsilon_k / k_B T} d\varepsilon_k \\ &= \sigma L \sqrt{\frac{8}{\pi \mu k_B T}} (k_B T + \varepsilon_a) e^{-\varepsilon_a / k_B T}. \end{aligned}$$

Note that we get something that looks like an Arrhenius equation, with a pre-exponential factor of

$$\begin{aligned} A &= \sigma L \sqrt{\frac{8}{\pi \mu k_B T}} (k_B T + \varepsilon_a) \\ &= \sigma L \sqrt{\frac{8 k_B T}{\pi \mu}} \left(1 + \frac{\varepsilon_a}{k_B T}\right) \\ &= \sigma \langle u_{\text{rel}} \rangle L \left(1 + \frac{\varepsilon_a}{k_B T}\right). \end{aligned}$$

Compare this with the simple collision theory result, equation (3.3). The reactive scattering theory with the assumption of an abrupt energy cutoff has given us a very similar result to simple collision theory, but with a correction that becomes smaller as T increases.

A hard cutoff of the total cross-section at the activation energy is probably not realistic. Instead, it is likely that the cross-section goes to zero smoothly as the kinetic energy approaches the activation energy. Thus, consider the following form for $\sigma_R(\varepsilon_k)$:

$$\sigma_R(\varepsilon_k) = \begin{cases} 0 & \text{if} \quad \varepsilon_k < \varepsilon_a, \\ \sigma^*(1 - \varepsilon_a/\varepsilon_k) & \text{if} \quad \varepsilon_k \geqslant \varepsilon_a. \end{cases} \tag{3.6}$$

If you have trouble picturing this function, it is shown in figure 3.9. Calculating k for this total cross-section, we have

$$\begin{aligned} k &= \frac{L}{k_B T} \sqrt{\frac{8}{\pi \mu k_B T}} \int_{\varepsilon_a}^{\infty} \varepsilon_k \sigma^*(1 - \varepsilon_a/\varepsilon_k) e^{-\varepsilon_k / k_B T} d\varepsilon_k \\ &= \sigma^* L \sqrt{\frac{8 k_B T}{\pi \mu}} e^{-\varepsilon_a / k_B T} \\ &= \sigma \langle u_{\text{rel}} \rangle L e^{-\varepsilon_a / k_B T}. \end{aligned}$$

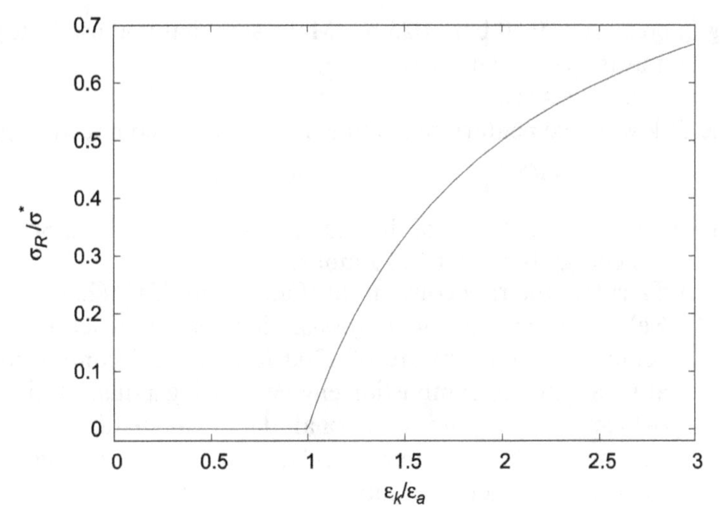

Figure 3.9. The total cross-section defined by equation (3.6). $\sigma_R/\sigma^* \to 1$ as $\varepsilon_k/\varepsilon_a \to \infty$.

This is *exactly* the result from simple collision theory! The two examples we have just seen reinforce Arrhenius' original insight that the activation energy is a threshold needed for a reaction to occur.

Reactive scattering theory can give us a variety of results, including the simple collision theory limit, depending on the form assumed for the total cross-section. Experimental chemical dynamics allows us to measure these cross-sections by controlling the kinetic energies of the reactants and observing the rate of formation of products. These experimental results then provide the raw data that allow us to verify various theories of chemical kinetics that predict cross-sections.

Exercise

3.1 As an intermediate result in deriving the equations of simple collision theory, we found that each molecule of A undergoes[3] $Z_A = \sigma\langle u_{rel}\rangle L[B]$ collisions per second with molecules of [B]. The mean free path is the distance traveled between collisions, which we can express as the distance an A molecule travels per second relative to the B molecules divided by the number of collisions per second, i.e.

$$l_{free} = \langle u_{rel}\rangle / Z_A.$$

The average time between collisions is just the inverse of the collision frequency: $\tau_{coll} = 1/Z_A$. Calculate the average time between collisions and the mean free path for molecules of $^{35}Cl_2$ colliding with ^{40}Ar atoms if the

[3] We actually had u_A rather than $\langle u_{rel}\rangle$ at this stage of the derivation and made the substitution for the relative velocity later, but we could have made this replacement here. Also, the formula in the text has n_B/V instead of [B].

argon pressure is 0.50 bar at 25 °C. Make sure that you use isotopic masses and not averaged molar masses.

Hint: ideal gas law!

3.2 The following elementary reaction is important in combustion processes:

$$\cdot OOH_{(g)} + \cdot OH_{(g)} \rightarrow H_2O_{(g)} + O_{2(g)}.$$

The pre-exponential factor of this reaction is 5.01×10^{10} L mol^{-1} s^{-1} and the activation energy is $E_a = 4.18$ kJ mol^{-1}.

(a) Calculate the rate constant at 'Fahrenheit 451'[4] (233 °C).

(b) Calculate the collisional cross-section for this reaction, assuming a combustion temperature of 2200 K (a typical combustion temperature in internal combustion engines). Using a quantitative argument (which may require additional data, calculations, or estimates) determine whether your calculated cross-section suggests a collision-limited reaction or not.

3.3

(a) The reaction

$$K_{(g)} + Br_{2(g)} \rightarrow KBr_{(g)} + Br_{(g)}$$

has an activation energy of zero so that the rate constant is equal to the pre-exponential factor. At 600 K, the rate constant is approximately 10^{12} L mol^{-1} s^{-1}. Assuming that the reaction occurs at the collision-limited rate, what is the collision cross-section? What does the size of the cross-section tell us about this reaction?

(b) We can estimate the change in energy in this reaction from the bond energies, which are as follows:

Species	Bond dissociation energy/kJ mol^{-1}
$Br_{2(g)}$	190.33
$KBr_{(g)}$	378.46

Using these data and information from part (a), sketch the reaction profile for this reaction, i.e. a graph of energy vs reaction coordinate.

3.4 We saw earlier that reactive scattering theory 'magically' gives the simple collision-theory rate constant if the cross-section takes the form of equation (3.6). You might have wondered whether this result is particular to this form of the total reactive cross-section. (It's the kind of thing that theoreticians

[4] For the Ray Bradbury fans.

stare up at the ceiling thinking about late at night.) Let's try a different form for σ_R, namely

$$\sigma_R(\varepsilon_k) = \begin{cases} 0 & \text{for} \quad \varepsilon_k < \varepsilon_a, \\ \sigma^* \arctan\left(\dfrac{\varepsilon_k - \varepsilon_a}{k_B T}\right) & \text{for} \quad \varepsilon_k \geqslant \varepsilon_a. \end{cases}$$

As in the case studied earlier, this reactive cross-section is a continuous function of ε_k, going to zero as $\varepsilon_k \to \varepsilon_a^+$.

Calculate the rate constant using reactive scattering theory for this 'arctan' reactive cross-section. Do you get a rate constant similar in form to that from simple collision theory?

The integrals you need cannot be evaluated analytically, but they can be numerically integrated[5]. Here are some useful integrals:

$$\int_0^\infty \arctan(x)e^{-x}dx = 0.6214.$$

$$\int_0^\infty x \arctan(x)e^{-x}dx = 0.9648.$$

The key to this problem is making a substitution such that you end up with an $\arctan(x)$ term in the integrand.

3.5 Suppose that we want to design a velocity selector that will be able to select speeds in the range 100–800 m s^{-1}. The beam will pass through the disks at a distance of 3.5 cm from the center. We have a motor that can operate reliably at rotational speeds of up to 400 Hz. The disks should not be thinner than 0.1 mm for mechanical stability. The disks should not be mounted more than 10 cm from each other, and they should be at least 1 cm apart. Provide a full set of design parameters for this velocity selector: the number of slits, the distance between disks, the thicknesses of the disks, and the widths of the slits.

3.6 The theory of reactive scattering studied in class is quite general and can be applied to a variety of rate processes. For example, it has been applied to processes involving neutron impacts with nuclei (neutron capture, induced fission, etc).

(a) The radius of the neutron is not directly relevant in estimating the cross-section. Because of the quantum mechanical effects associated with small particles such as neutrons, the de Broglie wavelength ($\lambda = h/p$) is a better estimate of the 'extent' of a neutron in space. Starting from the classical expression for the kinetic energy, obtain an equation relating the momentum to the kinetic energy, and then give the de Broglie wavelength as a function of the kinetic energy of the neutron.

[5] There are actually 'analytic' solutions, but they involve strange functions such as the so-called cosine integral, $\mathrm{Ci}(x) = -\int_x^\infty \cos t \, dt/t$.

(b) Estimate the dependence of the cross-section σ_R on the kinetic energy of the neutron using the expression for the cross-section found in simple collision theory, assuming a nucleus of radius R and a neutron of effective radius equal to its de Broglie wavelength.

(c) Typically, $m_n \ll m_2$, where m_n is the mass of the neutron and m_2 is the mass of the nucleus. What limit does the reduced mass approach in this case? Use this limit in subsequent calculations.

(d) Obtain an equation for the rate constant.

Note: If you have access to a symbolic algebra system, you may want to use it.

(e) ^{10}B is used in an experimental cancer treatment called neutron capture therapy: the patient is administered a drug containing this isotope that is preferentially taken up by the tumor. The tumor is then bombarded with thermal neutrons (neutrons whose kinetic energies obey a Maxwell–Boltzmann distribution with a T similar to room temperature). The neutrons are absorbed, triggering the following nuclear reaction:

$$^{10}_{5}\mathrm{B} + {}^{1}_{0}\mathrm{n} \rightarrow {}^{11}_{5}\mathrm{B} \rightarrow {}^{7}_{3}\mathrm{Li} + {}^{4}_{2}\alpha + \gamma.$$

Alpha particles are a particularly lethal form of radiation due to their mass. On the other hand, their large size prevents this radiation from getting very far, i.e. it tends to mostly damage the tumor where the drug is located and where the neutron irradiation has been applied. In this reaction, the lithium ion also gains a large kinetic energy and acts much like an alpha particle.

Based on the theory developed above, estimate the rate constant at 37 °C for neutron capture by ^{10}B. The radius of a ^{10}B nucleus is approximately 2.7 fm.

(f) The expression you developed for the rate constant behaves oddly at small and large temperatures. Describe the odd behavior. Why does this odd behavior occur? In other words, what assumption(s) made in our calculation are causing the odd behavior?

Reference

[1] Mantina M, Chamberlin A C, Valero R, Cramer C J and Truhlar D G 2009 Consistent van der Waals radii for the whole main group *J. Phys. Chem. A* **113** 5806–12

IOP Publishing

Foundations of Chemical Kinetics
A hands-on approach
Marc R Roussel

Chapter 4

Molecular properties from Gaussian calculations

Given reactants and products, can we predict a rate constant *ab initio*, i.e. from first principles, without the need for experimental data? This is the grand challenge of theoretical chemical kinetics. In the previous chapter, we obtained an equation for the rate constant that depends on the total cross-section at relative kinetic energy ε_{k} (equation (3.5)). One possible approach is to try to determine $\sigma_{\mathrm{R}}(\varepsilon_{\mathrm{k}})$ and then to compute the rate constant from this quantity. This involves computing the relative motions of atoms during a reaction at different kinetic energies and with different initial molecular positions and velocities. The subfield of kinetics in which such calculations are performed is called computational reaction dynamics. We will briefly mention this approach in the next chapter, but this is the domain of experts and would require its own book in order to pursue the hands-on approach I am aiming for in this book. For most readers, it would also pose significant problems from the point of view of acquiring sufficient computational resources to actually do the calculations. A much more accessible theory, transition-state theory (TST), also requires us to have access to some quantum mechanical details of a reaction. The data we need for TST calculations can in principle be obtained from experiments. However, transition-state spectroscopy, which is required to get properties of the transition state, is not a routine set of techniques, so the necessary data are in fact rarely available from experiments. Computational chemistry will fill this gap. I also happen to think that it's useful for a chemist to have a bit of basic computational chemistry in their bag of tricks.

Since this is not a textbook of computational chemistry, I'm going to be very focused on some basic skills. The calculations we will do will be helpful to illustrate the theory, but the state of the art is constantly changing, so if your ambition is to do publishable research in this area, you will need to study computational chemistry in greater depth than is possible here. Still, you will hopefully develop some sense of how the various pieces fit together, from computational chemistry to the calculation of a rate constant using TST.

doi:10.1088/978-0-7503-5321-2ch4

Gaussian [1] is the most popular computational chemistry program available today. There are lots of others, some free, some not. (Gaussian falls into the latter category.) The reason for using Gaussian in this book is its widespread availability on university campuses, but of course any modern computational chemistry software would do. I will also be focusing on operations using the graphical front end GaussView [2] to make the experience as painless as possible.

But first, a bit of background in computational chemistry ···

4.1 An overview of important ideas in computational chemistry

We will start out with a very rapid overview of concepts, many of which should have been encountered in previous courses.

Electrons, as you know, are much lighter than the nuclei of atoms. This allows a very important approximation called the **Born–Oppenheimer approximation** to be obtained. The very light electrons have much higher speeds than the heavier nuclei. From the point of view of solving the quantum-mechanical problem for the electrons, the nuclei move so slowly that they can be treated as quasi-static, in the sense that we can solve the electronic problem at fixed nuclear positions. One way to think about this is that the electronic wavefunction adjusts to the positions of the nuclei much faster than the nuclei move. We will see below that it is often useful to map out the energy of a molecule (or group of molecules) as a function of nuclear coordinates, but this can only be done by taking a set of 'snapshots' at different nuclear positions, a bit like stop-motion animation.

4.1.1 Molecular orbitals

You will recall that electrons occupy **molecular orbitals** (MOs), which are single-electron wavefunctions, either singly or in opposite-spin pairs. How can we represent an MO? A frequently used motif in applied mathematics is to expand a function in a set of **basis functions**. Although this may not be immediately obvious, you may actually have used this idea in your calculus courses, where it appears in the form of Taylor series expansions. In a Taylor series, we write

$$f(x) = a_0 + a_1x + a_2x^2 + \cdots$$

Our basis functions are $\{1, x, x^2, \ldots\}$. Taylor's theorem says that we can expand any smooth function (any function whose derivatives are all continuous) in this **basis set** (the set of basis functions), and that if we include enough terms, we can make this approximation as accurate as we want (provided the function has bounded derivatives over the domain where it is being approximated). If we can represent any function of a certain class to arbitrary accuracy by a linear combination of basis functions, we say that the basis set is **complete**. The monomial basis set used in Taylor expansions is thus complete with respect to the set of smooth functions of a single variable.

In your previous courses, you probably ran into LCAO-MO theory. LCAO stands for 'linear combination of atomic orbitals.' The MOs of a molecule are *not* just linear combinations of the atomic orbitals of the constituent atoms. However, it is possible to build a complete basis set out of functions that look like atomic orbitals but allowing a parameter corresponding to the nuclear charge to assume different values.

Formally, we need an infinite number of such functions for a complete basis set. Of course, in practice we want to use a finite number of basis functions. The fact that the simplest version of LCAO-MO theory already gives reasonable answers to many questions about bonding and molecular structure using only the valence atomic orbitals suggests that terms arising from added basis functions must in most cases be appearing as corrections rather than as dominant terms, so that we should be able to carry out accurate calculations with a finite set of basis functions.

There is another issue, which is the computational inefficiency[1] of the correct atomic orbitals, which include an $e^{-\zeta r/r_0}$ dependence on the distance from the nucleus, where ζ is an effective nuclear charge[2] and r_0 is a scale factor (typically the Bohr radius). To deal with this problem, we switch to a basis set that approximates the MOs but has a Gaussian dependence of $e^{-\zeta(r/r_0)^2}$ on the distance from the nucleus. Since Gaussians have a different shape than exponentials, we use several Gaussians to approximate one exponential. The advantage of using Gaussians, even if we have to use many more of them than of exact atomic orbitals, is that integrals involving Gaussians can be evaluated extremely quickly due to some nice properties of these functions.

So now suppose that our basis set of atomic orbitals approximated by Gaussians, $\{\chi_j\}$, is used to represent a molecular orbital, ϕ_i:

$$\phi_i = \sum_j c_{ij}\chi_j.$$

How do we pick the coefficients c_{ij}? We start by guessing the coefficients c_{ij} for every basis function used in every MO. We can calculate what the electronic energy would be if these guessed coefficients were in fact the correct coefficients of the MOs. We then use a key theorem called the **variational theorem** which, roughly stated, says that the best wavefunction is the one that minimizes the energy. This tells us that we just have to minimize the energy with respect to the coefficients c_{ij} to get the best representation of the wavefunction for the chosen basis set. Carrying out this minimization is the key procedure at the center of software that calculates MOs, such as Gaussian. If we want to get a more accurate solution, all we have to do is to expand the basis set, i.e. use more basis functions.

There is an art to choosing basis sets. What we want to do is to use the smallest number of basis functions that give us a sufficiently accurate solution to answer the particular question we are tackling. In most problems, our first priority would be to add functions that are particularly effective at representing chemical bonds and perhaps bond-making or bond-breaking events, which are often important in kinetic mechanisms.

The pure atomic problem is spherically symmetrical, but chemical bonds are directed in space. An efficient way to enhance the ability of a basis set to represent

[1] There are nevertheless basis sets that include the correct exponential dependence. These are known as Slater-type orbitals (STOs). Pure Slater-type orbitals are rarely used. Rather, STOs are approximated by linear combinations of Gaussian functions.

[2] The Greek letter ζ is pronounced 'zeta.'

the directionality of chemical bonds is to add some nonphysical orbitals with higher angular momentum than are available in the valence shell. For example, we can add p orbitals to hydrogen atoms and d orbitals to main-group elements from the second period down. These are called **polarization functions**. These added orbitals will use values of ζ chosen to give good results for a variety of bonding situations and not values that are directly related to the actual atomic charge. Adding polarization functions to a basis set is the most economical means of improving the accuracy of calculations and is therefore highly recommended.

Another issue has to do with long-range interactions, which are poorly modeled using basis sets tuned to model chemical bonds. This problem is addressed by adding **diffuse functions** to the basis set. Diffuse functions are unphysical functions that look like atomic orbitals but with a very low value of ζ. These functions therefore extend far into space. These are particularly important in modeling bond-making and bond-breaking events.

Basis sets are generally grouped into families based on their original designer as well as on the general philosophy for choosing basis functions. One of the earliest rationally designed families of basis sets is the Pople family. Pople basis sets have names like '6-31G' and '6-311G**.' The numerical part of the name tells us how many Gaussians are used to represent various atomic orbitals. The number before the hyphen is the number of Gaussians used for each inner-shell orbital, while the numbers after the hyphen tell us both how many ζ values are used in the valence shell, one for each digit, and how many Gaussians are used for each value of ζ. A 6-31G basis set, for example, uses six Gaussians per core orbital. It is a **split-valence basis set**, i.e. two different values of ζ are used to represent the valence atomic orbitals. Three Gaussians are used to represent an orbital-like function for one value of ζ, and one Gaussian is used for the second value of ζ. The values of ζ are chosen during the construction of the basis set to approximate the atomic orbital as accurately as possible. The basic name is then 'decorated' to indicate whether polarization or diffuse functions have been added to the basis set. In computational chemistry, a 'heavy atom' is anything with more electrons than helium. The decoration '*,' placed after the 'G,' is used to indicate that polarization functions were added to the heavy atoms, while '**' means that polarization functions were also added to hydrogen. A slightly different notation is often used, with a 'd' in parentheses indicating that polarization functions are being added to the heavy atoms and a 'p' indicating that polarization functions are being added to hydrogen[3]. Thus, 6-31G** and 6-31G(d,p) mean the same thing. A '+' before the 'G' similarly indicates the addition of diffuse functions to the heavy atoms, while '++' indicates that diffuse functions were also added to the hydrogen atoms. A basis set can have both polarization and diffuse functions. Thus, we can have a 6-31+G** basis set, for example.

The Dunning basis set family always includes polarization functions, so these don't need to be added. Dunning basis sets have names like cc-pVDZ and

[3] I'm assuming main-group elements here. Other types of polarization function can be added, particularly in the transition metals.

aug-cc-pVTZ. For these basis sets, you need to look at the character following the 'V,' which encodes how many functions with different ζ values are provided in the valence shell. 'D' tells you that you have a double-zeta basis set. In other words, instead of having one basis function made to represent the 1s orbital in a hydrogen atom, two different 1s-like functions with different ζ values are provided. Similarly, all valence orbitals in heavy atoms are 'doubled up.' A 'T' after the 'V' would indicate a triple-zeta basis set, a 'Q' a quadruple-zeta basis set, and a digit (5, 6, ...) would tell us that basis functions with this many different ζ values are provided per valence atomic orbital. The 'aug-' prefix indicates that the basis set was augmented with diffuse functions.

If you can afford the computer time, you generally can't go wrong by using a larger basis set. Most of us don't have access to a supercomputer, so we have to make some decisions. This means that we should learn a bit about the basis sets that are available and how to make wise use of them. As a rough guideline, the sizes of the most commonly used basis sets vary as follows:

$$\|6\text{-}31\text{G}\| < \|6\text{-}311\text{G}\| < \|\text{cc-pVDZ}\| \ll \|\text{cc-pVTZ}\|.$$

Note that the 'bare' cc-pVDZ basis set is larger than the 6-311G basis set, but the former automatically includes polarization functions. It would be fairer to compare the 6-311G** basis set to cc-pVDZ, and if we do that, the latter is only slightly larger. Again, we will let the available computer time, the size of the system, and the type of calculation dictate the basis set to be used, keeping in mind that we should probably add polarization functions and will often need diffuse functions as well.

4.1.2 Density functional theory

It may seem obvious to say this, but electrons repel each other. As a result, they tend to stay away from each other. This is called **electron correlation**. Electron correlation actually creates a serious problem for a straightforward application of MO theory because, if two opposite-spin electrons share a spatial wavefunction, there is nothing in our representation of the electronic wavefunction to keep the electrons away from each other. This results in the computed electronic energy being systematically too high.

In addition, a quantum-mechanical effect known as exchange antisymmetry causes same-spin electrons to similarly stay away from each other. (Exchange antisymmetry is at the root of the Pauli exclusion principle.) The traditional workhorse of quantum chemistry, the Hartree–Fock method, automatically constructs a many-electron wavefunction with the required antisymmetry. However, it does nothing to address the correlation problem. Correlation effects are accounted for by post-Hartree–Fock methods, which perform additional, typically highly compute-intensive work to estimate the correlation energy.

An alternative approach to solving for the electronic wavefunction known as density functional theory (DFT) tackles the correlation and exchange problems head-on. DFT uses a formulation of the quantum-mechanical equations that focuses on the electron density rather than the wavefunction. Correlation and exchange effects are included through semiempirical functionals. As a rule, they do a very reasonable job of estimating both the correlation and exchange energies. DFT methods also turn out to be extremely fast, so for much routine quantum-chemical work, they are now preferred to Hartree–Fock-based methods.

The catch—there is always a catch, isn't there?—is that there are *a lot* of DFT functionals, and choosing one is a real puzzle at times. The correlation and exchange functionals are semiempirical expressions with parameters that are either fit to different data or chosen to satisfy certain physical requirements. Consequently, each functional has its strengths and weaknesses. For our purposes, the B3LYP functional is a good general-purpose exchange–correlation functional. B3LYP is not one of the most modern functionals, but it gives surprisingly good results in a wide variety of problems. The M06-2X density functional, although not available directly from the GaussView menus, is also a useful, general-purpose functional that is particularly good at describing long-range forces between molecules.

Many density functionals are not particularly good at capturing long-range dispersion forces because they have the wrong long-range limiting behavior. In addition to M06-2X, there are functionals that are specifically designed to add dispersion interactions to exchange–correlation functionals. These should be used if long-range forces are important to a particular calculation. Two of these are directly accessible from GaussView, namely the APF-D and ωB97X-D functionals. (A suffix of -D often indicates a functional to which dispersion corrections have been added.)

4.1.3 Spin

In MO theory, we typically describe electron states using **spin-orbitals**. A spin-orbital is a spatial orbital with a spin function attached to it. The spin-up and spin-down spin functions are usually called α and β. Thus, the two spin-orbitals associated with a spatial orbital would be $\alpha\phi_i$ and $\beta\phi_i$. If we stick to the idea that a spatial orbital can hold two electrons, there ends the story.

However, it is sometimes convenient to let each spin-orbital have an independently computed spatial part. This leads to lower electronic energies, partly because it allows for some electron correlation, at the cost of losing the nice picture of a single orbital holding two electrons.

A calculation in which we force pairs of spin-orbitals to have the same spatial part is called a **restricted-spin** calculation. If we allow each spin-orbital to have a different spatial part, we say that the calculation is **unrestricted**. Because of the simpler interpretation of restricted spin-orbitals, I tend to use these preferentially. However, there are some calculations, especially those involving bond dissociation, that behave better using unrestricted wavefunctions. This is one of those computational issues about which it's best to keep an open mind.

One of the pieces of data we need to provide for an MO calculation is the **spin multiplicity**. This is calculated by

$$\text{multiplicity} = 2S + 1,$$

where S is the net spin. There is a bit of jargon here, summarized in table 4.1. The issue is that you need to have some idea in advance of how many unpaired electrons there will

Table 4.1. Unpaired electrons and multiplicity.

# Unpaired	S	Multiplicity	Spin state
0	0	1	Singlet
1	1/2	2	Doublet
2	1	3	Triplet

be. This is easy in some cases, trickier in others. Most molecules with an even number of electrons and a 'reasonable' Lewis diagram won't have unpaired electrons, so they have a singlet spin state, but there are exceptions, such as oxygen, whose two π electrons each go into a different π orbital according to Hund's rule, and which accordingly has a triplet ground state. Radicals with a single unpaired electron have a doublet spin state.

4.1.4 The potential energy surface

Solving the electronic problem gives us an electronic energy and, if we vary the nuclear coordinates, the dependence of the electronic energy on the nuclear coordinate vector $\mathbf{R}$. The electronic energy has three components, namely the electron kinetic energy K_e, the Coulombic electron–electron repulsion V_{ee}, and the electron–nuclear attraction V_{en}:

$$\varepsilon_e(\mathbf{R}) = K_e + V_{ee} + V_{en}.$$

All of the terms on the right-hand side depend on $\mathbf{R}$, not just V_{en}, because the nuclear positions affect the electron distribution, which in turn affects both the electron kinetic energy and electron–electron repulsion terms.

When the nuclei move, they 'drag along' the electrons in the Born–Oppenheimer picture. Consequently, the electronic energy acts as a potential energy term for the nuclei. Of course, the nuclei also repel each other, so the **effective potential**, V_{eff}, experienced by the nuclei is

$$V_{eff}(\mathbf{R}) = \varepsilon_e(\mathbf{R}) + V_{nn}(\mathbf{R}),$$

where V_{nn} is the nuclear–nuclear repulsion potential energy.

In the Born–Oppenheimer approximation, $V_{eff}(\mathbf{R})$ defines a potential energy surface for nuclear motion. There are many ways to use this potential energy surface, but almost all of them start with the location of one or more equilibrium geometries, which are minima on the potential energy surface. The calculation of a minimum-energy molecular structure is called **geometry optimization**. A very simple molecule, such as water, only has one minimum-energy structure. More complicated molecules may have a number of different conformations that are local minima on the potential energy surface. This may be important to us if these minima have similar energies, because there will then be a distribution of conformations, each of which might react at a different rate and which will thus have to be taken into account to compute a theoretical rate of reaction. If these processes are of interest, it will also be possible to calculate transition rates between neighboring local minima.

The calculation of a minimum-energy structure is based on a very simple strategy of 'walking downhill' on the potential energy surface. Given that this is the case, a single geometry optimization will only discover the minimum-energy structure immediately downhill from the starting point. In cases in which we want to look for multiple minima, we have to start optimizations from many different initial geometries. Depending on the molecule, this can be simple (e.g. in something like CH_2FCH_2F, where we really only need to consider a few rotamers) or a real challenge (e.g. in long alkanes, where there are many conformers).

4.2 Learning Gaussian

The calculations described in this section are intended to illustrate some features of Gaussian that will be particularly important to us. Gaussian is a general-purpose computational chemistry program with *many* capabilities. We will be particularly focusing on *ab initio* calculations, i.e. calculations in which we compute electronic wavefunctions based on the methods described above.

4.2.1 Our first MO calculation: the ·OH radical

We will start with a simple geometry optimization of the ·OH radical. This will allow you to learn both how to do the calculation and where to find the output we will need later.

Building the ·OH radical The strategy will be to start with water and then to delete a hydrogen atom. Initially building a molecule that is easy to make and then deleting or changing atoms is often the easiest way to build a molecule in GaussView.

- Start up GaussView.
- Click on the element fragment button (circled in figure 4.1). This will pop up a periodic table (figure 4.2).
- In the periodic table, click on O. At the bottom of the table, you will see a set of possible fragments, namely a bare oxygen atom, an oxygen atom with a double bond, and one with two single bonds (figure 4.2). Click on the oxygen with two single bonds.
- Click in the blank canvas. A water molecule appears.

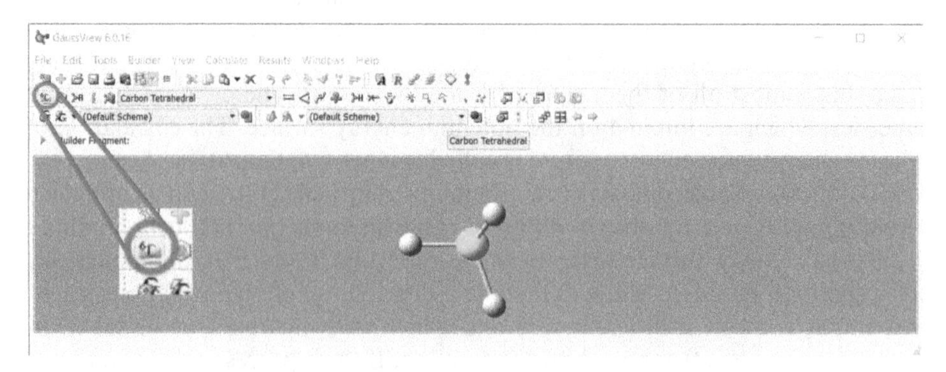

Figure 4.1. GaussView main window, with the element fragment button circled in red.

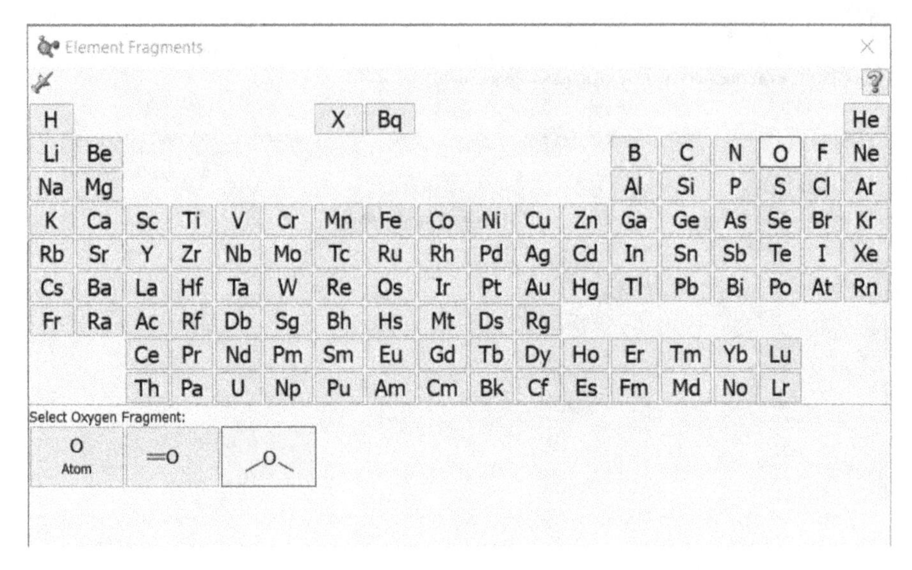

Figure 4.2. GaussView periodic table, with fragment selector under the periodic table.

- Right-click on one of the hydrogen atoms, then choose `Select Atom H?`. (The question mark will be an atom number, e.g. 'H2.')
- Hit the delete key on your keyboard. You now have an ·OH radical.

Setting up a geometry optimization We now want to optimize the geometry. Once that is done, we can collect some statistics about the molecule.

- In the main window's menu bar, click on `Calculate` → `Gaussian Calculation Setup`.
- In the `Job Type` tab, select `Opt+Freq`, which both optimizes the geometry and calculates the vibrational frequency. The defaults should be OK.
- In the `Method` tab, in the `Method` row, select `Ground State`, `DFT`, `Restricted-Open`, `B3LYP`. `Restricted-Open` calculates a wavefunction in which all of the electrons are paired in spatial orbitals, except for the one unpaired electron.

  In the `Basis set` row, select `6-31G`. Adding polarization functions is always a good idea, so also select `(d,p)`.

  GaussView will have recognized that ·OH is a radical and will already have guessed `Doublet` for the `Spin`. It's always a good idea to look at this before proceeding.

  Once you are done, you should have something that looks like the setup window shown in figure 4.3.

  There are many other options in Gaussian not offered by GaussView, which tries to simplify the process of creating and running Gaussian jobs. If you need a DFT method not offered in the GaussView menus (e.g. M06-2X), you can modify the Gaussian input file we will create in the next step using a text editor, as described in section 4.2.3.

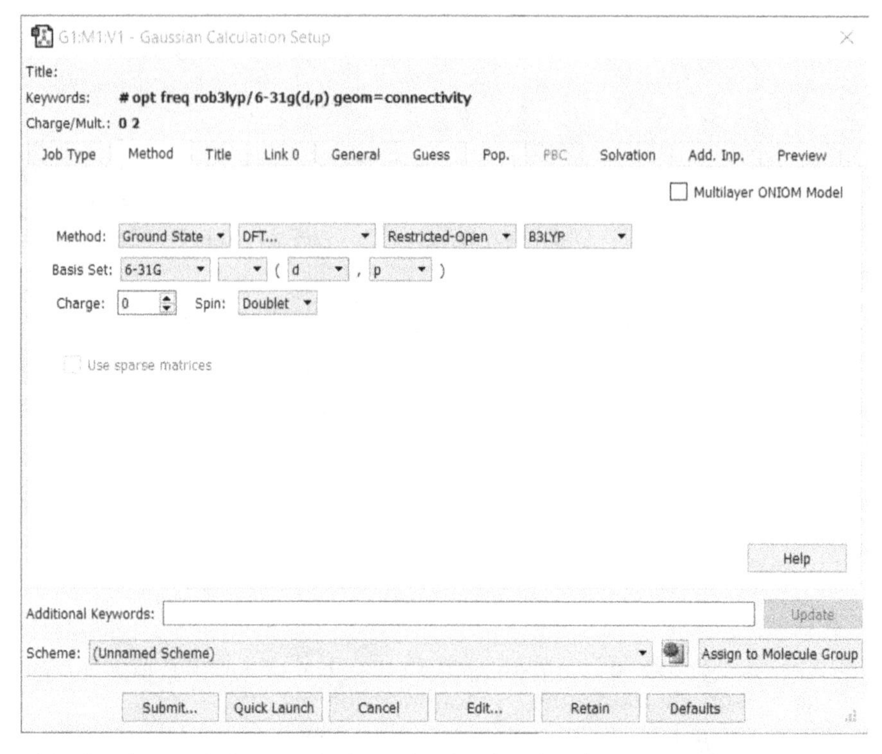

Figure 4.3. Gaussian calculation setup in GaussView for the calculation described in the text for the ·OH radical.

Running the job

- Click on `Submit`.
- GaussView will prompt you to save the input file. Gaussian input files have `.gjf` extensions.
- Gaussian will then run for a minute or so. As it is running, it will print information on what it is doing in an output window.
- When the job finishes, say `No` to closing the Gaussian window. (If you make a mistake and hit `Yes`, it's OK. All of this output went into a `.log` file with the same base name as the `.gjf` file. You can just open the log file with any text editor, or with GaussView for that matter.)

Reading the output Gaussian output files are best read backwards, unless you want to see the blow-by-blow of how Gaussian got to the final solution. If you start from the bottom, one of the first things you will see is a table that looks like this:

```
                           ----------------------------
                           !   Optimized Parameters   !
                           ! (Angstroms and Degrees)  !
 --------------------------                            --------------------------
 ! Name  Definition              Value          Derivative Info.                !
 --------------------------------------------------------------------------------
 ! R1    R(1,2)                  0.9794         -DE/DX =    0.0                 !
 --------------------------------------------------------------------------------
```

This table contains the optimized bond lengths and bond angles. In the case of a diatomic molecule, there is, of course, only a single bond length. The first column gives a shorthand name for the coordinate, in this case `R1`. All bonds have names that start with `R`. The second column tells us which atoms are connected by this bond. The numbers of the atoms are most easily retrieved by clicking on atoms in the optimized structure. We will get to that soon. The third column gives the value of the structural parameter, in this case 0.9794 Å. The fourth column should show a zero derivative (or very nearly so), since that's what determines whether the optimizer has reached a minimum.

Keep going backwards until you find a table that looks like the following:

```
                     E (Thermal)             CV                S
                      KCal/Mol        Cal/Mol-Kelvin    Cal/Mol-Kelvin
 Total                   6.770                4.968             42.598
 Electronic              0.000                0.000              1.377
 Translational           0.889                2.981             34.437
 Rotational              0.592                1.987              6.784
 Vibrational             5.289                0.000              0.000
                       Q            Log10(Q)             Ln(Q)
 Total Bot       0.817795D+04         3.912644          9.009196
 Total V=0       0.616095D+08         7.789648         17.936327
 Vib (Bot)       0.132738D-03        -3.877003         -8.927130
 Vib (V=0)       0.100000D+01         0.000000          0.000000
 Electronic      0.200000D+01         0.301030          0.693147
 Translational   0.275570D+07         6.440232         14.829183
 Rotational      0.111785D+02         1.048385          2.413997
```

Gaussian is calculating gas-phase thermodynamic properties for our molecules, namely the internal energy (`E`), the constant-volume heat capacity (`CV`), and the entropy (`S`). Note that values containing a '`D`' are in scientific notation. The use of `D` instead of the more familiar `E` is a quirk of Fortran, the computer language in which Gaussian was written. Each of the quantities reported is broken down into electronic, translational, rotational, and vibrational contributions. By default, these quantities are computed at 25 °C, although this can be changed.

The second part of the table contains the values of the partition functions, again at 25 °C. We first ran into partition functions in chapter 2, and we will return to them in chapter 6. These are computed in the approximation that the energy is separable into vibrational, electronic, translational, and rotational degrees of freedom. Two values are given for the overall ('`Total`') partition function as well as for the vibrational partition function. These two values differ by the zero point of the energy chosen. The '`V=0`' partition functions use the ground vibrational state as the zero of energy. Note that the vibrational partition function has a value of one in this case. This is because the vibrational spacing is sufficiently large that essentially all the molecules are in the ground vibrational state at 25 °C, in accord with the interpretation of the partition function discussed in section 2.2. The partition functions labeled '`Bot,`' on the other hand, use the bottom of the potential energy well as the reference energy. We will see later why this might be convenient.

While we're looking at the partition functions, notice that the electronic partition function has a value of two. It's a suspiciously round number, isn't it? The electronic states are typically very well separated, so that only the ground states are populated at room temperature. The value of the electronic partition function is therefore just the degeneracy of the ground state. This degeneracy is, in most cases, just the spin multiplicity, although there are occasionally some subtleties associated with spin–orbit coupling as well as coupling with nuclear spins.

Just above these thermodynamic data, you will find some information on the molecule's rotational energetics, which we will revisit in a later chapter.

Going back just a bit more, we find some vibrational data for ·OH in a little table that looks like this:

```
Harmonic frequencies (cm**-1), IR intensities (KM/Mole), Raman scattering
activities (A**4/AMU), depolarization ratios for plane and unpolarized
incident light, reduced masses (AMU), force constants (mDyne/A),
and normal coordinates:
                         1
                         SG
Frequencies --    3699.8422
Red. masses --       1.0671
Frc consts  --       8.6063
IR Inten    --       2.4553
```

Note the units in the header of the table. In particular, the 'frequency' is actually a wavenumber, the inverse of a wavelength. Since most infrared spectrometers output wavenumbers, if you are trying to compare the results of such a calculation to experimental data, this is probably the most convenient unit. Incidentally, the experimental ground-state vibrational frequency of a hydroxyl radical is 3737.76 cm^{-1}, so our calculation did a decent job of estimating this frequency.

You can now close the Gaussian window. At this point, GaussView will offer to open a `.chk` file. Go ahead. The `.chk` file contains your optimized molecule, which GaussView will now display.

Geometric parameters from the GUI You should see your molecule in a new window, except that you will be looking at it end-on due to Gaussian's default coordinate system orientation. Click and drag to rotate the molecule into an orientation in which you can see both atoms. To measure the geometric parameters, it's now just a matter of clicking on atoms.

- Click on one of the two atoms. GaussView will show you the number it associates with this atom.
- Now click on the second atom. Again, you will see the number of this atom, but even better, at the bottom of the screen, GaussView will show you the bond length. You should see something like

```
B = 0.979 37 (O1 H2)
```

indicating that the bond (B) between the two atoms is 0.979 37 Å long. Given that we asked Gaussian to perform a geometry optimization, this is the equilibrium bond length, i.e. the bond length that gives the minimum energy.

- For more complex molecules, clicking on three atoms in succession gives a bond angle.
- Clicking on four atoms (if you had that many) would give a dihedral angle.

4.2.2 Coordinate scanning

There are times when we don't just want an equilibrium geometry. We want to see the whole potential energy surface, or at least as much of it as we can visualize given that, in general, this is a high-dimensional object. To do this, we need to scan the molecule's internal coordinates (bond lengths, bond angles, dihedral angles). Let's start with a particularly simple example, namely our ·OH radical, which has just one internal coordinate, the bond length. Varying the bond length will trace out the potential energy curve of this diatomic molecule.

Before you start a parameter scan, you should have an equilibrium geometry. This will give you an idea of the range of values to scan (e.g. the range of bond lengths). It's a good idea to do a very coarse scan first to get a sense of whether the range of values you have chosen will show you the features you are trying to visualize and also to figure out whether there will be any issues with the method chosen. So let's start with that.

In order to set up a scan, you either need to go back to your original `.gjf` file or work from your `.chk` file. Since the `.chk` file is already open, I'm going to work with that.

The first thing we need to do is to set the bond length to either the minimum or maximum value we want to consider. Here, we know that the equilibrium bond length is about 1 Å, so I'm going to start at 0.5 Å. To set the bond length to a specific value, proceed as follows:

- Right-click on the canvas. From the pop-up menu, select `View` → `Builder`.
  (The Builder tools are all available in the main GaussView window. However, the icons are much smaller there, so at first, you may find the Builder easier to use. These tools are also accessible directly from the `Builder` item in the right-click pop-up.)
- In the Builder pop-up, click on `Modify bond` (diatomic molecule with double-ended arrow over it; figure 4.4).
- Now select the two atoms in your ·OH radical by clicking on each one in turn. This will open a window that allows you to modify bond properties. Change the bond length to approximately 0.5 either by using the slider or by typing in a value.
- In the main GaussView window, click on `Tools` → `Redundant Coordinates`.
- In the window that opens up, click on `Add`. In the drop-down menu about halfway down the page, replace `Unidentified` by `Bond`. Then, in the canvas, click on each of the two atoms. The two boxes labeled `Coordinate` should be filled with the numbers one and two (not necessarily in this order).

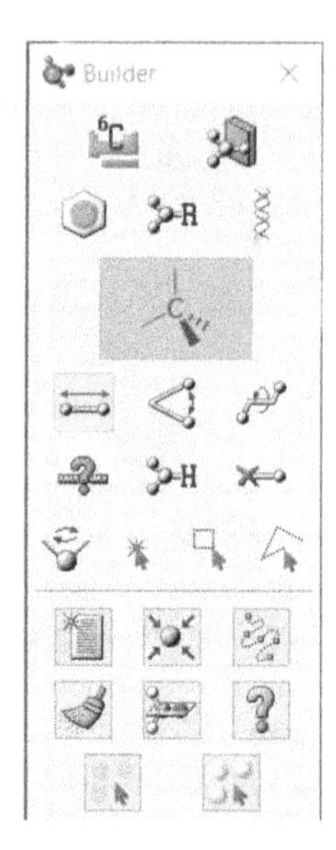

Figure 4.4. The 'Builder' pop-up from Gaussian, with the 'Modify bond' option selected (blue).

- In the next drop-down, select `Scan Coordinate`.
- We now need to select the number and size of steps to take in our scan. We need to include large values of the bond length to approach the separated atom limit. I will target a maximum bond length of 5 Å. This is a guess of how far apart the atoms will have to be in order to approach the dissociation limit. For a quick scan, I will use steps of 0.5 Å. Since 5 − 0.5 Å = 4.5 Å and 4.5 Å/0.5 Å = 9, I will choose nine steps of 0.5 Å each.
- Your Redundant Coordinate Editor window should now be filled in as shown in figure 4.5. Click on `OK` to confirm these selections.
- Click on `Calculate` → `Gaussian Calculation Setup`.
- `Job Type` is automatically set to `Scan Relaxed (Redundant Coord)` by GaussView once you have set up a scan with the Redundant Coordinate Editor.
- We are going to leave the method as it was previously set up (`Ground State`, `DFT`, `Restricted-Open`, `B3LYP`), but we're going to add diffuse functions to the basis set, which should now read `6-31G++(d,p)`. The diffuse functions, if you remember, are important to properly capture intermolecular forces, and in this case we're going to pull the molecule apart until, in effect, the only forces left are long-range interatomic forces. Arguably, we should use a density functional with better long-range behavior than B3LYP. We won't worry about this detail for now.
- Click on `Submit`. Save the job with a new file name.

If you do everything exactly the way I did it, Gaussian will fail, telling you that 'The processing of the last link ended abnormally.' If you now look at the end of the log file, you will see the message 'SCF has not converged.' There can be many

Figure 4.5. The Redundant Coordinate Editor window, showing the setup for a coarse scan of the ·OH bond length.

reasons for this message, but in the case of a radical, the issue is often that you need to switch to an unrestricted wavefunction.

- Go back to the `Gaussian Calculation Setup` and to the `Method` tab, and change `Restricted-Open` to `Unrestricted`. Also make sure that the `Job Type` is still set to `Scan Relaxed (Redundant Coord)`. Submit the job.
- Once the calculation ends, say `Yes` to closing the Gaussian window and open the `.log` file (**not** the `.chk` file). If all went well, you should have a canvas containing your ·OH radical with a counter at the top, like this:

This counter allows you to step through the structures generated during your scan, while the green dot runs an animation. The drop-down next to the green dot lets you adjust the parameters of the animation. Go ahead and give it a try!

- What we really want to do is to look at the effective potential. To do this, click `Results` → `Scan`. This will give you the graph shown in figure 4.6. Note that the 'Total Energy' is really the effective potential and that the 'Scan Coordinate' is the bond length in angstroms, even though neither is labeled that way by Gaussian.
- We now know that our chosen computational method will work across the full range of bond lengths, and that we have chosen a reasonable range of

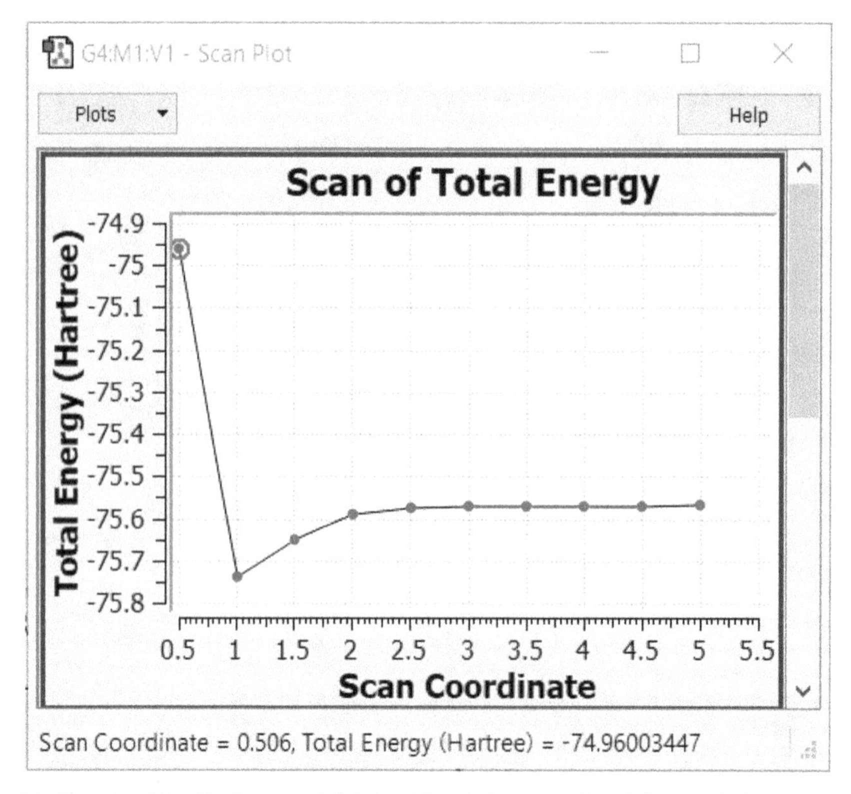

Figure 4.6. A graph of the effective potential vs bond length (scan coordinate) from an initial, coarse scan of the ·OH bond length.

bond lengths: we can see the steeply rising part of the potential energy at small distances, and the flat part where the atoms no longer have strong bonding interactions. If we did not see these features, we could make another coarse scan to find a better range of bond lengths.

If we wanted to emphasize the area around the minimum a bit more, we could start at a slightly larger value of the bond length, say 0.6 Å. It also looks as though the potential energy is quite flat long before we reach 5 Å, so we could save a bit of computational time by limiting the maximum bond length considered to, say, 4 Å.

- We now need to repeat all of the steps of setting up the scan, starting with fixing the initial bond length, this time at 0.6 Å. (We could still start at 0.5 Å if we really wanted to.) We can take 34 steps of 0.1 Å each, since 34(0.1 Å) + 0.6 Å = 4 Å.
- Now go get yourself a drink. This will take a few minutes.
- When the job is done, have a look at the plot. There is just one more thing to do. We don't want the effective potential plotted vs an unnamed 'Scan Coordinate.' We want the axis to be properly labeled. And although hartrees, the atomic unit of energy used in computational chemistry calculations, are fine units, we probably want the energy in more familiar units.

Open up the `Plots` drop-down menu, and select `Properties`, then proceed as follows:

- In the y axis box, choose units of '`KCal/Mol`'[4] instead of hartrees.
- Now click on the `Title` tab. Since my figure will have a caption, I'm going to delete the default plot title from this tab. I also set the x axis title to[5] 'R (Å)' and the y axis title to 'Effective potential.' Click on `Ok`.
- If you want to export your graph, right-click on the graph, and choose `Save Picture`. You can also choose `Save Data` to save the raw data to a text file, in case you want to use a better plotting program, or want to work with the data outside of GaussView. Another use for `Save Data` is that, if you realize you want more points, you can keep the ones you have already calculated by saving them, and then go and collect more points in some region of interest, following which you save the new points and put the two data sets together using a text editor or spreadsheet.

Your final result should look like figure 4.7. Note the description of the calculation in the caption. These details should always be given for a quantum chemical calculation, whether in the 'Methods' section of a paper or in the figure caption if you will be using different computational methods for different calculations.

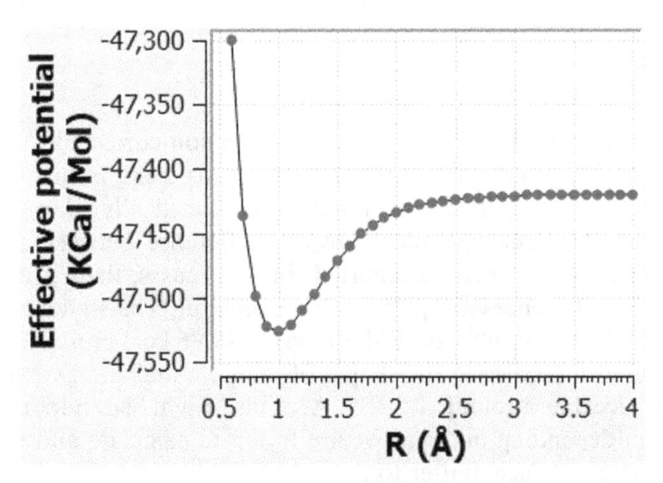

Figure 4.7. The effective potential energy curve of the ·OH radical obtained using Gaussian 16 [1]. The DFT calculation used the B3LYP exchange–correlation functional and unrestricted wavefunctions, with a 6-31G++(d,p) basis set.

[4] If you're picky (like me), you won't be very happy at the capitalization in GaussView, especially since 'Cal' is a commonly used abbreviation for kilocalorie, and of course, using a capital K for kilo is an abomination.

[5] The simplest way to get an Angstrom symbol in Windows is to pick it from the Character Map utility. On a Mac, Option-shift-A gives you this character.

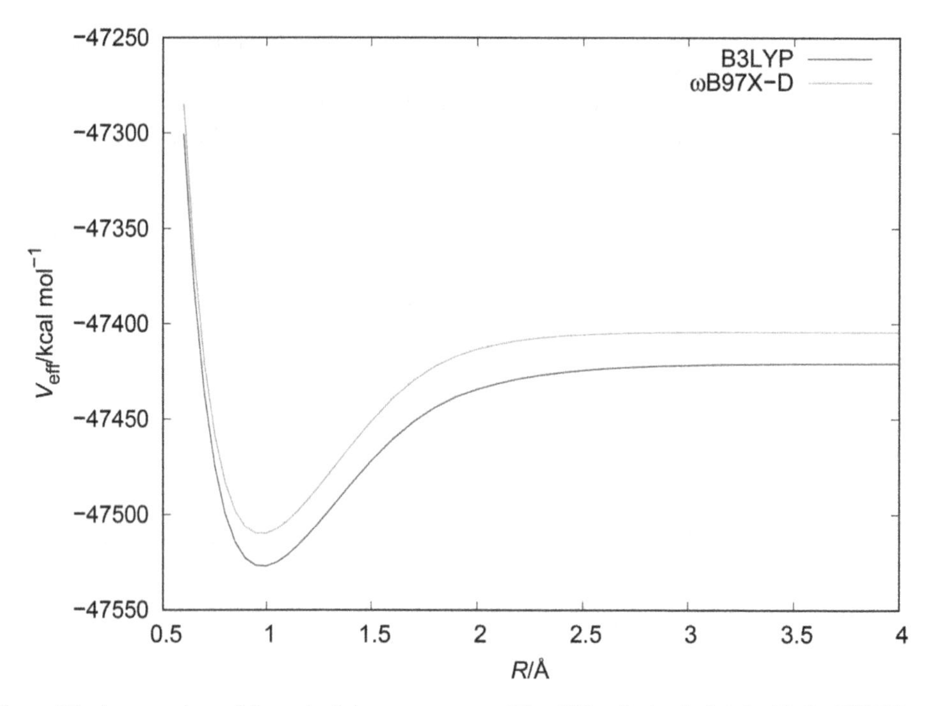

Figure 4.8. A comparison of the potential energy curves of the ·OH radical calculated with the B3LYP and ωB97X-D density functionals. Other than the functionals, all computational parameters were set as in figure 4.7.

There is the issue of whether we needed a dispersion-corrected functional to do this calculation. Figure 4.8 shows the results of scans using both the B3LYP and ωB97X-D functionals, the latter being a functional specifically adapted to deal with dispersion forces. The first and most obvious difference between the two is the difference in the absolute energies reported. In most cases, these are unimportant, since only energy differences are physically meaningful. The well depths are about the same (-120.31 kcal mol^{-1} for B3LYP vs -119.96 kcal mol^{-1} for ωB97X-D). There is a bit of a difference in the shapes of the potential energy curves, which is particularly noticeable around 1.5–2.5 Å. This might be important in some situations, again depending on what we are trying to calculate and which parts of the potential energy surface matter to us.

4.2.3 Editing a Gaussian input file

Gaussian provides a large selection of functionals, but only some of them are available directly in GaussView. The full list of Gaussian functionals is available on Gaussian's web site (https://gaussian.com/dft). On occasion, you may want to use a functional that is not available in the GaussView menus. What do you do then?

The `.gjf` file has a reasonably straightforward format, and can be edited with any convenient text editor. In the file we used to carry out a geometry optimization for ·OH, the second line reads:

```
# opt freq rob3lyp/6-31g(d,p) geom=connectivity
```

This line tells Gaussian what we want it to do. The first two keywords, `opt freq`, tell Gaussian that we want to optimize the geometry and calculate the vibrational frequencies. The next part specifies the quantum-mechanical calculation method and basis set. `ro` means 'restricted open.' If we had chosen an unrestricted wavefunction, these two letters would have been replaced by `u`. We then have the name of the functional: `b3lyp`. The basis set appears after the slash. You can easily replace the name of the functional by anything you want, say `m062x`, to get the M06-2X density functional. You could then open the file in GaussView and start the calculation from the `Calculate` menu by choosing `Gaussian Quick Launch` → `Current Input File`.

You can also edit files used to scan an internal coordinate. The file we created to scan the ·OH bond length has the following contents:

```
%chk=C:\Users\mrous\Documents\courses\C4000\foundations\Gaussian\scanOH.chk
# opt=modredundant ub3lyp/6-31++g(d,p) geom=connectivity

Title Card Required

0 2
 O                  -0.00000000    -0.00000000    -0.08086564
 H                   0.00000000    -0.00000000    -0.68086564

 1 2 1.0
 2

B 1 2 S 34 0.100000
```

Note the use of `opt=modredundant` to indicate that all of the coordinates except those listed later will be optimized. The last line tells us which coordinate is not to be optimized and in this case is instead to be scanned systematically. It's a bond (`B`) between atoms one and two. Thirty-four steps of 0.1 Å will be taken. The initial coordinates of the two atoms specify the initial bond length indirectly; in this case, the z coordinates differ by 0.6 Å. If we were generally happy with this file except for one or two details, we could easily go in and tweak those details by hand. Depending on what we want to do, this may or may not be easier than using the GaussView interface.

4.2.4 Gaussian documentation

Gaussian has extensive documentation available online: https://gaussian.com/man.

Exercise

4.1 GaussView has the capability to scan two coordinates, generating three-dimensional plots of the effective potential against the scanned coordinates. Determine and plot the potential energy of HCN as a function of both the H-C bond length and the bond angle. Make sure to report all of the details of your calculation (in the style of figure 4.7).

References

[1] Frisch M J *et al* 2016 *Gaussian Inc 16 Revision B.01* (Wallingford, CT: Gaussian Inc)

[2] Dennington R, Keith T A and Millam J M 2019 *GaussView Version 6* (Shawnee Mission, KS: Semichem Inc)

IOP Publishing

Foundations of Chemical Kinetics

A hands-on approach

Marc R Roussel

Chapter 5

Potential energy surfaces

In the last chapter, we learned how to do Gaussian calculations and how to calculate potential energy curves associated with changing one internal molecular coordinate as well as the potential energy surfaces (PESs) generated by varying two coordinates of a molecule (exercise 4.4). We can also calculate PESs for systems consisting of the reactants and products of a reaction. The lowest-energy path from the reactants to products is of particular interest and allows us to obtain a clear definition of the transition state.

5.1 The potential energy surfaces of chemical reactions

Suppose we have a system consisting of N atoms. In their initial configuration, these atoms may be bonded to form one or more molecules. In the interesting cases, it will be possible, through some relatively simple motions, to rearrange the atoms into different chemical species, which we will think of as the reactants and products of a chemical reaction. In other words, there is a rearrangement of the atoms that would constitute an elementary reaction.

The potential energy of a set of atoms depends on their relative positions. We need some internal coordinates to describe these. If a system (a molecule or a group of molecules) has N atoms, then $3N - 6$ internal coordinates are sufficient to describe the relative positions of all the atoms[1]. Let's think specifically about a three-atom reaction $A + BC \rightarrow AB + C$. For this system, $3N - 6 = 3$. There are many choices for the internal coordinates, but a convenient set of coordinates consists of R_{AB}, R_{BC} (the distances between atoms A and B and between B and C, respectively) and θ, the A–B–C angle. This would make the potential energy $V_{eff}(R_{AB}, R_{BC}, \theta)$ a

[1] You may have learned in earlier courses that linear molecules have $3N - 5$ vibrational degrees of freedom. The question we are asking here is slightly different. If we take a linear molecule such as CO_2 as an example, two of its vibrational modes are bends. From the point of view of the potential energy, it hardly matters in which direction we are bending the molecule, so the number of coordinates we need to describe the *relative* positions of the atoms is $3N - 6$, even for a linear molecule.

doi:10.1088/978-0-7503-5321-2ch5

function of three variables, i.e. a hypersurface in a four-dimensional space. This is hard to visualize. One possible solution to this problem is to study cuts through this hypersurface, usually taken at fixed angles. We could therefore look at a series of two-dimensional surfaces depending on R_{AB} and R_{BC}, each of which would be drawn for a different value of θ.

We will start by thinking about what happens in a collinear reaction ($\theta = 180°$). Furthermore, we will think about the simplest possible A + BC reaction, namely $H + H_2 \rightarrow H_2 + H$. We can compute the potential energy surface for this reaction at any given, fixed angle using Gaussian. A new trick or two will be useful.

The calculation starts by placing three H atoms roughly in a line using GaussView. Note that we don't have to put any bonds between them. Bonds are a convenience for creating reasonable starting geometries, but they don't actually mean anything to the Gaussian computational engine, except insofar as they may suggest how many electrons and lone pairs a given system has. Instead of dropping three separate, unbonded atoms onto the canvas, we could equally well put down an H_2 molecule and an H atom. This would have no effect on the results.

Once the atoms are placed, use the Builder to create an initial geometry with an H–H–H angle of 180°. We also want to place the atoms close together (corresponding to a corner of the surface to be scanned), say 0.6 Å apart. The problem is that if I change one H–H distance, it will change the other. GaussView has a nice mechanism for handling this. Choose the `Modify Bond` tool of the Builder. Select (say) the atom on the left and the middle atom. If you select them in this order, the leftmost atom will be numbered 1 and the middle atom will be 2. (If you selected them in the other order, just swap the labels in my instructions.) In the `Bond` box, from the drop-down menu for Atom 1, choose `Translate atom` (or `Translate group`, which has the same effect here, given that Atom 1 isn't attached to anything else), but for Atom 2 choose `Fixed`. This has the effect of changing the bond length by moving Atom 1 only. Set the distance between these two atoms at 0.6 Å. Now repeat this procedure with the rightmost and middle atoms. If you select the rightmost atom first, the drop-downs will be set in the appropriate way. Otherwise, again, you can reverse the drop-down selections so that the middle atom is the one that doesn't move.

There are often problems with redundant coordinate scans and linear molecules, essentially due to an internal coordinate transformation that only this particular approach to scanning coordinates suffers from. We're going to use a different technique to set up this scan.

- Click on `Tools →Atom List`.
- If you don't see columns with the titles `Opt 1`, `Opt 2`, and `Opt 3`, click on the '`O`' (`Show Optimization Flag Column`) button in the row of buttons just below the Atom List Editor window's menus. You may also want to get rid of the display of Cartesian coordinates if they appear by clicking on the '`C`' button in order to simplify the display.
- The way to think about this screen (figure 5.1) is that the columns `Bond`, `Angle`, and Dihedral are paired with the `Opt 1`, `Opt 2`, and `Opt 3` columns in a 1:1 fashion: notice that with one exception, the row four–`Opt 3` cell, there

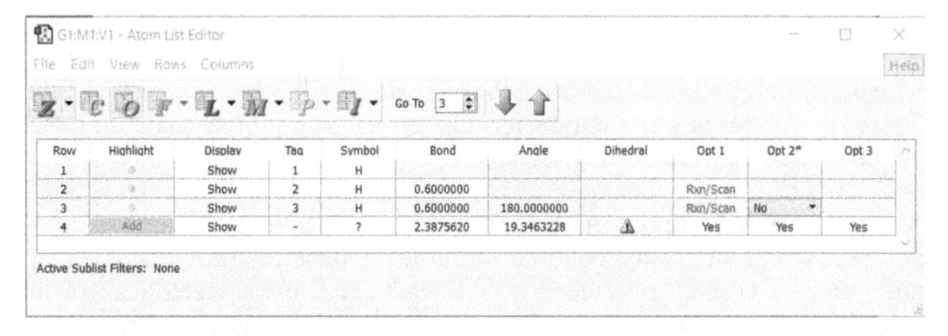

Figure 5.1. The atom list editor set up to scan the two interatomic distances of the H_3 system at a fixed angle.

is an entry in the `Opt` cells only if there is an entry in the corresponding cells of the geometric parameter columns. For example, in row two of the figure, the bond length corresponds to the `Opt 1` column; in row three, the bond length again corresponds to the `Opt 1` column and the angle to `Opt 2`.

- Rows two and three should show the 'bond lengths' (interatomic distances) in your initial structure. For each of these, select `Rxn/Scan` from the `Opt 1` column. This will do exactly what we were previously doing with redundant coordinate scans, namely set us up to scan these two interatomic distances.
- To model a collinear reaction, we also want to prevent the H–H–H angle from changing. You should see a bond angle of 180° in row three. Set the corresponding cell of the `Opt 2` column to `No`. This will keep this angle fixed during the optimization. Your atom list editor window should look like figure 5.1. You can now close the atom list editor window.
- Now click on `Calculate →Gaussian Calculation Setup`.
- In the `Job Type` tab, select `Scan` and `Rigid`. `Scan, Relaxed (Z-matrix)` would have the same effect in this case because there are no other coordinates than the fixed angle and the two interatomic distances to be scanned. As we will see later, a `Relaxed` optimization would allow the energy to be minimized by optimizing all degrees of freedom other than those that are explicitly fixed or varied.
- In the same tab, you can set up the number of steps you want to take, the size of these steps, and the start and end values. If you specify three out of four of these, the other is calculated. You could, for example, set up a coarse scan using just three steps over the range from 0.6 to 2 Å. You might do this to check that the calculation works over the full ranges of the two variables scanned but also to figure out the run time per point and therefore what would be feasible given the computational facilities available to you and your time constraints. Note that you need to independently set the scan parameters for the two coordinates.
- In the `Method` tab, choose `DFT`, `Unrestricted`, `B3LYP` and a 6-31G++(d,p) basis set. The 'd' doesn't do anything because there are no heavy atoms in this calculation, but Gaussian gets confused and throws a syntax error if you don't select something in the first position of the parenthesis.

- You can then submit the job, view the results, and if everything seems to have worked, you can then go back and set up a scan on a denser mesh. Note that you should check the atom list editor to make sure nothing has changed there before you set the scan parameters and rerun. Don't forget that you need to open the `.log` file to view the scan.

My final potential energy surface appears in figure 5.2. We get a bit of a sense of the shape of the PES this way, but because of the rapid increase in energy as the two bonds shrink or expand in tandem, it is difficult (even if we spend a lot of time rotating it around) to get a real sense of important features of the PES. We will come back to this problem in a moment. There are some things we can clearly see. There are two valleys, one near $R_{AB} \sim 0.8$ Å and the other near $R_{BC} \sim 0.8$ Å, corresponding to an H_2 molecule and, at larger distances along either of these valleys, a separate H atom. These two valleys are clearly connected, but it's hard to see exactly how they are connected.

A useful way to visualize surfaces in three dimensions is to use a contour plot. In a contour plot, curves connect points with the same function value (height above or below the coordinate plane), in our case points of equal potential energy. This reduces the three-dimensional surface to a two-dimensional picture.

We can generate a contour plot using Matlab. The free Matlab clone Octave can also be used. Generally, the instructions are the same for both software packages. The few differences that crop up revolve around the final adjustments to the figure and are discussed below.

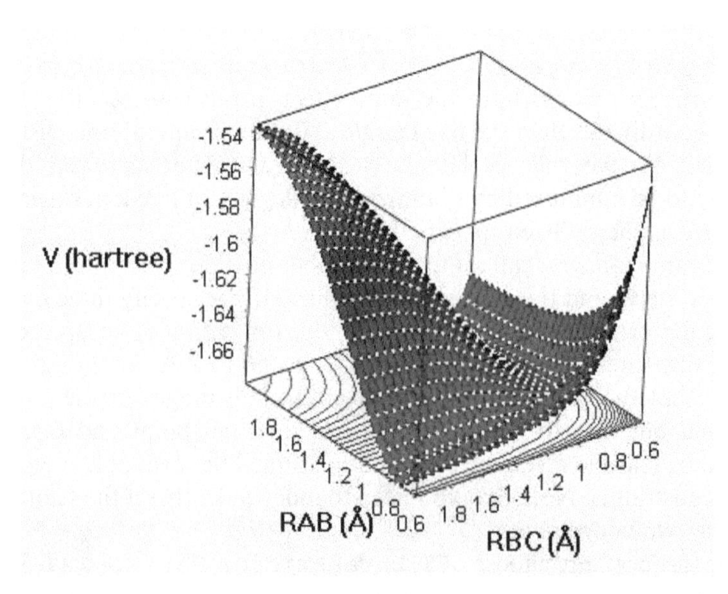

Figure 5.2. The PES for the collinear H + H_2 reaction computed on a grid with a 0.05 Å spacing at the following level of theory: B3LYP exchange–correlation functional, unrestricted wavefunction, 6-31G++(d,p) basis set.

When a range expression has two colons, the middle value is interpreted as a step size. The square brackets create a vector (in general, a matrix) containing the values created by the range operator.

We can use a similar command to create RBC, or we can use the fact that the same values were used for R_{AB} and R_{BC}:

```
>> RBC = RAB
```

Incidentally, Matlab ignores spaces around operators (equal signs, arithmetic operators), so feel free to insert spaces (as I did) to make your code more readable.

- We now have everything we need to generate a contour plot. We can generate a simple contour plot with

```
>> contour(RAB,RBC,m)
```

The problem is that the contour plot you get isn't very detailed in the region we really want to look at closely. To fix that, we just need to use more contours. We can get more contours by telling Matlab directly how many we want:

```
>> contour(RAB,RBC,m,25)
```

I found that 25 contours gave me the desired amount of detail by trial-and-error.

- You probably want to know the values that correspond to each contour. There are two ways to do this. You can use either one, or both.
  1. You can directly label the contours:

```
>> contour(RAB,RBC,m,25,'ShowText','On')
```

  This will put the contour value on most of the contours. It makes the plot a little busy, but it's probably easier to interpret plots labeled this way.

  2. You can add a color bar that shows how the colors correspond to values. To do this, you add the colorbar() command *after* the contour plotting command:

```
>> contour(RAB,RBC,m,25)
>> colorbar()
```

- You will definitely want to label your axes. This is done with the xlabel() and ylabel() functions. These functions allow fancy markup using a built-in LaTeX interpreter. For example

```
>> xlabel('$R_{\rm AB}$ (\AA)','Interpreter','latex')
```

generates the x-axis label 'R_{AB} (Å).' The dollar signs indicate math mode, which we need for subscripts. The underscore creates a subscript. (A caret ^ would create a superscript.) Finally, \AA is LaTeX markup for the angstrom symbol.

- There is one more detail. Since the two axes are both in angstroms and run over the same range, we want to make sure that the plot is square. We do this by using

```
>> axis equal
```

after generating the plot.

- Start by saving the data to a text file. (Right-click on the 3D plot window, a use `Save Data`.)
- We're going to do some surgery on this text file, so make a copy in case y make a mistake and need to back out.
- Open the copy of the data file with a text editor. Find the lines that start follows:

```
#
# Table Format
        X/Y             6.0000000000E-01        6.5000000000E-01
```

(The '`X/Y`' line may start with different numbers depending on how tight you mad the computational mesh.) Delete everything from the top of the file up to an including this line. Save the file.

- Start up Matlab. Look for an area labeled `Command Window`, which should contain a prompt that looks like this: `>>`
- At the prompt, type

  ```
  >> m = load('myfile.txt')
  ```

  with `myfile.txt` replaced by the name of your edited data file.
- If all went well, in the `Workspace` area, you should see something like this:

Workspace	
Name ▲	Value
m	29x30 double

This tells you that `m` is a 29 × 30 matrix (29 rows, 30 columns). The exact dimensions depend on how many steps you took in your computation. If you took n steps in each direction, you should have $n + 1$ rows and $n + 2$ columns. If this isn't the case, you made a mistake in editing the data file. Make another copy from the original and try again.

- The reason that there are more columns than rows is that the first column contains the R_{AB} values. We need to get rid of that. Use the following command, adjusting according to the size of your matrix:

  ```
  >> m = m(:,2:30)
  ```

  Here, we are using Matlab's range operator, the colon. The colon before the comma says to take all of the rows. After the comma, we are telling Matlab to take columns 2:30 of `m`. We store the result back in `m`. The matrix `m` should now have dimensions of 29 × 29 (in my example). This matrix now contains the V_{eff} values only.
- We now need to create vectors of R_{AB} and R_{BC} values, which will be the x and y coordinates of our contour plot. The range operator will help us out again. I calculated the effective potential for R_{AB} values from 0.6 to 2 Å in steps of 0.05 Å. The following will create a vector of the R_{AB} values:

  ```
  >> RAB=[0.6:0.05:2]
  ```

A reasonable contour plot of our data would therefore be generated by

```
>> contour(RAB,RBC,m,25,'ShowText','On')
>> axis equal
>> xlabel('$R_{AB}$ (\AA)','Interpreter','latex')
>> ylabel('$R_{BC}$ (\AA)','Interpreter','latex')
```

You could optionally add `colorbar()` to this sequence of commands if you wish. The result is shown in figure 5.3 (along with some embellishments to be discussed later).

If you are using Octave instead of Matlab, the above instructions will generally work, but only the simpler TEX interpreter is available for the axis labels. Thus, you would type the following to label your axes:

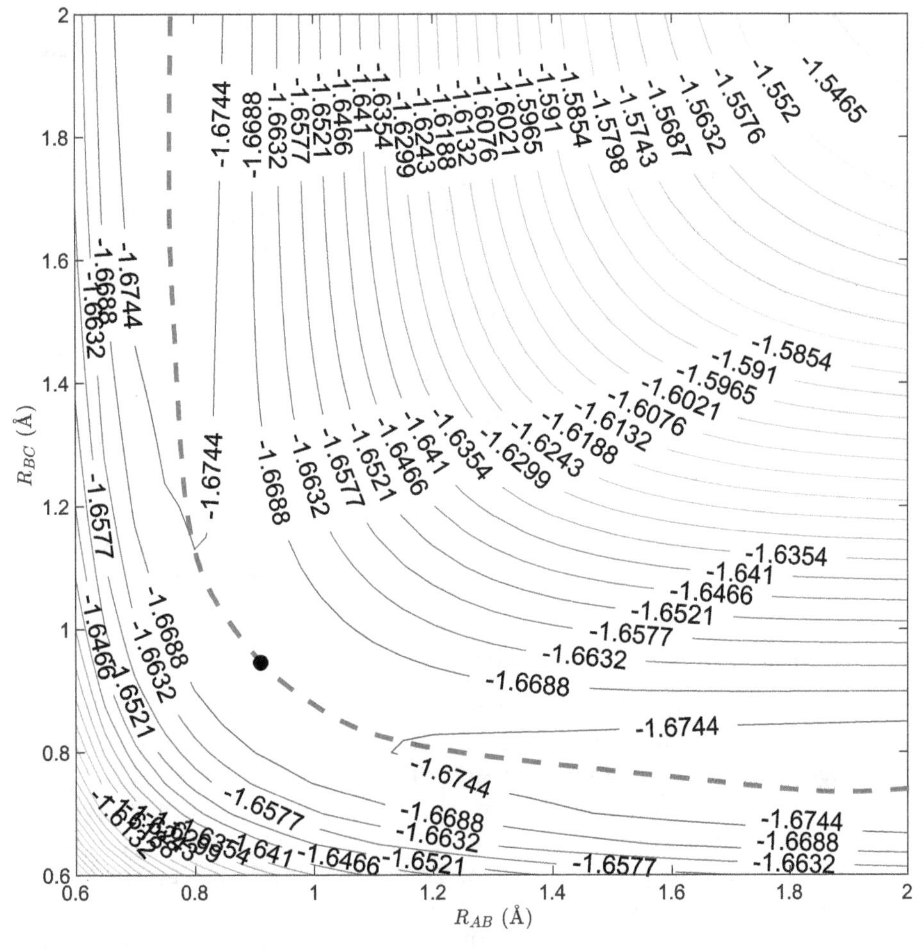

Figure 5.3. A contour plot of the data from figure 5.2. A hand-sketched minimum-energy path connecting reactants to products (dashed curve) has been added as well as a dot representing the transition state.

```
>> xlabel('R_{AB} (angstrom)','interpreter','tex')
>> ylabel('R_{BC} (angstrom)','Interpreter','tex')
```

Now let's think about what this contour plot is trying to tell us. From the values of the contours, we see that the energy goes down as we approach the valleys whose outer contours appear at −1.6744 hartree, whether we approach the valley from one side or the other. However, the valleys' contours are not connected. This can only be true if we go uphill again when crossing from one valley to the other, i.e. the potential energy increases between the two valleys. The curved path shown in the figure is a hand-drawn approximation to the lowest-energy path from one valley to the other. The dot marks the highest point along this path, which is a minimum if we approach it along the bottom-left to top-right diagonal. A point that is a maximum in one direction and a minimum in another is called a **saddle point**. This saddle point is the **transition state** of the reaction, which we had previously talked about as a simple maximum in section 2.5. In general, a transition state is not a simple maximum but a saddle point on the PES. One possible interpretation of the **reaction coordinate** shown in one-dimensional sketches of a reaction profile would be as the distance traveled along the minimum-energy path connecting reactants to products. Along this path, the potential energy would vary roughly as the usual one-dimensional reaction profiles suggest, but figure 5.3 clearly shows us that these one-dimensional profiles represent a huge simplification. It is also clear from these considerations that the reaction coordinate is not a Cartesian coordinate with a simple relationship to the internal coordinates of the set of atoms involved.

Computing a PES is an example of what computer scientists call an 'embarrassingly parallel' calculation. Why? Because each point on the PES is an independent calculation. This means that, if you have appropriate resources, you can share the task of computing a PES across multiple CPUs in a single machine, or even across multiple machines on a network. Gaussian has facilities for distributing a calculation across CPUs or across a network, which you can learn about from the Gaussian manual if you ever need to do this (https://gaussian.com/running). Calculating a PES for a larger system would definitely be an example of a calculation for which you would want to find out about Gaussian's parallel computing capabilities.

5.2 The importance of the potential energy surface

Why do we care about the PES? The minimum-energy path from reactants to products is sometimes called the **reaction path**. We will see later that there is a bit more to this concept than the classical minimum-energy path. However, the true reaction path is generally close to the minimum-energy path. This being the case, studying the minimum-energy path on the PES tells us *how* the reactants are converted into products, both from the point of view of the motions of the atoms during the reaction and from the point of view of the way in which bonding changes as the reaction proceeds.

If we have a PES, we can easily compute classical trajectories, since the force acting along a particular coordinate is just $F_i = -\partial V_{\text{eff}}/\partial q_i$, where q_i is a generalized coordinate. This would give us exactly the kind of information described above, namely a movie of the transformation of reactants into products. These calculations would allow us to sample different initial velocities, angular momenta, relative orientations, etc.

We can also solve the time-dependent Schrödinger equation for the nuclei on a PES. This would give us a more realistic view of the course of the reaction, although at a significant computational and conceptual cost.

If we run enough trajectories on a PES, whether those trajectories are the results of classical or quantum calculations, we can collect statistics about the outcome of a reaction. This is a direct computational counterpart of the molecular beam experiments described in section 3.2 and would provide the same kind of information. This could then be used to estimate the cross-sections that appear in reactive scattering theory.

5.3 Avoided crossings

Throughout this chapter, we have implicitly assumed that a reaction occurs on a single potential energy surface. However, this is not always the case.

The potential energy curves of the electronic states of a diatomic molecule with the same orbital symmetry cannot cross. They may, however, come very close to each other, as shown in figure 5.4. This is called an **avoided crossing**. In the case of

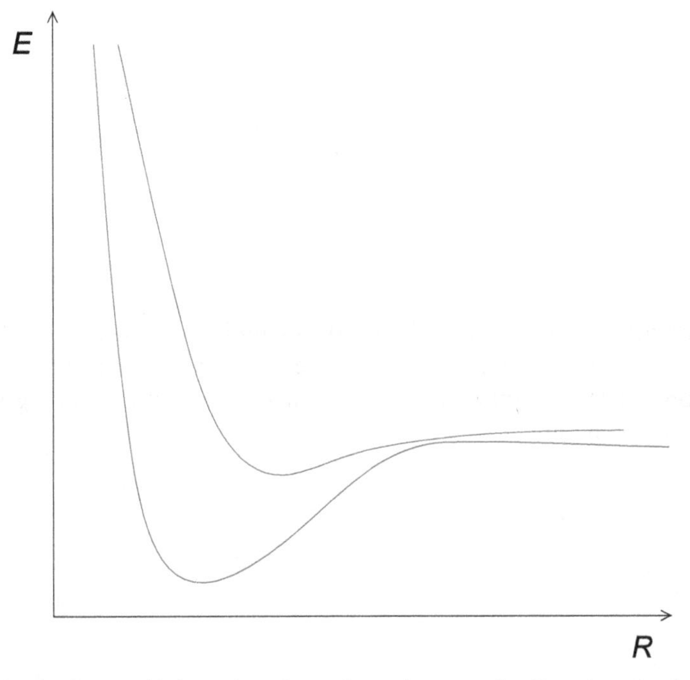

Figure 5.4. A sketch of an avoided crossing of two electronic states of a diatomic molecule with the same orbital symmetry.

polyatomic molecules, the potential energy surfaces can touch, but they still cannot cross each other.

In one case or the other, near an avoided crossing, systems can cross from one potential energy surface to the other. Thus, we not only need to think about the motion on a PES, we also have to consider the quantum mechanics of transitions between PESs. This is only a problem if the electronic states are not well separated, which is, fortunately, only an issue in some systems. While the dynamics across multiple potential energy surfaces is an extremely interesting topic, studying it would require a substantial enlargement of this book and a considerably deeper quantum mechanical treatment. We therefore leave it to more advanced courses in quantum dynamics.

Exercise

5.1 The following is a PES for a collinear reaction between an oxygen atom and a hydrogen molecule:

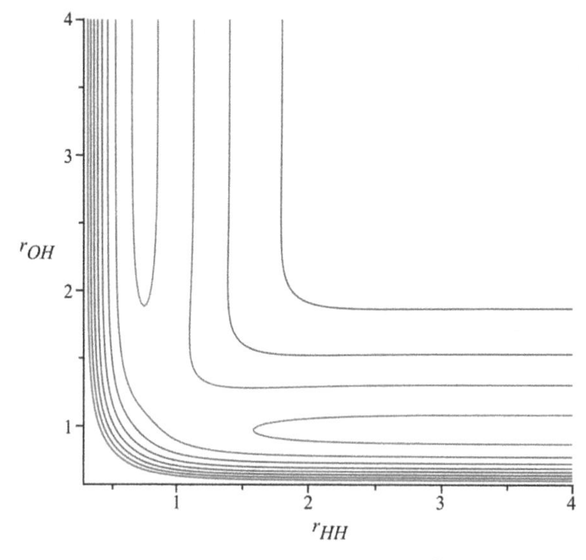

The contours are drawn at intervals of 100 kJ mol^{-1}. Based on the shape of this potential energy surface, would you expect the product OH to carry about the same amount, more, or less vibrational energy than H_2 had before passing through the transition state? Explain.

IOP Publishing

Foundations of Chemical Kinetics
A hands-on approach
Marc R Roussel

Chapter 6

The statistical treatment of equilibrium

6.1 A brief review of molecular energy levels

Molecules are quantum mechanical entities, and it turns out to be important to know something about the various ways they can store energy, which we have briefly touched on previously. Here, we will review some equations and concepts that you should have previously run into. We only need to know a little in order to move ahead, so the treatment will be brief.

6.1.1 Translational kinetic energy

To treat the translational kinetic energy, it is simplest to assume that a molecule occupies a rectangular box. Qualitatively, the shape of the box changes nothing provided the box is much larger than the molecular dimensions. The energy levels of a particle in a box are given by

$$\varepsilon_{n_x,n_y,n_z} = \frac{h^2}{8m}\left(\frac{n_x^2}{L_x^2} + \frac{n_y^2}{L_y^2} + \frac{n_z^2}{L_z^2}\right). \tag{6.1}$$

In this equation, h is Planck's constant, m is the mass of the particle (e.g. a molecule), L_ξ is the length of the box in the corresponding dimension, and n_ξ is a quantum number, in this case a positive whole number: 1, 2, 3, …

The translational energy levels are incredibly close together. You may have learned in a previous course that the average translational kinetic energy of the molecules in a gas is $\frac{3}{2}k_BT$, or $\frac{1}{2}k_BT$ per dimension, with k_B being the Boltzmann constant. At room temperature (20 °C), this works out to about 2×10^{-21} J per dimension. For a molecule of $^{14}N_2$ with a mass of 4.65×10^{-26} kg, the value of n_x corresponding to this average kinetic energy in a container that has a length of 1 cm is about 4×10^8. For this value of n_x, the spacing between the energy levels (i.e. the change in $\varepsilon_{n_x,n_y,n_z}$ when n_x is increased by one) is about 10^{-29} J, which is a tiny fraction of the average energy.

6.1.2 Rotational energy

This section and the next will discuss the rotational and vibrational energy levels of molecules—both associated with nuclear-motion degrees of freedom—as separate quantities. While rotation and vibration (and the electronic state, for that matter) are coupled, treating them as separate degrees of freedom is often a good approximation.

The general treatment of rotation starts with a matrix called the inertia tensor[1]. An important theorem of mechanics states that it is always possible to choose a set of axes such that the inertia tensor is diagonal. In this special coordinate system, we only have to concern ourselves with three moments of inertia, computed from the masses and positions of the atoms in the molecule. For example,

$$I_{xx} = \sum_{i=1}^{N} m_i \left[(y_i - y_{\text{cm}})^2 + (z_i - z_{\text{cm}})^2 \right]. \tag{6.2}$$

In this equation, the subscript i denotes one particular atom of the N atoms in the molecule, and the subscript 'cm' denotes the center of mass. By permuting the coordinates, we obtain similar equations for the remaining two moments of inertia, I_{yy} and I_{zz}. Let us rename these three moments of inertia I_{A}, I_{B} and I_{C}, ordered so that $I_{\text{A}} \leqslant I_{\text{B}} \leqslant I_{\text{C}}$. We can also define the rotational constants A, B and C as follows:

$$A = \frac{\hbar^2}{2I_{\text{A}}},$$

and B and C are defined analogously. A, B, and C have units of energy, and $\hbar = h/2\pi$.

In general, the quantum mechanical treatment of rotation is difficult, but some cases give simple equations for the energy. For a linear molecule, with the convention that the bond lies along the z axis, $I_{zz} = 0$ and $I_{xx} = I_{yy} = I$. The rotational energy is

$$\varepsilon_J = BJ(J+1), \tag{6.3}$$

where J is a quantum number that can assume any of the values 0, 1, 2, ... You may recall from your previous courses that angular momentum has to be measured relative to an axis in space, which for a linear molecule can be any axis perpendicular to the bond axis and passing through the center of mass. This introduces a new quantum number K whose value is chosen from the set $0, \pm1, \pm2, \dots, \pm J$. There are therefore $2J + 1$ possible values, meaning that the energy level ε_J has a degeneracy of

$$g_J = 2J + 1, \tag{6.4}$$

i.e. there are $2J + 1$ distinct quantum states corresponding to ε_J.

---

[1] A tensor is an object that defines a transformation between two spaces. Although tensors can always be represented by matrices or their higher-dimensional generalizations, the concept of a tensor is much more general.

Since the nuclei lie along the z axis for a linear molecule, equation (6.2) simplifies to

$$I = \sum_{i=1}^{N} m_i (z_i - z_{\mathrm{cm}})^2.$$

If the molecule is diatomic, this equation simplifies further to

$$I = \mu R^2,$$

where μ is the reduced mass (equation (2.9)) and R is the bond length.

A spherical top is a molecule for which $A = B = C$. These are highly symmetric molecules such as CH_4 and SF_6. Spherical tops also obey equation (6.3). However, the degeneracy of each level is

$$g_J = (2J + 1)^2.$$

Symmetric tops have two identical moments of inertia. If $I_A = I_B \leqslant I_C$, then we say that the symmetric top is oblate, and the rotational energy is given by

$$\varepsilon_{J,K} = BJ(J + 1) + (C - B)K^2,$$

where again, $J = 0, 1, 2, \ldots$ and K is a new quantum number satisfying $K = 0, \pm 1, \pm 2, \ldots, \pm J$. Note that $\varepsilon_{J,K} = \varepsilon_{J,-K}$. Accordingly, with the exception of $\varepsilon(J, 0)$, which is nondegenerate, the rotational energy level $\varepsilon_{J,K}$ is twofold degenerate.

There are also prolate symmetric tops. These have $I_A \leqslant I_B = I_C$. The rotational energy of a prolate symmetric top is given by

$$\varepsilon_{J,K} = BJ(J + 1) + (A - B)K^2.$$

K satisfies the same rule as for oblate symmetric tops, and again we have a nondegenerate $E(J, 0)$ energy level and twofold degenerate $\varepsilon_{J,K}$ levels for $K \neq 0$.

It is again useful to do a simple calculation. For the $^{14}N_2$ molecule previously considered, $\mu = 1.16 \times 10^{-26}$ kg, and the bond length is 109.76 pm. This gives $I = 1.40 \times 10^{-46}$ kg m^2. Thus, $B = 3.97 \times 10^{-23}$ J. We still expect to have about 2×10^{-21} J of energy stored in a rotational mode at room temperature, on average. Solving for J at this energy, we get $J \approx 7$. The difference between ε_{J+1} and ε_J is $2(J + 1)B \approx 6 \times 10^{-22}$ J, which is somewhat smaller than the average energy. As for the translational kinetic energy, the rotational levels are packed fairly close together relative to the average energy at room temperature.

6.1.3 Vibrational energy

In contrast to the treatment of rotational motion, the quantum mechanical treatment of vibrational motion, with suitable approximations, is relatively simple. A molecule with N atoms has $3N$ total degrees of freedom, i.e. different ways that it can move. Three of those degrees of freedom correspond to the translational motion of the molecule as a whole. A nonlinear molecule has three degrees of freedom that correspond to rigid rotations of the molecule around its three principal axes.

The remaining $3N - 6$ degrees of freedom are vibrational modes. A linear molecule only has two rotational degrees of freedom, so it has $3N - 5$ vibrational modes.

The potential energy associated with the vibrational motion generally has a complicated dependence on the relative positions of the atoms. However, if we take a multivariate Taylor series of this potential energy around a minimum and keep only quadratic terms, it turns out that we can transform to a special coordinate system in which only the diagonal terms (terms in q_i^2, where q_i is one of the special coordinates) remain, and all the cross-terms in $q_i q_j$ vanish. These coordinates are called normal-mode coordinates. In other words, in the normal-mode coordinate system, the potential energy is a sum of simple quadratic terms $U_i = k_i q_i^2$, where U_i is the potential energy associated with the ith normal mode and k_i is a constant, sometimes called a Hooke's law constant. Writing the vibrational energy of a mode as a quadratic function of a generalized coordinate is called the harmonic-oscillator approximation. The Hooke's law constant is a property of the potential energy surface, specifically

$$k_i = \frac{\partial^2 V_{\text{eff}}}{\partial q_i^2}. \tag{6.5}$$

To the extent that we can represent the potential energy surface as a quadratic form, i.e. ignoring higher-order terms in the Taylor series, the normal modes behave as if they were independent. The quantum mechanical energy of a normal mode satisfies the equation

$$\varepsilon_{v_i} = \hbar\omega_0^{(i)}\left(v_i + \frac{1}{2}\right). \tag{6.6}$$

The quantum number v_i can take any of the values 1, 2, ... Each normal mode has its own natural frequency $\omega_0^{(i)}$ as well as its own quantum number v_i. The natural frequency is connected to k_i by

$$\omega_0^{(i)} = \sqrt{k_i/\mu_i}, \tag{6.7}$$

where μ_i is a generalized reduced mass associated with the ith vibrational mode.

Molecules can have a wide range of vibrational frequencies. Bond stretches tend to have large frequencies, while torsional motions tend to have small frequencies.

6.1.4 Electronic energy

There are no simple equations for electronic energies. They can be measured spectroscopically or computed using an *ab initio* program such as Gaussian.

6.2 Partition functions

6.2.1 The translational partition function

Equation (6.1) gives the translational energy levels for a particle in a box. Note that this equation can be separated into a sum of contributions corresponding to each coordinate axis, each of which can be treated separately. The partition function associated with motion along the x-axis is

$$q_x = \sum_{n_x=1}^{\infty} \exp(-\varepsilon_{n_x}/k_B T), \tag{6.8}$$

where

$$\varepsilon_{n_x} = \frac{n_x^2 h^2}{8mL_x^2}.$$

We previously calculated that changing the quantum number by one unit has a negligible effect on the energy for typical room-temperature energies. The same is true of the exponential in equation (6.8). From your calculus course, you may recall that you can approximate an integral using a Riemann sum:

$$\int_a^b f(x)\,\mathrm{d}x \approx \sum_{i=1}^{n} f(x_i)\Delta x,$$

where we have divided the interval from a to b into n subintervals of width $\Delta x = (b - a)/n$, and x_i is a point in subinterval i. There are various ways of choosing x_i (left endpoint of the subinterval, right endpoint, midpoint), but in the limit as $\Delta x \to 0$ ($n \to \infty$), they all converge to the value of the integral provided the integrand is continuous. We can think of equation (6.8) as a Riemann sum if we take $\Delta n_x = 1$ and tell ourselves that the integrand is being evaluated at the right endpoint of each subinterval. We then get

$$q_x \approx \int_0^{\infty} \exp\left(-\frac{n_x^2 h^2}{8mL_x^2 k_B T}\right) \mathrm{d}n_x.$$

This is a known integral[2]:

$$q_x = \frac{L_x}{h}\sqrt{2\pi m k_B T}. \tag{6.9}$$

We could repeat this calculation for the y and z directions, but of course the result would be the same, except for the axis labels. The translational partition function is therefore

$$q_{tr} = q_x q_y q_z = \frac{L_x L_y L_z}{h^3}(2\pi m k_B T)^{3/2}.$$

Note that $L_x L_y L_z$ is the volume (V) of the container. Thus,

$$q_{tr} = \frac{V}{h^3}(2\pi m k_B T)^{3/2}.$$

[2] Recall that $\int_{-\infty}^{\infty} e^{-x^2}\mathrm{d}x = \sqrt{\pi}$ from section 2.6.

This final formula does not depend on the shape of the container, provided the container is convex, i.e. provided that a straight line connecting any two points inside the container does not cross the container's walls.

It is also convenient to define the volumic translational partition function[3]:

$$\mathfrak{q}_{\text{tr}} = \frac{q_{\text{tr}}}{V} = \frac{(2\pi m k_{\text{B}} T)^{3/2}}{h^3}.$$

6.2.2 The vibrational partition function

The vibrational energy is given by equation (6.6). Since we are allowed to set the zero of the energy scale anywhere we want, let us first pick the ground vibrational state as our zero point. The vibrational energy then reduces to

$$\varepsilon_{v_i} = \hbar\omega_0^{(i)} v_i.$$

The partition function for vibrational mode i is then

$$\begin{aligned} q_{\text{vib}}^{(i)} &= \sum_{v_i=0}^{\infty} \exp\left(-\frac{\hbar\omega_0^{(i)} v_i}{k_{\text{B}}T}\right) \\ &= \sum_{v_i=0}^{\infty} \left[\exp\left(-\frac{\hbar\omega_0^{(i)}}{k_{\text{B}}T}\right)\right]^{v_i}. \end{aligned} \tag{6.10}$$

This is a geometric series. You may recall from your previous mathematical studies that

$$\sum_{k=0}^{\infty} x^k = (1 + x)^{-1}$$

provided $x < 1$. The quantity in square brackets in equation (6.10), being the exponential of a negative number, is necessarily smaller than one. Thus,

$$q_{\text{vib}}^{(i)} = \left[1 - \exp\left(-\frac{\hbar\omega_0^{(i)}}{k_{\text{B}}T}\right)\right]^{-1}. \tag{6.11}$$

It turns out that using the ground vibrational state as our reference energy isn't all that convenient. At the cost of losing the nice interpretation of the partition function as counting accessible vibrational states, we can use the bottom of the potential well as the reference energy. This corresponds to adding back the zero-point energy $\hbar\omega_0^{(i)}/2$. It was briefly mentioned in section 2.2 that adding a constant to the energy

[3] The adjective 'volumic' means 'per unit volume' [1, p. 6]. Admittedly, there are a lot of different q's floating around. I'm using $\mathcal{q}$ for generalized coordinates, q for partition functions, and $\mathfrak{q}$ for volumic partition functions. Hopefully the visual difference, added to the context, is clear enough.

has a simple effect on the partition function. Specifically, if we add an amount $\Delta\varepsilon$ to all of the energies, we get:

$$\begin{aligned} q_{\text{new}} &= \sum_i g_i \exp\left(\frac{-(\varepsilon_i + \Delta\varepsilon)}{k_B T}\right) \\ &= \exp\left(\frac{-\Delta\varepsilon}{k_B T}\right) \sum_i g_i \exp\left(\frac{-\varepsilon_i}{k_B T}\right) \\ &= \exp\left(\frac{-\Delta\varepsilon}{k_B T}\right) q_{\text{old}}. \end{aligned}$$

In these equations, q_{old} is the partition function based on the original energy scale, and q_{new} is the partition function for the energy scale shifted by $\Delta\varepsilon$. If we use the bottom of the potential energy well to evaluate the vibrational partition function, q_{old} is given by equation (6.11), and $\Delta\varepsilon = \hbar\omega_0^{(i)}/2$. The 'new' partition function is therefore

$$q_{\text{vib}}^{(i)} = \frac{\exp\left(-\frac{\hbar\omega_0^{(i)}}{2k_B T}\right)}{1 - \exp\left(-\frac{\hbar\omega_0^{(i)}}{k_B T}\right)}. \tag{6.12}$$

Note that the two exponentials are slightly different; the exponential in the numerator contains a factor of $\frac{1}{2}$ that is absent in the denominator.

The harmonic approximation allows us to write the vibrational energy as a sum of terms for individual modes. The vibrational partition function is therefore a product:

$$q_{\text{vib}} = \prod_{i=1}^{n_{\text{vib}}} q_{\text{vib}}^{(i)},$$

where n_{vib} is the number of vibrational modes of the molecule.

6.2.3 The rotational partition function

As you may have gathered from section 6.1.2, the treatment of rotational motion is more complicated than the treatment of other types of motion. We will treat the case of a linear molecule in detail and then skip straight to the result for nonlinear molecules.

Equations (6.3) and (6.4) give the rotational energy and degeneracy of a linear molecule, respectively. The rotational partition function is therefore

$$q_{\text{rot}} = \sum_{J=0}^{\infty} (2J+1) \exp\left(-\frac{BJ(J+1)}{k_B T}\right). \tag{6.13}$$

At low temperatures ($k_B T \lesssim B$), we have no option but to evaluate this sum numerically, but that's acceptable because the terms shrink quickly with increasing J, so we only need a few to get an accurate value of the partition function. On the other hand, at high temperatures ($k_BT \gg B$), many terms contribute, since $\exp[-BJ(J+1)/k_BT] \sim 1$ unless J is very large. Using the Euler–Maclaurin formula, which relates the value of a sum to the corresponding integral in the case that the sum is not a good approximation to the integral, it is possible to show that (6.13) is well approximated by

$$q_{\text{rot}} \approx \frac{2Ik_BT}{\hbar^2}.$$

This isn't quite the end of the story, because of the symmetry requirements of quantum mechanics for indistinguishable particles. The correct rotational partition functions are

$$\begin{aligned} q_{\text{rot}} &= \frac{1}{\sigma}\sum_{J=0}^{\infty}(2J+1)\exp\left(-\frac{BJ(J+1)}{k_BT}\right) \\ &\approx \frac{2Ik_BT}{\sigma\hbar^2} \qquad (k_BT \gg B), \end{aligned}$$

where σ is a **symmetry number**. The symmetry number counts the number of different but indistinguishable ways in which you can orient the molecule in space. For a heteronuclear diatomic molecule, $\sigma = 1$, while for a homonuclear diatomic molecule, $\sigma = 2$, since we can rotate the molecule 180° around an axis perpendicular to the bond and obtain an indistinguishable configuration of atoms.

For any nonlinear molecule, the rotational partition function works out to

$$q_{\text{rot}} = \frac{\pi^{1/2}}{\sigma}\left(\frac{2I_Ak_BT}{\hbar^2}\right)^{1/2}\left(\frac{2I_Bk_BT}{\hbar^2}\right)^{1/2}\left(\frac{2I_Ck_BT}{\hbar^2}\right)^{1/2},$$

where I_A, I_B, and I_C are the moments of inertia around the molecule's three principal rotational axes. We again encounter the symmetry number. This is one for molecules lacking any rotational symmetries. For molecules with rotational symmetries, we have to imagine using these symmetries to rotate the molecule into different, indistinguishable positions. The procedure is illustrated in figure 6.1 using a methane molecule, with colored balls representing the different, but identical, hydrogen atoms. In each row, we rotate a third of a turn (120°) around the vertical axis to obtain three indistinguishable positions of the molecule. Going from the last position in row one to the first position in row two, we rotate the molecule 120° around the carbon-purple axis. To go from the end of row two to the first position in row three, we rotate around the carbon-purple axis again. Finally, to get from the last position in the third row to the first position in the last row, we rotate around the carbon-white axis. In total, there are 12 indistinguishable positions, given that the colors were just there for our convenience, so for methane, $\sigma = 12$.

If you have studied group theory, you might guess that the number of equivalent positions has something to do with the rotational subgroup of the molecule. This is, in fact, correct. Methane, for example, has four C_3 axes and six independent C_2 axes. We used the four C_3 axes to generate all of the $4 \times 3 = 12$ positions in figure 6.1. We could also have used the C_2 rotations; note that $6 \times 2 = 12$.

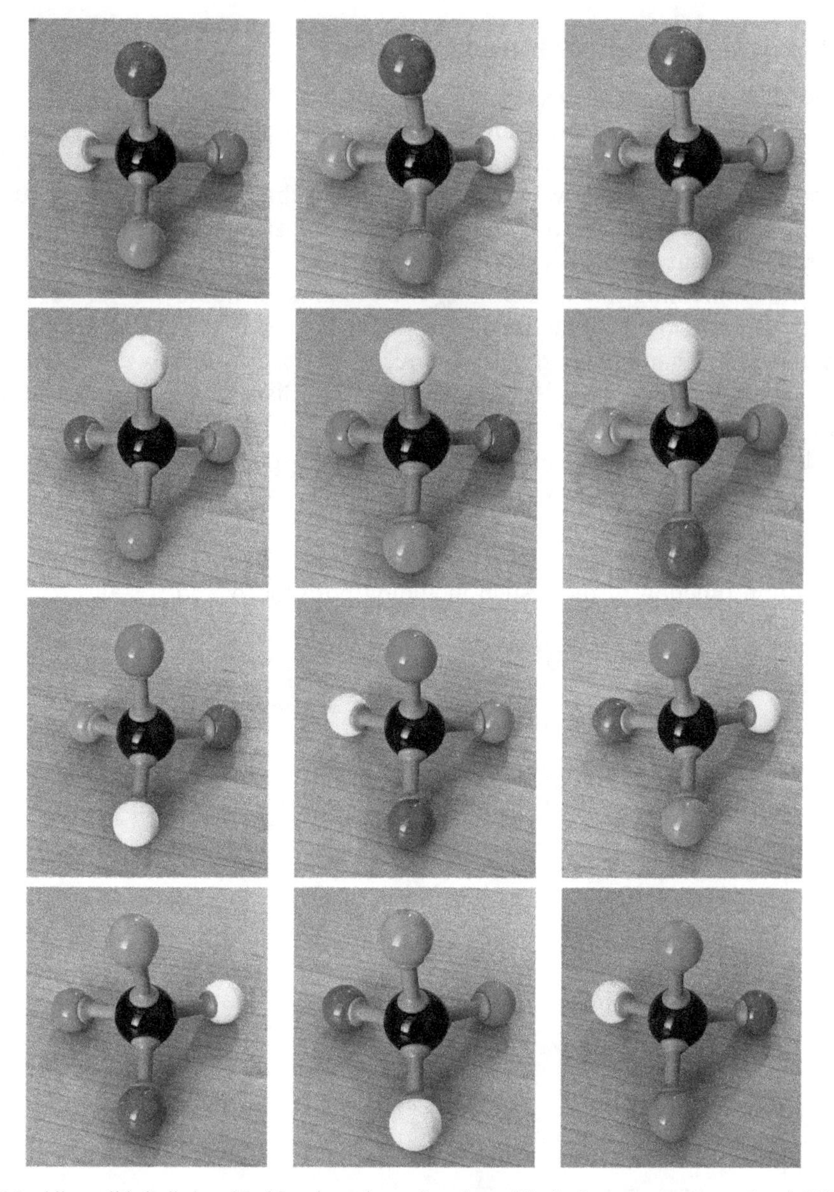

Figure 6.1. All possible indistinguishable orientations of methane. The hydrogen atoms are colored differently to allow us to see how the atoms are interchanged by the rotations.

6.2.4 The electronic partition function

Because electronic energy level separations are usually large, the value of the electronic partition function is usually the degeneracy of the ground state. This turns out to be the spin multiplicity, which we encountered previously.

On occasion, the electronic energy levels are sufficiently close that we need to consider two (or a few) of them. In this case, we explicitly evaluate the partition function as a sum over the accessible levels.

6.3 A statistical approach to equilibrium

Suppose that we want to discuss an isomerization equilibrium, $A \rightleftharpoons B$, from a *statistical* perspective. For an isomerization, we can simply think of A and B as being two different states of the system consisting of all of the atoms that make up these molecules. However, if we are talking about isomers, A and B are presumably stable entities separated by an energy barrier (as illustrated in figure 2.3). We can therefore talk about the energy levels of A and B separately. If we do so, we might end up with something like the leftmost two columns of figure 6.2. However, if we think of the two isomers as a single system, we can merge their energy levels as shown in the right-hand column of the figure. We can then ask how molecules will be distributed among these levels at equilibrium. This distribution is governed by the Boltzmann distribution. As a final step, we can count how many molecules are in energy levels that correspond to the A isomer versus those that correspond to the B isomer. The ratio of these two numbers should give us the equilibrium constant.

For simplicity in evaluating partition functions in this section, we're going to sum over the quantum states of the two molecules rather than over their energy levels. The partition function of the system including both the A and B states is

$$q_{\text{sys}} = \sum_{i \in \{a,b\}} \exp\left(-\frac{\varepsilon_i}{k_B T}\right),$$

Figure 6.2. The energy levels of a molecule A and its isomer B. The lowest energy state of B is $\Delta\varepsilon_{gs}$ above the lowest energy state of A. (It could be below it, of course. We would then have a negative value of $\Delta\varepsilon_{gs}$.) The column on the right merges the energy levels of A and B into a single set of levels.

where a and b represent, respectively, the set of possible states of A and B. In other words, we are summing over *all* the states in the right-hand column of figure 6.2. We could also separate the partition function into two sums, one for the A states and one for the B states, and add them up:

$$q_{\mathrm{sys}} = \sum_{i \in a} \exp\left(-\frac{\varepsilon_i}{k_{\mathrm{B}}T}\right) + \sum_{i \in b} \exp\left(-\frac{\varepsilon_i}{k_{\mathrm{B}}T}\right).$$

Of course, the two sums on the right-hand side of this equation are just the partition functions of A and B:

$$q_{\mathrm{sys}} = q_{\mathrm{A}} + q_{\mathrm{B}}.$$

Note that these equations assume that we are measuring the energies in the individual partition functions relative to a common zero.

If we have many copies of this molecular system (i.e. many molecules), then the probability that any given molecule is in a particular state corresponding to isomer A is just

$$P(i) = \frac{\exp\left(-\frac{\varepsilon_i}{k_{\mathrm{B}}T}\right)}{q_{\mathrm{sys}}}$$

for any desired value of $i \in a$. If we want to know the probability that a molecule is in *any* of the A states, i.e. that it is the A isomer, we just have to sum over all the A states:

$$P(\mathrm{A}) = \frac{\sum_{i \in a} \exp\left(-\frac{\varepsilon_i}{k_{\mathrm{B}}T}\right)}{q_{\mathrm{sys}}} = \frac{q_{\mathrm{A}}}{q_{\mathrm{sys}}}.$$

Now suppose that a macroscopic system contains N molecules. The expected number of A molecules in this system, once the system has come to equilibrium, is

$$N_{\mathrm{A}} = NP(\mathrm{A}) = N\frac{q_{\mathrm{A}}}{q_{\mathrm{sys}}}.$$

An analogous equation would hold for B provided that we measured all of the energies relative to the same zero, say the lowest energy level of A. More likely, because it's usually easier to do so, we calculated the partition function of B relative to its lowest energy level. This means that we have to add $\Delta\varepsilon_{\mathrm{gs}}$, the difference in energy between the ground states of A and B (figure 6.2), to all of the energy levels of B. As we saw previously, this multiplies the partition function of B by a factor of $\exp(-\Delta\varepsilon_{\mathrm{gs}}/k_{\mathrm{B}}T)$. Thus,

$$N_{\mathrm{B}} = NP(\mathrm{B}) = N\frac{q_{\mathrm{B}} \exp(-\Delta\varepsilon_{\mathrm{gs}}/k_{\mathrm{B}}T)}{q_{\mathrm{sys}}}.$$

The equilibrium constant is N_B/N_A, or

$$K = \frac{q_B}{q_A} \exp\left(-\frac{\Delta\varepsilon_{gs}}{k_B T}\right).$$

In the foregoing, we assumed that the partition functions were measured from the lowest available energy level, but this is not in fact necessary, provided we use a consistent zero of energy. Figure 6.3 shows two possible reference levels and the energy adjustments that must be made to use either of these. The figure shows the energy profile along a reaction path for an isomerization reaction. The states belonging to A and B are colored blue and red, respectively. In order for the theory we have developed to work, there has to be negligible occupation of the states colored black, which belong to neither isomer. In other words, the wells need to be sufficiently deep that the occupied states at the temperature of interest are overwhelmingly those that can be assigned to one of the isomers. This condition coincidentally makes the harmonic approximation accurate and thus allows us to use the vibrational partition functions we obtained earlier. Supposing that is so, we could decide to use the lowest vibrational state of A as our reference. We would then use equation (6.11), which assumes the lowest vibrational state as a reference, to calculate the vibrational partition functions of both A and B and apply the correction shown on the figure as $\Delta\varepsilon_{v=0}$ to the energy levels of B. *Ab initio* programs such as Gaussian calculate the optimized energy in the Born–Oppenheimer

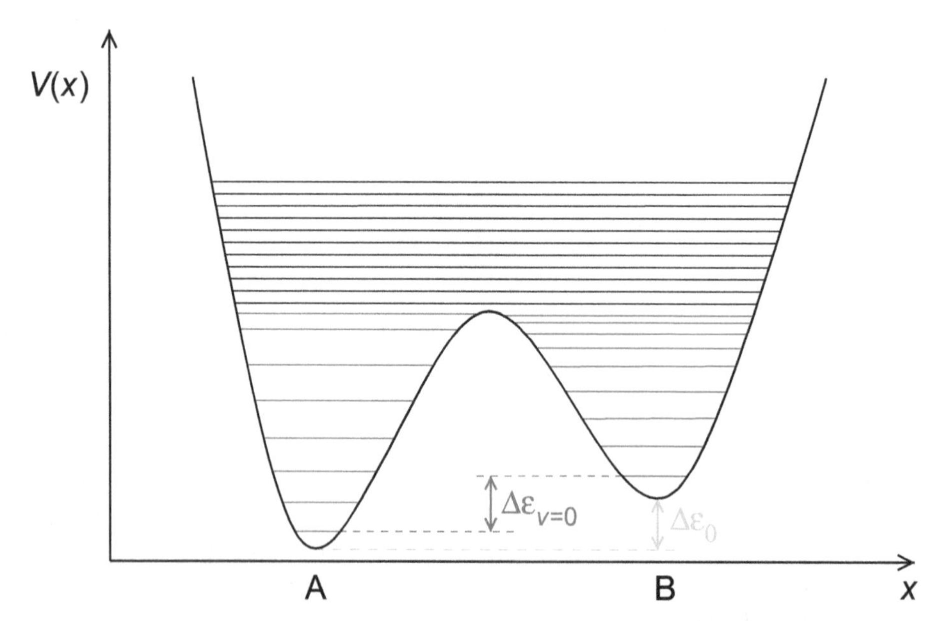

Figure 6.3. Two possible choices for the reference energy in an equilibrium calculation. We could use the lowest energy level of isomer A (its ground vibrational state), which would require an adjustment $\Delta\varepsilon_{v=0}$, the difference in energy between the ground vibrational states of the two isomers, to the vibrational energies of B. Alternatively, we could use the bottom of the A potential well and adjust the energies of B by $\Delta\varepsilon_0$, the difference between the minima of the two potential energy surfaces. See text for additional details.

approximation, in which the nuclei are frozen in place, i.e. they give the energy at the very bottom of the potential well. Using the lowest vibrational level as our reference is thus a little inconvenient because it forces us to manually adjust for the zero-point vibrational energies.

The alternative is to use the bottom of the well as a reference. We would use the corresponding vibrational partition functions, equation (6.12). All we have to do then is to apply the correction $\Delta\varepsilon_0$ to the energy levels of B, but that is just a difference of the energies at the equilibrium geometries calculated by Gaussian.

The treatment given above can be generalized to any desired stoichiometry. Consider, for example, the general case with two distinct reactants and two products:

$$a\mathrm{A} + b\mathrm{B} \rightleftharpoons c\mathrm{C} + d\mathrm{D}.$$

The equilibrium constant in this case takes the form

$$K = \frac{q_{\mathrm{C}}^{\mathrm{c}} q_{\mathrm{D}}^{\mathrm{d}}}{q_{\mathrm{A}}^{\mathrm{a}} q_{\mathrm{B}}^{\mathrm{b}}} N^{-\Delta n} \exp\left(-\frac{\Delta\varepsilon_0}{k_{\mathrm{B}}T}\right), \tag{6.14}$$

where

$$\Delta\varepsilon_0 = c\varepsilon_0^{(\mathrm{C})} + d\varepsilon_0^{(\mathrm{D})} - (a\varepsilon_0^{(\mathrm{A})} + b\varepsilon_0^{(\mathrm{B})})$$

is the difference in the potential energy minima of the appropriate stoichiometric mixtures of reactants and products, $\Delta n = c + d - (a + b)$ is just the difference in the stoichiometric coefficients of the reactants and the products, and N is a number of molecules chosen (along with the V appearing in the translational partition functions) to fix the standard state.

Equation (6.14) is not the most convenient form of the equilibrium constant. Recall that, provided we can write the energy as a sum of contributions, the molecular partition function can be written

$$\begin{aligned} q_X &= q_{\mathrm{tr}} q_{\mathrm{vib}} q_{\mathrm{rot}} q_{\mathrm{elec}} \\ &= V \mathfrak{q}_{\mathrm{tr}} q_{\mathrm{vib}} q_{\mathrm{rot}} q_{\mathrm{elec}}. \end{aligned}$$

Note that only q_{tr} depends on V. We can define the volumic molecular partition function analogously to the volumic translational partition function:

$$\begin{aligned} \mathfrak{q}_X &= q_X / V \\ &= \mathfrak{q}_{\mathrm{tr}} q_{\mathrm{vib}} q_{\mathrm{rot}} q_{\mathrm{elec}}. \end{aligned}$$

We can substitute $V\mathfrak{q}_X$ into equation (6.14) for each species:

$$\begin{aligned} K &= \frac{q_{\mathrm{C}}^{\mathrm{c}} q_{\mathrm{D}}^{\mathrm{d}}}{q_{\mathrm{A}}^{\mathrm{a}} q_{\mathrm{B}}^{\mathrm{b}}} N^{-\Delta n} \exp\left(-\frac{\Delta\varepsilon_0}{k_{\mathrm{B}}T}\right) \\ &= \frac{(V\mathfrak{q}_{\mathrm{C}})^{\mathrm{c}} (V\mathfrak{q}_{\mathrm{D}})^{\mathrm{d}}}{(V\mathfrak{q}_{\mathrm{A}})^{\mathrm{a}} (V\mathfrak{q}_{\mathrm{B}})^{\mathrm{b}}} N^{-\Delta n} \exp\left(-\frac{\Delta\varepsilon_0}{k_{\mathrm{B}}T}\right) \\ &= \frac{\mathfrak{q}_{\mathrm{C}}^{\mathrm{c}} \mathfrak{q}_{\mathrm{D}}^{\mathrm{d}}}{\mathfrak{q}_{\mathrm{A}}^{\mathrm{a}} \mathfrak{q}_{\mathrm{B}}^{\mathrm{b}}} \left(\frac{N}{V}\right)^{-\Delta n} \exp\left(-\frac{\Delta\varepsilon_0}{k_{\mathrm{B}}T}\right). \end{aligned}$$

Now note that, from the ideal gas law, $V/N = k_B T/p \equiv \hat{V}$, where $\hat{V}$ is the volume per molecule. We want to evaluate K relative to the standard state. The standard state used most often in thermodynamics corresponds to a pressure of $p° = 1$ bar. Thus,

$$K = \frac{\mathsf{q}_C^c \mathsf{q}_D^d}{\mathsf{q}_A^a \mathsf{q}_B^b} \hat{V}^{\Delta n} \exp\left(-\frac{\Delta\varepsilon_0}{k_B T}\right). \tag{6.15}$$

Equation (6.15) is the desired equation for the equilibrium constant for a gas-phase reaction. There is, however, another equivalent expression that is more convenient for use with Gaussian. Note that

$$\hat{V}^{\Delta n} = \frac{\hat{V}^c \hat{V}^d}{\hat{V}^a \hat{V}^b}.$$

We can combine the factors of $\hat{V}$ with the volumic partition functions to define a new kind of partition function:

$$\hat{q} = \hat{V}\mathsf{q} = \hat{V}\mathsf{q}_{tr} q_{vib} q_{rot} q_{elec}.$$

We then have

$$K = \frac{\hat{q}_C^c \hat{q}_D^d}{\hat{q}_A^a \hat{q}_B^b} \exp\left(-\frac{\Delta\varepsilon_0}{k_B T}\right). \tag{6.16}$$

When asked to calculate vibrational frequencies, Gaussian also calculates partition functions. As it turns out, Gaussian calculates a translational partition function corresponding to $\hat{q}_{tr} = \hat{V}\mathsf{q}_{tr}$. To calculate equilibrium constants using Gaussian, it is therefore easiest to use equation (6.16). Note, however, that $\hat{q}$ is not the translational partition function in any real system because the volume $\hat{V}$ is not the volume of any container.

Example 6.1. *We're going to use Gaussian to get all of the data necessary to calculate an equilibrium constant for the reaction* $2H_2O_{(g)} \rightleftharpoons H_3O^+_{(g)} + OH^-_{(g)}$ *at* 25 °C.

To calculate this equilibrium constant, we will set up a Gaussian `Opt+Freq` *calculation in GaussView. Generally, this is straightforward, except that building an initial geometry for the* H_3O^+ *ion requires a little bit of craft. Make a water molecule, then rotate it so that you can place a simple hydrogen atom at about the right place in terms of both bonding distance and bond angles. Use the measurement tool to measure the O–H bond length for one of the bonds from the original water molecule, then use the* `Modify bond` *tool to make a bond between the oxygen and the extra hydrogen atom of the same length. Finally, use the* `Modify angle` *tool to start off with an approximately tetrahedral bond angle of* 109°.

For these calculations, I used the B3LYP density functional with a 6-31G(d,p) basis set. I used restricted-spin wavefunctions. By default, Gaussian uses a standard-state pressure of 1 atm, which is quaintly old-fashioned[4]. *Gaussian gives us the option to change this using the* `pressure` *keyword. To use this, type the following into the* `Additional Keywords` *box at the bottom of the* `Gaussian Calculation Setup` *window in GaussView:* `Pressure=0.986 923`. *(The pressure has to be in atm, and 1 bar = 0.986 923 atm.)*

It is always a good idea to check ab initio *calculations against experimental data. Table 6.1 compares my calculations against experimental results. With the exception of the* OH^- *bond length, the agreement is very good. The long bond length calculated for* OH^- *is a sign that we could do better, probably by using a larger basis set, but we will carry on and see what effect this has on the final result.*

Table 6.2 gives the data computed by Gaussian for all the species in this reaction. Note that $\hat{q}$ *is what Gaussian calls the* `Total Bot` *value of the partition function. We can now start to put the pieces together:*

$$\begin{aligned}\Delta\varepsilon_0 &= \varepsilon_0(H_3O^+) + \varepsilon_0(OH^-) - 2\varepsilon_0(H_2O)\\ &= -76.7056 + (-75.7263) - 2(-76.4197)\text{ hartree} = 0.4075\text{ hartree}\\ &\equiv (0.4075\text{ hartree})(2625.500\text{ kJ mol}^{-1}\text{hartree}^{-1}) = 1070\text{ kJ mol}^{-1}.\end{aligned}$$

$$\begin{aligned}K &= \frac{(\hat{q}_{H_3O^+})(\hat{q}_{OH}^-)}{\hat{q}_{H_2O}^2}e^{-\Delta\varepsilon_0/RT}\\ &= \frac{(9.531\times10^{-8})(6.803\times10^{3})}{(1.985\times10^{-2})^2}\exp\left(\frac{-1070\times10^{3}\text{ Jmol}^{-1}}{(8.314\,462\,618\text{ JK}^{-1}\text{mol}^{-1})(298.15\text{ K})}\right)\\ &= 5.76\times10^{-188}.\end{aligned}$$

We can also calculate this equilibrium constant from the standard free energies of formation available from the NIST-JANAF tables (https://janaf.nist.gov). These tables give $\Delta_r G^\circ = 925.083$ kJ mol^{-1}, *from which we calculate an equilibrium constant of* 8.56×10^{-163}. *If the error in our calculation came entirely from the energies, which*

Table 6.1. A comparison of experimental and calculated properties of the reactants and products in the reaction $2H_2O_{(g)} \rightleftharpoons H_3O^+_{(g)} + OH^-_{(g)}$. Method: B3LYP with the 6-31G(d,p) basis set and restricted electronic wavefunctions as implemented in Gaussian [2].

	Calculated		Experimental		
Species	Bond length/Å	Bond angle/°	Bond length/Å	Bond angle/°	Reference
H_2O	0.965	103.7	0.958	104.5	[3]
H_3O^+	0.982	113.4	0.974	113.6	[4]
OH^-	0.983		0.964		[5]

[4] IUPAC recommended a change to a standard pressure of 1 bar in 1982. The difference between 1 atm and 1 bar is admittedly small and would not have a very large effect on the results. Am I being fussy by insisting on this correction? Yes.

Table 6.2. Data calculated using Gaussian as described in the text for the reaction $2H_2O_{(g)} \rightleftharpoons H_3O^+_{(g)} + OH^-_{(g)}$. For the partition functions, the `Total Bot` values are used.

Species	$\hat{q}$	ε_0/hartree
H_2O	1.985×10^{-2}	−76.4197
H_3O^+	9.531×10^{-8}	−76.7056
OH^-	6.803×10^{3}	−75.7263

it doesn't, the discrepancy would correspond to an error of about 144 kJ mol^{-1} *in* $\Delta\varepsilon_0$. *The B3LYP density functional generally isn't bad for these kinds of problems, but we probably needed a larger basis set.*

In any event, our calculation and the thermodynamic data both predict a tiny equilibrium constant for this reaction, which you may find surprising. After all, the equilibrium constant for the analogous reaction in an aqueous solution is 10^{-14}. *This goes to show the critical role of the solvent in stabilizing some of the species involved in reactions in solution. Another lesson to draw is that we don't expect proton transfer between water molecules to be a significant reaction in gas-phase reactions.*

Exercise

6.1

(a) Calculate the zero-point energy of an electron in a 1 nm box.

(b) Since there is no potential energy, the energy of a particle in a box is all kinetic. Calculate the speed of the electron.

(c) Calculate the absolute value of the momentum of the electron.

6.2 How many normal modes would cyclobutane (C_4H_8) have?

6.3

(a) Roughly how many translational states are accessible to a $^1H_2$ molecule at 20 °C in a 1.050 L container?

(b) Without doing detailed calculations, estimate the number of translational states available to a molecule of $^2H_2$ under the conditions of question (a).

6.4 Determine the symmetry number of benzene.

Hint: Draw the carbon skeleton of benzene, and number the carbon atoms. Then think about how you can rotate the molecule to get back to an indistinguishable position. What happens if you flip the molecule over?

6.5 $^{35}Cl_2$ has a vibrational wavenumber of 559.71 cm^{-1}. Calculate the vibrational partition function of $^{35}Cl_2$ with respect to the bottom of its potential energy well at 350 °C.

6.6

(a) Vibrational frequencies are often given as a wavenumber, $\tilde{\nu}_0$, in units of cm^{-1}. The wavenumber is the inverse of the wavelength. From elementary quantum mechanics, recall that for photons $E = hc/\lambda$, so we have $E = hc\tilde{\nu}$. Correspondingly, for a harmonic oscillator, we could write $E_v = hc\tilde{\nu}_0\left(v + \frac{1}{2}\right)$. Obtain an equation relating $\tilde{\nu}_0$ to ω_0.

(b) For $^1H^{35}Cl$, $\tilde{\nu}_0 = 2990.95\ \text{cm}^{-1}$. What is ω_0?

(c) What is the probability that a molecule of HCl is in the ground vibrational state at 20 °C?

(d) Plot the vibrational partition function vs temperature. Around what temperature would the excited states become significantly populated?

6.7 Choose one of the following diatomic molecules for this question: HF, HCl, LiH, LiF, LiCl, ClF, OH^-, Li_2, Cl_2, F_2.

(a) Choose a basis set that you consider to be reasonable for your molecule and for the task of computing both equilibrium properties and a potential energy curve. Explain briefly how you chose this basis set.

(b) Carry out a geometry optimization. Report both the optimized geometry and corresponding energy.

Note: The easiest way to find the energy after the geometry optimization is to open up the `.chk` or `.log` file, then click on `Results→Summary`. The '`Electronic Energy`' is misnamed. This is really the value of the effective potential (which includes nuclear–nuclear repulsion) at the computed geometry. The value is given in hartree. The Wikipedia page for the hartree (https://en.wikipedia.org/wiki/Hartree) gives conversion factors to many other units of energy.

(c) Obtain and graph the effective potential energy curve for your diatomic molecule.

(d) Carry out a vibrational analysis (frequency calculation) using Gaussian. Use data from your geometry optimization and vibrational analysis to calculate the partition functions using the formulas from section 6.2, and compare the values you obtain to those given in the `.log` file.

Note that Gaussian uses the old standard pressure of 1 atm, instead of the modern value of 1 bar. Make sure to use the same pressure in your Gaussian calculations as in your manual calculations.

(e) Calculate the equilibrium constant at 25 °C for the dissociation of your molecule using the results of Gaussian calculations.

Note: A table may be a good way to present the data extracted from Gaussian for each relevant species.

References

[1] Cohen E R *et al* 2008 *Quantities, Units and Symbols in Physical Chemistry: IUPAC Green Book* 3rd edn (Cambridge: IUPAC & RSC Publishing)

[2] Frisch M J *et al* 2016 *Gaussian Inc 16 Revision B.01* (Wallingford, CT: Gaussian Inc)

[3] Hoy A R and Bunker P R 1979 A precise solution of the rotation bending schrödinger equation for a triatomic molecule with application to the water macromolecules *J. Mol. Spectrosc.* **74** 1–8

[4] Tang J and Oka T 1999 Infrared spectroscopy of H_3O^+: the ν_1 fundamental band *J. Mol. Spectrosc.* **196** 120–30

[5] Rosenbaum N H, Owrutsky J C, Tack L M and Saykally R J 1986 Velocity modulation laser spectroscopy of negative ions: the infrared spectrum of hydroxide (OH^- *J. Chem. Phys.* **84** 5308–13

IOP Publishing

Foundations of Chemical Kinetics
A hands-on approach
Marc R Roussel

Chapter 7

Transition-state theory

Transition-state theory (TST) is one of the great theories of chemical kinetics. Like all great theories, it can be understood at a number of levels. Most chemists will eventually be exposed to the thermodynamic formalism, which will be our starting point. We will develop this formalism into a computational method using ideas from the statistical theory of equilibrium developed in the last chapter. We will briefly survey some modern developments of the theory in section 7.4. Finally, we will consider tunneling corrections in transition-state theory.

On a historical note, transition-state theory has gone by a few different names over time. If you go digging through the older literature and see papers on 'activated complex theory' or on the 'theory of absolute rates,' these are in fact papers on TST.

7.1 The thermodynamic formalism

The fundamental approximation of transition-state theory is that we can break down an elementary reaction, reactant(s) → product(s), into two steps: (i) reaching the transition state and then (ii) going across the transition state to the product(s). As the title of this section hints, we will be introducing thermodynamic considerations, in which the choice of standard state matters. This affects the form of the equations for first- and second-order reactions slightly differently, so we will develop both cases in parallel. We consider the elementary reactions

$$R \overset{k^{(1)}}{\rightarrow} \text{product(s)}, \tag{7.1a}$$

and

$$X + Y \overset{k^{(2)}}{\rightarrow} \text{product(s)}, \tag{7.1b}$$

whose rates are given by the law of mass action as follows:

$$v^{(1)} = k^{(1)}[\mathrm{R}] \tag{7.2a}$$

and

$$v^{(2)} = k^{(2)}[\mathrm{X}][\mathrm{Y}], \tag{7.2b}$$

respectively.

The fundamental assumption of TST, expressed as a mechanism, can be written as

$$R \overset{K_{\mathrm{tot}}^{\ddagger}}{\rightleftharpoons} \mathrm{TS} \overset{k^{\ddagger}}{\rightarrow} \text{product(s)} \tag{7.3a}$$

or

$$X + Y \overset{K_{\mathrm{tot}}^{\ddagger}}{\rightleftharpoons} \mathrm{TS} \overset{k^{\ddagger}}{\rightarrow} \text{product(s)}, \tag{7.3b}$$

respectively, for first- or second-order reactions. We assume that the reactant(s) and transition state (TS) are in equilibrium with an equilibrium constant $K_{\mathrm{tot}}^{\ddagger}$. The meaning of the subscript 'tot' (for total) will become clear later. The rate constant $k^{\ddagger}$ then gives the specific rate for proceeding from the TS to the product(s). If the reactants and transition state are in equilibrium, then we have

$$K_{\mathrm{tot}}^{\ddagger} = \frac{a_{\mathrm{TS}}}{a_{\mathrm{R}}}$$

or

$$K_{\mathrm{tot}}^{\ddagger} = \frac{a_{\mathrm{TS}}}{a_{\mathrm{X}} a_{\mathrm{Y}}}.$$

In these equations, a_i denotes the activity of chemical species i. For the moment, we will adopt a concentration-based activity standard with standard concentration $c°$. Because we're really thinking about gas-phase reactions in this chapter, $c° = n/V = p°/RT$. The advantage of using concentration-based activity is that it will dovetail easily with the most common usage of the law of mass action, in which we express rates in terms of concentrations.

The concentration-based activity of a substance i is $a_i = \gamma_i c_i / c°$, where γ_i is the activity coefficient of i. Thus,

$$K_{\mathrm{tot}}^{\ddagger} = \frac{\gamma_{\mathrm{TS}}}{\gamma_{\mathrm{R}}} \frac{[\mathrm{TS}]}{[\mathrm{R}]} \tag{7.4a}$$

or

$$K_{\mathrm{tot}}^{\ddagger} = \frac{\gamma_{\mathrm{TS}}}{\gamma_{\mathrm{X}} \gamma_{\mathrm{Y}}} \frac{c°[\mathrm{TS}]}{[\mathrm{X}][\mathrm{Y}]}, \tag{7.4b}$$

where again the first equation of this pair applies to first-order reactions, and the second equation to second-order reactions. Equations (7.4) are the basis for understanding many nonideal effects in chemical kinetics via the activity coefficients γ_i. For now though, we are focused on the gas phase. Except at very high pressures, gases tend to behave reasonably ideally, so we assume that $\gamma_i = 1$ for all species appearing in equations (7.4). Thus,

$$[\mathrm{TS}] = K_{\mathrm{tot}}^{\ddagger}[\mathrm{R}]$$

or

$$[\mathrm{TS}] = K_{\mathrm{tot}}^{\ddagger}[\mathrm{X}][\mathrm{Y}]/c^{\circ}.$$

From the first- and second-order pseudo-mechanisms (7.3), we have $v = k^{\ddagger}[\mathrm{TS}]$. Thus

$$v^{(1)} = k^{\ddagger}K_{\mathrm{tot}}^{\ddagger}[\mathrm{R}]$$

or

$$v^{(2)} = \frac{k^{\ddagger}K_{\mathrm{tot}}^{\ddagger}}{c^{\circ}}[\mathrm{X}][\mathrm{Y}].$$

However, for the elementary reactions (7.1), $v^{(1)} = k^{(1)}[\mathrm{R}]$ and $v^{(2)} = k^{(2)}[\mathrm{X}][\mathrm{Y}]$, respectively. Thus,

$$k^{(1)} = k^{\ddagger}K_{\mathrm{tot}}^{\ddagger} \tag{7.5a}$$

and

$$k^{(2)} = \frac{k^{\ddagger}K_{\mathrm{tot}}^{\ddagger}}{c^{\circ}}. \tag{7.5b}$$

To go further, we have to think about the treatment of the reactive mode. To my knowledge, there are at least three different ways to deal with this, all of which give the same final equation. I will present two of them in this section, starting with one in which we treat the reactive mode as a translational mode, which is perhaps the most intuitive treatment.

The reaction coordinate as a translational mode The equilibrium constant $K_{\mathrm{tot}}^{\ddagger}$ appearing in all of the equations we have written is the complete ('total') partition function for the transition state. This includes the mode of motion across the barrier. From section 6.3, we know that the equilibrium constant contains a factor that is a ratio of partition functions (e.g. equation (6.16)). For the equilibria in the pseudo-mechanisms (7.3), the numerator of this fraction is the partition function of the transition state. But this is a strange equilibrium constant, since the reactive mode cannot be in equilibrium: as long as there is a deficiency of products relative to reactants given the equilibrium constant of the reaction (which would typically be the case during most of a kinetics experiment), the population of transition states is mainly sustained by molecules climbing the energy ladder from the reactant side. We don't have a balance between this process and molecules coming up to the transition state from the product side, which would be required for equilibrium. Our first move is therefore to pull the reactive mode out of $K_{\mathrm{tot}}^{\ddagger}$:

$$K_{\text{tot}}^{\ddagger} = q_{\text{r}}^{\ddagger} K^{\ddagger}, \tag{7.6}$$

where $q_{\text{r}}^{\ddagger}$ is the partition function for the reactive mode, and $K^{\ddagger}$ is the rest of the equilibrium constant after we have removed this reactive mode.

It makes sense to think of the reactive mode as a translational mode, since motion along this mode is free, at least in the vicinity of the transition state. We have previously obtained the partition function for a one-dimensional translation, equation (6.9). The partition function depends on a mass. In the normal-mode coordinates, the reactive coordinate, q_{r}, is associated with a reduced mass μ_{r}, which is the appropriate mass to use here. This partition function also depends on the size of the region 'containing' the motion. Broadening our definition of the transition state slightly, we can say that it is, roughly speaking, the set of quantum states above the barrier separating reactants from products. Let's say that this barrier has a width of δ. The partition function for the transition state is then

$$q_{\text{r}}^{\ddagger} = \frac{\delta}{h}\sqrt{2\pi\mu_{\text{r}} k_{\text{B}} T}. \tag{7.7}$$

As for $k^{\ddagger}$, it is the rate at which molecules cross the transition state from reactants to products. If molecules are traveling across the barrier at an average speed of $\langle u_{\text{r}} \rangle$, then the rate of crossing (the number of crossings per unit time) is just

$$k^{\ddagger} = \langle u_{\text{r}} \rangle / \delta. \tag{7.8}$$

Earlier, we derived the velocity distribution in one dimension, equation (2.6). If we set up our coordinate system so that the reactant-to-product direction is the direction of increasing q_{r}, we are interested in the average velocity considering only the molecules traveling in the positive direction. Taking into account that we must use the reduced mass, the average speed of travel across the top of the barrier from reactants to products is

$$\begin{aligned}
\langle u_{\text{r}} \rangle &= \int_0^{\infty} u_{\text{r}}\, p(u_{\text{r}})\, du_{\text{r}} \\
&= \sqrt{\frac{\mu_{\text{r}}}{2\pi k_{\text{B}} T}} \int_0^{\infty} u_{\text{r}} \exp\left(\frac{-\mu_{\text{r}} u_{\text{r}}^2}{2k_{\text{B}} T}\right) du_i \\
&= \sqrt{\frac{\mu_{\text{r}}}{2\pi k_{\text{B}} T}} \frac{k_{\text{B}} T}{\mu} = \sqrt{\frac{k_{\text{B}} T}{2\pi\mu_{\text{r}}}}.
\end{aligned} \tag{7.9}$$

If we now combine equations (7.5) to (7.9), we get the surprisingly simple results

$$k^{(1)} = \frac{k_{\text{B}} T}{h} K^{\ddagger} \tag{7.10a}$$

and

$$k^{(2)} = \frac{k_{\text{B}} T}{hc^{\circ}} K^{\ddagger}. \tag{7.10b}$$

Note that all of the details associated with the potential energy surface (PES) that we needed to consider on the way to these equations, in particular the width of the barrier (for which we would have been hard pressed to state a precise number) and the reduced mass associated with the reactive mode, have canceled out. The only part of the expression for the rate constant that depends on the PES is $K^{\ddagger}$, and it only depends on the properties of the transition state and reactants. If we can identify the transition state and compute its partition function and the height of the potential energy barrier separating the reactants from the products, we have everything we need to know to calculate a rate constant. We will look at the details of this calculation in the next section.

The reaction coordinate as a vibrational mode Although this is intuitively a bit less obvious, we can treat the reactive mode as a 'defective vibration,' for lack of a better term. The appeal of this approach is that we don't need to bring in the artificial parameter δ, which doesn't have a clear definition. We start from equation (7.6), but now we assume that $q_{\mathrm{r}}^{\ddagger}$ is a vibrational mode with a partition function given by equation (6.12). Here, it is convenient to use the frequency in Hz rather than the angular frequency, so we replace $\hbar\omega_{\mathrm{r}}$ by $h\nu_{\mathrm{r}}$, where the subscript r refers to the reaction coordinate. We therefore have

$$q_{\mathrm{r}}^{\ddagger} = \frac{\exp\left(-\frac{h\nu_{\mathrm{r}}}{2k_{\mathrm{B}}T}\right)}{1 - \exp\left(-\frac{h\nu_{\mathrm{r}}}{k_{\mathrm{B}}T}\right)}.$$

We now need a bit of math. Taylor's theorem says that a function can be approximated as follows:

$$f(x) \approx f(a) + \left.\frac{\mathrm{d}f}{\mathrm{d}x}\right|_{x=a}\left(x - a\right) + \left.\frac{1}{2!}\frac{\mathrm{d}^2 f}{\mathrm{d}x^2}\right|_{x=a}(x - a)^2 + \cdots$$
$$+ \left.\frac{1}{k!}\frac{\mathrm{d}^k f}{\mathrm{d}x^k}\right|_{x=a}(x - a)^k,$$

where $x = a$ is a point around which we want to approximate $f(x)$. If we only wanted to look at small deviations from $x = a$, we would typically keep the first nonzero term, and throw out the rest on the basis that if $(x - a)^k$ is small, then $(x - a)^{k+1}$ is smaller still.

The Hooke's law constant defined by equation (6.5) decreases as the second derivative decreases, i.e. as the potential energy becomes less strongly confining. A reactive coordinate is an extreme case of a weakly confined mode, so we expect the associated frequency (equation (6.7)) to be small. Let $x = h\nu_{\mathrm{r}}/k_{\mathrm{B}}T$. Using Taylor's theorem, and stopping at the first nonzero term, we have

$$\mathrm{e}^{-x/2} \approx 1,$$

and

$$1 - \mathrm{e}^{-x} \approx x.$$

Thus,

$$q_r^\ddagger \approx \frac{1}{x} = \frac{k_B T}{h\nu_r}.$$

A vibration doesn't intrinsically have a direction (unlike translational motion), so now we have to be careful about the fact that the equilibrium assumption we started with also doesn't distinguish between transition states traveling from left to right or from right to left across the top of the barrier. There is a 50% probability that a molecule, having reached the top of the barrier, will have a reactive-mode momentum that points towards products rather than back towards reactants. Thus, in this derivation, we should only count half of the transition states as proceeding to products. Consequently,

$$k^{(1)} = \frac{1}{2}k^\ddagger \frac{k_B T}{h\nu_r} K^\ddagger$$

and

$$k^{(2)} = \frac{1}{2}\frac{k^\ddagger}{c^\circ}\frac{k_B T}{h\nu_r} K^\ddagger.$$

Recall that $k^\ddagger$ represents the rate at which complexes with the correct energy cross the transition state. If the frequency of 'vibration' is ν_r, then a complete back-and-forth cycle is completed in time ν_r^{-1}. We are only interested in half of this cycle, i.e. in moving across the transition state from reactant to product, which will take half of this time, i.e. $\nu_r^{-1}/2 = (2\nu_r)^{-1}$. Since the rate is the inverse of the average time taken for a motion, we have $k^\ddagger = 2\nu_r$. Substituting this into our expressions for the rate constant, we get an exact match for equations (7.10)!

Transition-state theory is an odd duck. We can derive its fundamental equations either by treating motion across the transition state as a translation or as a half-vibration. Either way, the bits that depend on the path taken to get these equations cancel out, and the data we need in order to calculate rate constants ends up being exactly the same. Either transition-state theory is profoundly correct, or this is one of those cosmic coincidences that it's best not to think too hard about.

Whether we come to it one way or another, let's now try to think of equations (7.10) in thermodynamic terms. Suppose that we now define a standard free energy of activation $\Delta^\ddagger G^\circ$ as follows:

$$\Delta^\ddagger G^\circ = -RT \ln K^\ddagger.$$

This is a strange kind of free energy change because $K^\ddagger$ is an 'equilibrium constant' from which one mode of motion of the transition state has been removed. Nevertheless, let's pursue this idea. For first-order reactions,

$$\begin{aligned} k^{(1)} &= \frac{k_B T}{h} e^{-\Delta^\ddagger G^\circ/RT} \qquad (7.11a) \\ &= \frac{k_B T}{h} e^{-(\Delta^\ddagger H^\circ - T\Delta^\ddagger S^\circ)/RT} \end{aligned}$$

$$\therefore \frac{hk^{(1)}}{k_BT} = e^{\Delta^{\ddagger}S^{\circ}/R}e^{-\Delta^{\ddagger}H^{\circ}/RT}$$

$$\therefore \ln\left(\frac{hk^{(1)}}{k_BT}\right) = \frac{\Delta^{\ddagger}S^{\circ}}{R} - \frac{\Delta^{\ddagger}H^{\circ}}{RT}, \tag{7.11b}$$

or, for second-order reactions,

$$\ln\left(\frac{hk^{(2)}c^{\circ}}{k_BT}\right) = \frac{\Delta^{\ddagger}S^{\circ}}{R} - \frac{\Delta^{\ddagger}H^{\circ}}{RT}. \tag{7.11c}$$

The latter two equations are called the **Eyring equations**[1]. One way to think about them is that they *define* an entropy and enthalpy of activation. These can be retrieved by plotting $\ln(hk^{(1)}/k_BT)$ [or $\ln(hk^{(2)}c^{\circ}/k_BT)$] vs T^{-1}. Such a plot should have a slope of $-\Delta^{\ddagger}H^{\circ}/R$ and an intercept of $\Delta^{\ddagger}S^{\circ}/R$.

$\Delta^{\ddagger}H^{\circ}$ and $\Delta^{\ddagger}S^{\circ}$ should be thought of as empirical parameters. If we carry out a series of experiments varying, say, a substituent, then comparing the enthalpies and entropies of activation can be meaningful. But these quantities are often abused. The derivation of these parameters assumes an elementary reaction, but the Eyring equation is often applied to complex reactions, which makes it hard to know exactly what $\Delta^{\ddagger}H^{\circ}$ and $\Delta^{\ddagger}S^{\circ}$ mean [1]. In some simple cases, we can interpret the activation enthalpy and entropy as telling us the changes in enthalpy and entropy, respectively, going from the reactants to the last transition state (see [2], section 17.3.1). But it's easy to write down mechanisms where this interpretation would not hold.

Example 7.1. *Bromomethane is used extensively as a fumigant to kill a variety of pests. It is both a greenhouse gas and an ozone-depleting chemical. Larin and coworkers have studied the gas-phase reaction* $Cl^{\bullet} + CH_3Br \rightarrow CH_2Br^{\bullet} + HCl$ *[3]. The rate constant for this reaction varies with temperature as follows:*

T/K	294	306	315	328	337	350	360
$k/10^5\ m^3mol^{-1}s^{-1}$	2.70	3.01	3.35	3.72	4.14	4.61	5.14

We would like to use the Eyring equation to extract the enthalpy and entropy of activation. Because this is a second-order reaction, we will use equation (7.11c). As noted earlier, in the gas phase, $c^{\circ} = p^{\circ}/RT$. *Because this depends on the temperature,*

[1] Henry J Eyring, whose contributions were central to the development of transition-state theory, was also a devout member of the Church of Latter-Day Saints (LDS). Because of the way the LDS Church spread in Western North America, there happens to be a large community of LDS members in Southern Alberta, where Lethbridge is located. When the Eyring equation comes up in class, I will often be asked if the originator of this equation is the same Henry Eyring who is an important leader in the Church. And the answer is no, but they are closely related: the chemist Henry J Eyring was the father of the Church leader Henry B Eyring. Incidentally, Henry B Eyring has a degree in physics.

we have to calculate $c°(T)$ for each experimental data point. The easiest way to handle this is in a spreadsheet. Here is what mine looks like:

	A	B	C	D	E	F	G	H
1	Analysis of the data of Larin et al., *Kinet. Catal.* **59**, 11 (2018)							
2	for the reaction Cl + CH_3Br → CH_2Br + HCl							
3								
4	p^0:	1.00E+05 Pa						
5	k_B:	1.380649E-23 J/K						
6	h:	6.626070E-34 J s						
7	R:	8.314462618 J $K^{-1}mol^{-1}$						
8								
9	T/K	294	306	315	328	337	350	360
10	$k/m^3mol^{-1}s^{-1}$	2.70E+05	3.01E+05	3.35E+05	3.72E+05	4.14E+05	4.61E+05	5.14E+05
11	c^0/mol m^{-3}	40.91	39.30	38.18	36.67	35.69	34.36	33.41
12	$\ln(hkc^0/k_BT)$	-13.22603115	-13.197354	-13.148308	-13.124427	-13.071593	-13.039762	-12.987278
13	T^{-1}/K^{-1}	3.40E-03	3.27E-03	3.17E-03	3.05E-03	2.97E-03	2.86E-03	2.78E-03
14								

Once you have calculated all of the relevant quantities in a spreadsheet, you can plot the results to check for linearity. If the graph is nonlinear, you can't do an Eyring analysis. In this case, we get a straight line with a bit of scatter (figure 7.1), so we can carry on.

The slope of the graph is -379 K, *and since the slope is* $-\Delta^{\ddagger}H°/R$, *the enthalpy of activation is* $\Delta^{\ddagger}H° = 3.1$ kJ mol^{-1}. *Similarly, the intercept is* $\Delta^{\ddagger}S°/R = -11.95$, *giving an entropy of activation of* $\Delta^{\ddagger}S° = -99$ J K^{-1}mol^{-1}. *Entropies of activation of the order of* -100 J K^{-1}mol^{-1} *are typically seen in bimolecular reactions and can be thought of as the entropy loss when the two reactants are brought together to form a transition state. Large deviations from this typical value indicate that there are additional intra-molecular events contributing to the formation of the transition state.*

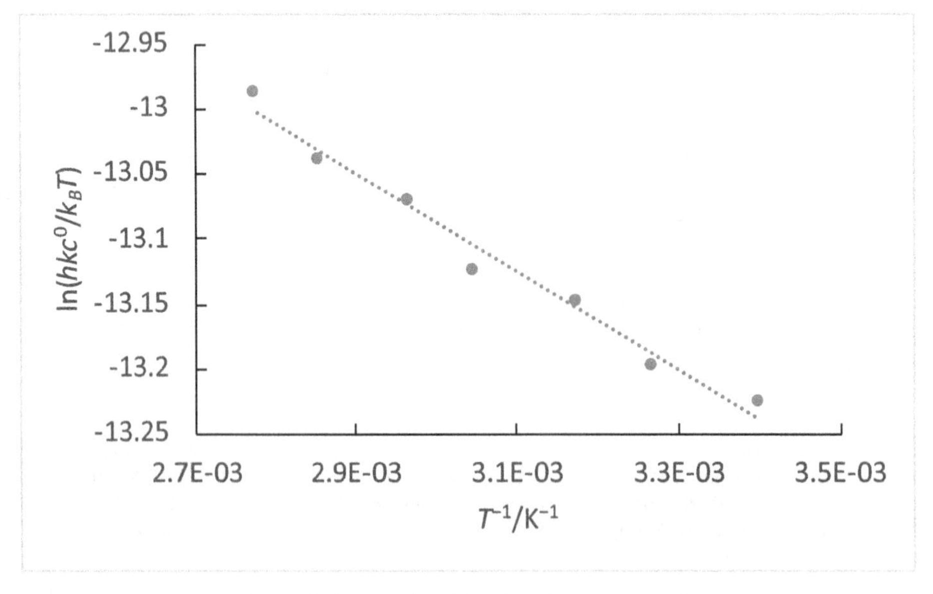

Figure 7.1. Eyring plot of the data from example 7.1.

7.2 Calculating rate constants in TST

The Eyring thermodynamic formalism lets us calculate the enthalpies and entropies of activation from experimental data, but it is not a prescription for computing a rate constant *ab initio*. For that, we go back to equations (7.10). We can use the techniques of section 6.3 to calculate the equilibrium constant. Any of the expressions for the equilibrium constant from section 6.3 would do, but we turn specifically to equation (6.16), which allows us to directly use the data generated by Gaussian. Using this expression, equations (7.10) become

$$k^{(1)} = \frac{k_B T}{h} \frac{\hat{q}^{\ddagger}}{\hat{q}_R} e^{-\Delta^{\ddagger}\varepsilon/k_B T} \tag{7.12a}$$

and

$$k^{(2)} = \frac{k_B T}{hc^{\circ}} \frac{\hat{q}^{\ddagger}}{\hat{q}_X \hat{q}_Y} e^{-\Delta^{\ddagger}\varepsilon/k_B T} \tag{7.12b}$$

respectively for the first- and second-order cases. In these equations, $\Delta^{\ddagger}\varepsilon$ represents the difference in energy between the transition state and the reactant(s). We have seen that the most convenient reference point from the point of view of computational chemistry is the effective potential energy at the local minimum. For the transition state, we want the effective potential energy at the saddle point. The partition functions of the reactants are easily obtained, as we discussed in the previous chapter. For the transition state, we need to recall that the part of the partition function associated with the reaction coordinate was previously removed (equation (7.6)). The partition function $q^{\ddagger}$ therefore only considers $3N - 7$ vibrational modes for a nonlinear molecule, or $3N - 6$ vibrational modes for a linear molecule.

It is worth spending a bit of time thinking about the units of equations (7.12). In equation (7.12a), since the partition functions are dimensionless, $k^{(1)}$ has units of s^{-1}, which is exactly what we expect for a first-order rate constant. The situation for equation (7.12b) is slightly more complicated due to the appearance of c°. Suppose that we use $c^{\circ} = p^{\circ}/RT$, where p° is the standard pressure (1 bar in the typical convention for gas-phase reactions). Working in SI units, c° would have units of mol m^{-3}, so $k^{(2)}$ would have units of $\text{m}^3\text{mol}^{-1}\text{s}^{-1}$, which are correct units for a second-order rate constant. Gas-phase kineticists often use units of $\text{cm}^3\text{molecule}^{-1}\text{s}^{-1}$. We can get something like these units if we use $c^{\circ} = N/V = p^{\circ}/k_B T$, which has units of molecules m^{-3}. With this c°, $k^{(2)}$ comes out in units of $\text{m}^3\text{molecule}^{-1}\text{s}^{-1}$. Confusingly, because 'molecule' is dimensionless, you will sometimes see units of cm^3s^{-1}, which is the same thing as $\text{cm}^3\text{molecule}^{-1}\text{s}^{-1}$. Another option offers itself if, in the rate laws we started with, we used the pressures of the reactants and transition state instead of concentrations. In this case, c° would have been replaced by p°, and the units of $k^{(2)}$ could be either $\text{bar}^{-1}\text{s}^{-1}$ or $\text{Pa}^{-1}\text{s}^{-1}$, depending on the units chosen for the standard pressure. Again, these units are not unusual in gas-phase kinetics.

In order to use equations (7.12), we have to be able to locate the transition state. Fortunately for us, Gaussian can do this and can also calculate the necessary partition functions at the transition state. It is important, however, to have some idea of how Gaussian does this in order to have a chance of success. Gaussian actually provides two quite different techniques for finding the transition state. We will focus on 'Berny' searches (named for Berny Schlegel, who developed the algorithm). In a Berny search, whose basic idea is illustrated in figure 7.2, Gaussian uses the local derivatives of the effective potential to walk uphill on the PES while staying at the minimum in the transverse directions. The difficulty with this approach is that you have to start somewhere near the reaction path in order for this to work. At a minimum in the PES, there isn't a distinct climbing direction. If you start in a strange geometry far from the reaction path, the algorithm might wander away to a physically irrelevant transition state or not find a transition state at all. At least for simple reactions, it is generally possible to guess at the transition-state geometry. Even in more complicated cases, roughly locating the transition state by doing a scan of the PES will generally give us an idea of where to start. You only have to nudge the starting geometry in the right direction in order for a Berny search to work. It isn't necessary to know exactly where the transition state is located.

Let us put what we have learned into practice. We are going to calculate the rate constant for the $H + H_2$ reaction at 1000 K. To use equation (7.12b), we need partition functions and the energy difference between the reactants and the transition state.

An aside about calculating $\Delta^{\ddagger}\varepsilon$: this is the most delicate quantity to calculate. Since it appears in an exponential, small errors in the energy can result in large errors in the equilibrium constant. One of the main sources of error in *ab initio* calculations is the size of the basis set. These errors cancel somewhat if we are careful about using the same basis set for all of the species involved. In the transition state, all three atoms are in relatively close proximity. In this situation, there is a sharing of basis functions between adjacent atoms. When we calculate the energies of an H atom and

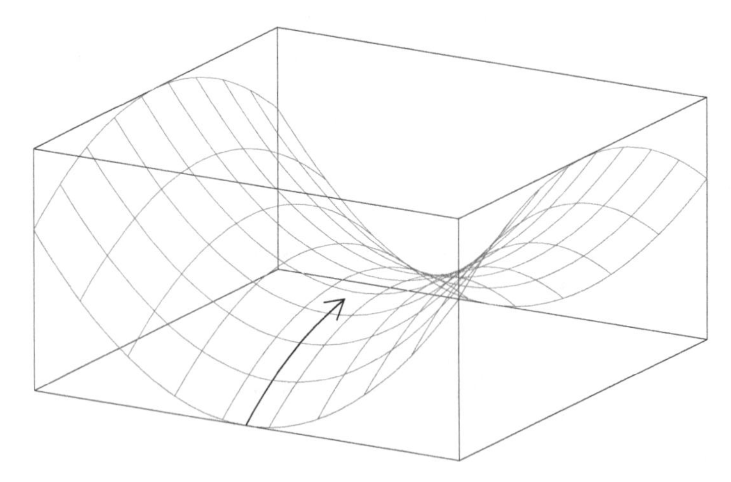

Figure 7.2. A schematic of a transition-state search with the Berny algorithm. The algorithm tries to walk uphill while staying in the middle of the minimum energy path.

an H_2 molecule separated by a large distance, this sharing of basis functions is lost. Effectively, we are working with a smaller basis set. The error associated with this issue is called the basis-set superposition error (BSSE). Dealing with basis-set superposition errors in transition states is extremely tricky. If you have heard about 'ghost atoms' or 'counterpoise corrections' (different names for the same thing), we can't easily use them here. There is also some question as to whether this technique is all that effective. Ideally, we would either use an enormous basis set or extrapolate results to the infinite basis-set limit, since the BSSE vanishes in these cases. Since this is not always practical, the best advice at this point is the following:

- Density functional theory (DFT) methods are less sensitive to the size of the basis set than Hartree–Fock calculations, so DFT methods are strongly recommended.
- It's still a good idea to use the biggest basis set you can afford.
- If possible, use an appropriately designed sequence of basis sets (e.g. cc-PVDZ, cc-PVTZ, cc-PVQZ, …) and extrapolate to the infinite basis-set limit. However, I will leave this topic to your computational chemistry course.

In the transition state, the atoms are farther apart than the normal bonding distances, so dispersion forces may be important. Accordingly, I used the ωB97X-D density functional. I chose the aug-cc-pVQZ basis set, which has diffuse wave-functions, for the same reason. Finally, given the odd number of electrons, I chose an unrestricted spin wavefunction.

When you set up your calculations in GaussView, you will note an '`Additional Keywords`' box near the bottom of the window. Since we want to use the standard pressure of 1 bar (0.986 923 atm) and a temperature of 1000 K, type the following into this box:

```
pressure=0.986 923 temperature=1000
```

Note that you will need to do this for *each* of the three necessary calculations. Otherwise, the calculations for H and H_2 are entirely straightforward.

We will need a guess for the transition-state geometry. It doesn't need to be a very accurate guess, but as noted above, it should suggest to Gaussian in what direction to search for a transition state. In section 5.1, we computed the PES for this reaction, obtaining the contour plot shown in figure 5.2. From this contour plot, we see that the saddle point for a collinear collision is somewhere near $R_{AB} = R_{BC} = 0.95$ Å. (We can peek into the data file to confirm this.) We can't be sure at this point that the transition state is collinear. If I set up an exactly collinear initial geometry, then the forces perpendicular to the molecular axis exactly cancel out, and only a collinear transition state can be found. Accordingly, my strategy here is to lay down three atoms approximately 0.95 Å apart with at least a slight bend and see what Gaussian does in the process of searching for a transition state.

Once you have created the initial geometry, set up an `Opt+Freq` calculation, but this time, select `Optimize to a TS (Berny)`. Also choose `Force Constants: Calculate at all points`. This has to do with how Gaussian calculates which direction is uphill vs the transverse directions along which it must still minimize the potential energy.

When I ran this calculation, I ended up with a nearly linear structure, bent at an angle of 179.7°. You may get a slightly different angle, depending on your initial geometry. Gaussian has a stopping rule that has to do, roughly, with the magnitude of the forces. At a saddle point, the forces acting on the atoms are exactly zero. Near the saddle point, they are small, and Gaussian may decide that the forces are small enough and stop before it gets to the exact saddle point. In this case, it is clear that Gaussian is moving towards a linear geometry. We really need to get this right because linear molecules are a special case, with an extra vibrational mode and one less rotational mode than a nonlinear molecule. This difference has a very large effect on the values of the partition functions. After seeing this, I straightened out the molecule obtained as a result of my initial search for the transition state (using the '`Modify angle`' tool), and then restarted the calculation. Although this is not conclusive, I obtained a slightly lower energy for the exactly linear transition state than for the bent transition state (−1.6652 vs −1.6650 hartree)[2]. This suggests that the bent configuration was slightly displaced from the saddle point, i.e. 'up the side' of the saddle. The strictly linear geometry is therefore our saddle point.

Gaussian calculates the partition function of the transition state in exactly the way you would want, i.e. leaving out the reactive mode. You can therefore use both the energy and the partition function calculated by Gaussian without any adjustments. My results are summarized in table 7.1. Given these data, calculating the rate constant is now straightforward:

$$\begin{aligned}\Delta^{\ddagger}\varepsilon &= \varepsilon_{TS} - (\varepsilon_{H_2} + \varepsilon_{H}) \\ &= -1.6652 - [-1.1771 + (-0.5029)]\text{ hartree} = 0.0148\text{ hartree} \\ &\equiv (0.0148\text{ hartree})(4.359\,744 \times 10^{-18}\text{ J hartree}^{-1}) \\ &= 6.45 \times 10^{-20}\text{ J}.\end{aligned}$$

Table 7.1. Thermodynamic data from the calculations described in the text for the H + H_2 reaction. The partition functions are computed relative to the energy minima.

Species	ε/hartree	$\hat{q}$
H	−0.5029	1.660×10^{6}
H_2	−1.1771	5.568×10^{5}
TS	−1.6652	3.833×10^{7}

[2] In theory, there could be a transition state of even lower energy at some other angle, which is why I say that this calculation is not conclusive. However, it is not obvious why that should be the case for this simple reaction.

We will want equilibrium constants in cm^3 molecule^{-1}s^{-1}. Therefore, we use $c° = p°/k_BT$, as discussed above. Substituting this into equation (7.12b), we get

$$\begin{aligned}
k^{(2)} &= \frac{(k_BT)^2}{hp°}\frac{\hat{q}^{\ddagger}}{\hat{q}_X\hat{q}_Y}e^{-\Delta^{\ddagger}\varepsilon/k_BT} \\
&= \frac{[(1.380\,649 \times 10^{-23}\ \mathrm{J\ K^{-1}})(1000\ \mathrm{K})]^2}{(10^5\ \mathrm{Pa})(6.626\,070 \times 10^{-34}\ \mathrm{J\ Hz^{-1}})} \\
&\quad \times \frac{3.833 \times 10^7}{(5.568 \times 10^5)(1.660 \times 10^6)} \\
&\quad \times \exp\left(\frac{-6.45 \times 10^{-20}\ \mathrm{J}}{(1.380\,649 \times 10^{-23}\ \mathrm{J\ K^{-1}})(1000\ \mathrm{K})}\right) \\
&= 1.11 \times 10^{-18}\ \mathrm{m^3\ molecule^{-1}}s^{-1} \\
&\equiv (1.11 \times 10^{-18}\ \mathrm{m^3\ molecule^{-1}}s^{-1})(100\ \mathrm{cm\ m^{-1}})^3 \\
&= 1.11 \times 10^{-12}\ \mathrm{cm^3 molecule^{-1}s^{-1}}.
\end{aligned}$$

It is a useful exercise to convince yourself that the units of $(k_BT)^2/hp°$ are m^3 molecule$^{-1}s^{-1}$.

The experimental value of this rate constant is $(2.1 \pm 0.6) \times 10^{-12}$ cm^{3}molecule^{-1}s^{-1}. Given the simplicity of this theory, we did fairly well. However, our estimate is clearly somewhat too small. The main source of error for this reaction is tunneling, i.e. the transmission of probability through the barrier rather than over the barrier. Tunneling is particularly significant for lighter particles and is generally worth taking into account when a reaction involves the motion of a hydrogen atom. We will shortly consider tunneling corrections in TST.

7.2.1 Isotope effects

Replacing one isotope by another often has a significant effect on the value of the rate constant of a chemical reaction. Given what we know about the calculation of rate constants in TST, we can begin to see why that might be the case: the mass affects the translational, vibrational, and rotational partition functions.

As an example, let's look at the $H_2 + D \rightarrow H + HD$ reaction. To change the mass of an atom in GaussView, we use the Atom List Editor found in the Tools menu. Click on the `Show Mass Columns` icon, which looks like this: 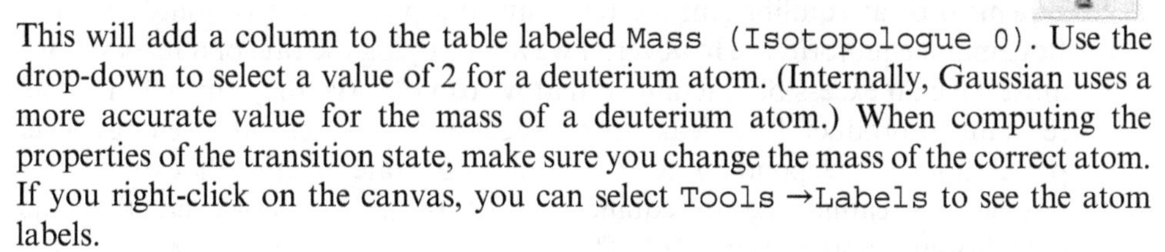.

This will add a column to the table labeled `Mass (Isotopologue 0)`. Use the drop-down to select a value of 2 for a deuterium atom. (Internally, Gaussian uses a more accurate value for the mass of a deuterium atom.) When computing the properties of the transition state, make sure you change the mass of the correct atom. If you right-click on the canvas, you can select `Tools` →`Labels` to see the atom labels.

We already have the partition function of H_2 from table 7.1. Table 7.2 gives the additional data required for this calculation. Notice that the change in mass doesn't affect the energy. This is because the energy of a molecule in any given nuclear

Table 7.2. Thermodynamic data for the H_2 + D reaction.

Species	ε/hartree	$\hat{q}$
D	−0.5029	4.691×10^6
TS	−1.6652	1.128×10^8

configuration depends only on electrostatics and not on the masses of the nuclei. However, we do need to recompute the partition function of the transition state because this depends on the masses of the atoms.

We calculate $k^{(2)} = 1.16 \times 10^{-12}$ cm^{3}molecule^{-1}s^{-1} for the H_2 + D reaction. For this particular reaction, the isotope effect predicted by classical TST is small. However, for hydrogen isotopes in particular, tunneling can have a much larger effect.

7.3 Strong and weak points of TST

At this point, it is worthwhile mentioning some of the strong and weak points of conventional transition-state theory, starting with the strong points:

- It is a relatively easy theory to use to predict rate constants. All of the required quantities are easily computed.
- The thermodynamic formalism of transition-state theory connects well to other parts of physical chemistry, making the theory conceptually useful.
- For simple reactions, transition-state theory often gives values in reasonable agreement with experiment.
- Even when the theory doesn't do a great job of predicting the value of a rate constant, it often correctly predicts the relative effects of various factors on the rate constant, e.g. isotope effects.
- It's an important starting point for understanding reactions under nonideal conditions, e.g. the effect of ionic strength. We will come back to this point in chapter 12.

There are some significant weaknesses to this theory as well:

- The equilibrium approximation at the heart of transition-state theory is dubious. To understand the problem, consider figure 6.2. In order for a system to be at equilibrium, the reactants and products must jointly be in a Boltzmann equilibrium. However, if we are studying the rate of a reaction, we must have an excess of reactant so that we can observe the conversion of the reactant to product. The 'leak' from reactant to product out of energy level populations corresponding to the transition state means that the latter populations cannot be in equilibrium. It turns out, however, that the equilibrium approximation of TST is reasonably accurate provided the potential energy barrier is sufficiently high. For reactions with low or no barriers, we will need a different approach.

Turning to figure 6.3, we see that it's also difficult to ascribe a precise meaning to the phrase 'the energy levels of the transition state.' Above the barrier, quantum mechanics says that the state of the molecule is delocalized and includes values of the nuclear coordinates that we would associate with both the reactant and product configurations. TST finagles this problem by fixing the reaction coordinate at the transition-state geometry, and then we can talk about the vibrational modes of the TS *excluding* the reactive mode of motion. Strictly speaking, though, we are imposing a classical picture on quantum mechanical entities, namely the molecules involved in the reaction. So what exactly do we mean by 'equilibrium' if the species that is supposed to be in equilibrium with the reactants is merely a point on the PES at which no forces constrain the molecule to remain? The somewhat unsatisfying answer must be that we mean an equilibrium of the populations of the various reactant energy levels with the populations of the energy levels above the barrier, especially those that are just above the energy of the transition-state structure. But the TS cannot, strictly speaking, have energy levels that unambiguously belong to this structure.

- In most reactions, there are specific conditions on the quantum states of the reactants that must be met in order for a reaction to occur, e.g. one specific vibrational mode might have to store enough energy to enable a rearrangement. This tends to create a non-equilibrium (i.e. non-Boltzmann) distribution of energy among the unreacted molecules because the populations of molecules in a specific set of quantum states are depleted by reaction. If this effect is sufficiently strong, the equilibrium approximation might not be valid, regardless of the height of the barrier.
- Transition-state theory often badly overestimates rate constants. This results from trajectories that cross the saddle point and then turn around and come back, which are not properly accounted for in the theory. This is sometimes fixed by multiplying transition-state rate constants by a **transmission coefficient** (κ). For example, we would write, for a second-order rate constant,

$$k^{(2)} = \kappa \frac{k_{\mathrm{B}}T}{hc^{\circ}} \frac{\hat{q}^{\ddagger}}{\hat{q}_{\mathrm{X}}\hat{q}_{\mathrm{Y}}} \mathrm{e}^{-\Delta^{\ddagger}\varepsilon/k_{\mathrm{B}}T}.$$

  The problem then becomes one of developing theories to calculate κ.
- The TST assumption that systems must pass through the transition state along a reaction coordinate in transiting from reactants to products is an irreducibly classical mechanical idea, as mentioned above. This idea leads to a number of additional difficulties for the theory:
  - What exactly do we mean by passage 'through' the transition state? Is it good enough to pass 'close' to the TS?
  - As mentioned above, tunneling is important in a variety of reactions, particularly but not exclusively [4] for reactions involving the motions of light particles (electron-transfer reactions, proton-transfer reactions, etc). A tunneling correction can be calculated and incorporated into the transmission coefficient.

- The separability of the reaction coordinate is not guaranteed. This means that energy stored in other molecular motions can affect whether or not a molecule crosses the TS and whether it will stay in the product well once it has crossed. This is again accounted for via the transmission coefficient.

Transition-state theory, while not flawless, is both conceptually and theoretically useful, so it is well worth mastering the foundations of this theory despite its flaws. And, as the following sections discuss, it can be improved upon.

7.4 Variational transition-state theory

The literature on transition-state theory is enormous, and it has spawned many interesting variations. In this section, we briefly examine one particularly important descendant of classical TST known as variational transition-state theory (VTST). But first, we should think a little about what we are doing in classical TST.

The saddle point we focus on in TST is a saddle in the conformational space of the atoms, i.e. in a space whose axes are the relative coordinates of the atoms making up the transition state. The partition functions appearing in the equations for the TST rate constant have the effect of averaging all of the quantum states of complexes with the transition-state conformation, weighting the states by the Boltzmann distribution. We could generalize this idea slightly by defining a transition-state dividing surface (TSDS), which would be a surface that divides the reactants from the products passing through the transition state orthogonally to the reaction path (figure 7.3). If we could calculate the probability that the trajectories from reactants to products would pass through any given point in the TSDS, which would involve

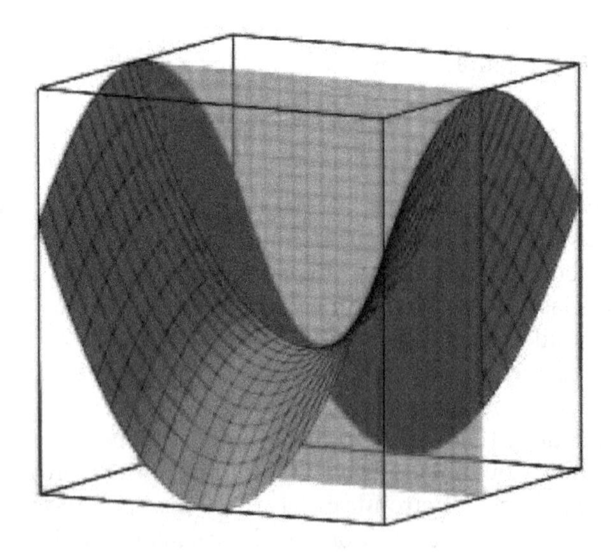

Figure 7.3. The transition-state dividing surface (TSDS), a surface separating reactants from products passing through the transition state orthogonally to the reaction path.

calculating the tunneling probabilities for points with energy below the PES, we could again take the Boltzmann average over all the points in the TSDS, and this would allow for the possibility that the system does not exactly pass through the saddle on the way to products.

In practice, this more complicated version of TST doesn't improve the results much. The problem is the recrossing issue mentioned briefly in the last section. Recrossings can have many causes. Just to fix an image in our heads, consider caroming, which happens when the system has excess energy in modes transverse to the reaction coordinate. The system will go up one side of the PES, slide down, and then go up the other side, and it may, depending on the exact distribution of energy among the various modes, just turn around after crossing the TSDS and go back the way it came (figure 7.4). A trajectory such as this one doesn't result in products, so it shouldn't count, but since it crosses the TSDS once in the reactant-to-product direction, it is counted in a TST calculation[3].

Figure 7.5 shows some possible trajectories that cross the TSDS in the reactant-to-product direction, along with some made-up statistics for the sake of discussion. These statistics are percentages of trajectories with the same qualitative appearance. Thus, in this example, 40% of trajectories would go straight through the TSDS and form products, 20% of trajectories that cross in the reactant-to-product direction in

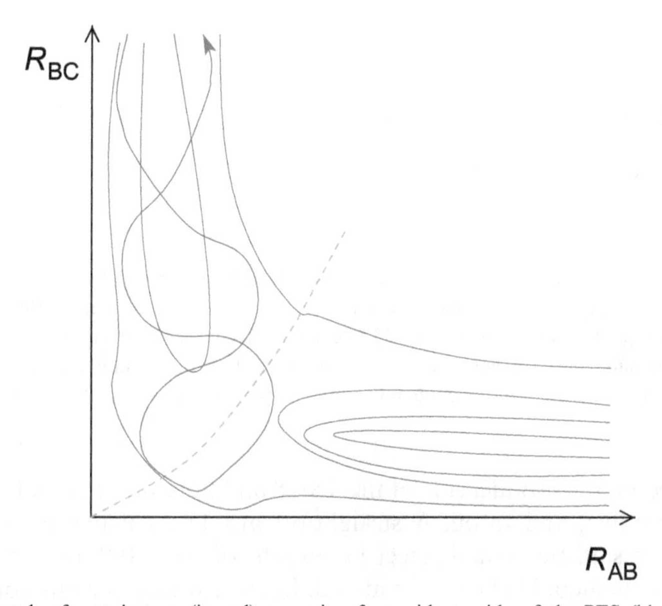

Figure 7.4. A sketch of a trajectory (in red) caroming from side to side of the PES (blue contours) and returning to the reactant valley after crossing the TSDS, shown as a dashed green curve.

[3] You may wonder, how TST 'counts' anything? We just calculate partition functions! The point is that certain quantum states are less likely to lead to products than their weight in the partition function would imply due to the kinds of dynamical effects described here. Somehow, we have to reduce the weight we accord to states in which products are less likely to form.

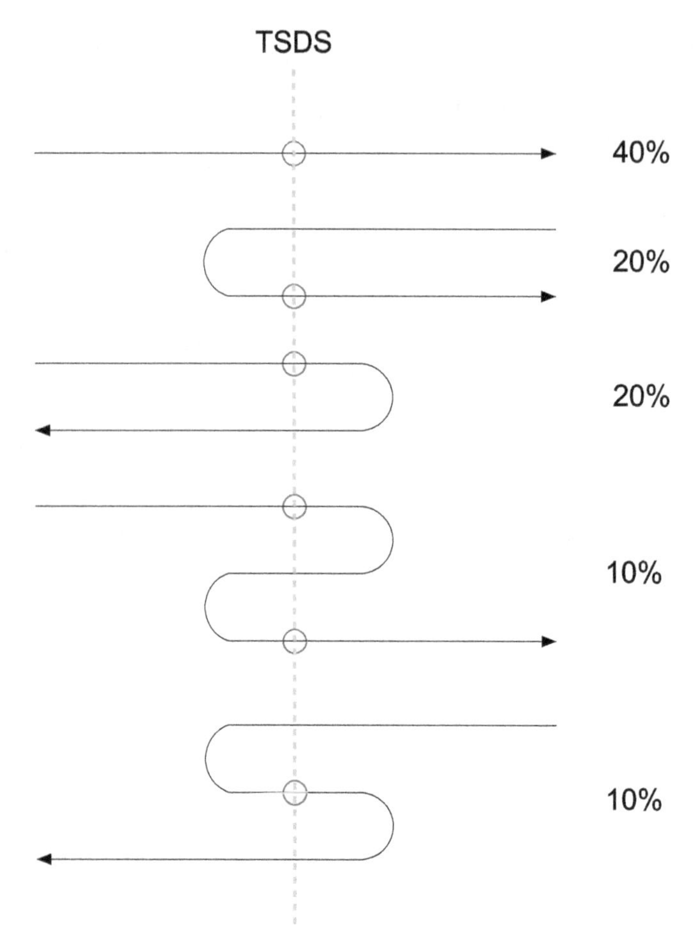

Figure 7.5. A schematic illustration of trajectories crossing the conventional TSDS, shown as a dashed line. The circles represent events that would be counted in a conventional TST calculation. Blue circles represent events that result in product formation and should be counted. Red circles represent events that should not be counted, either because they do not result in product formation or because they are redundant. The percentages are for discussion only and do not represent statistics from a real reaction. Adapted from Truhlar and Garrett [5].

fact originated on the product side of the TSDS and thus are not events converting a reactant to product, and so on. A straightforward TST calculation with a system displaying these statistics would detect 110 events for every 100 trajectories, but only 50 of these events should have been counted. In other words, the rate constant would be overestimated by a factor of 2.2.

Part of the problem is that the conventional TSDS is chosen without regard to the momenta. These clearly contribute to dynamics in various ways, including, for instance, the caroming effect discussed above. Moreover, we might be able to avoid some of the nonproductive trajectories shown in figure 7.5 by moving the TSDS. The basic idea of variational TST is that if we avoid all of the recrossings, the computed rate constant is minimized. The VTST calculation therefore involves moving or shaping the TSDS in order to minimize the value of the rate constant, hence the use of the word 'variational' in the name of this technique. In theory, if we could set the TSDS in the full phase space, i.e. the space of both coordinates and momenta, it would be possible to find a TSDS that is never recrossed. In practice, this is very hard to do, so VTST is generally applied to a TSDS in the conformational space. Even though this may not eliminate all recrossings, in practice this procedure gives good results, i.e. it eliminates most recrossings. After applying the variational TST method, we might end up with statistics along the lines of the illustration in figure 7.6. Counting fewer TSDS crossings as reactive events, i.e. minimizing the rate

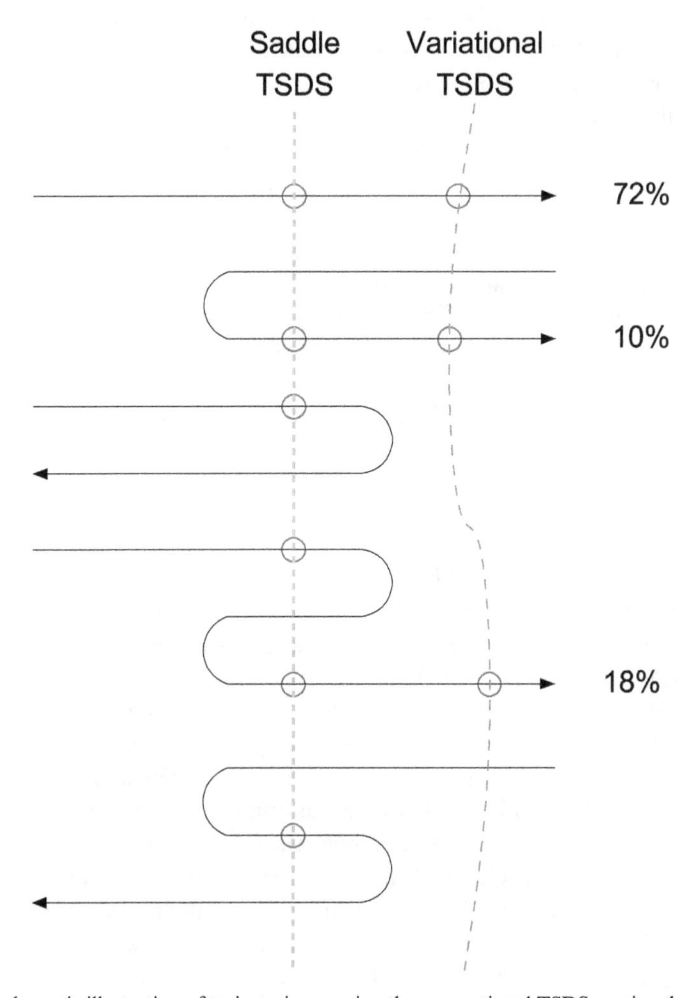

Figure 7.6. A schematic illustration of trajectories crossing the conventional TSDS passing through the saddle point in conformation space (short dashes) and the variational TSDS (long dashes). The symbolism is as in figure 7.5, except that now we show both events that would be counted in conventional TST where the trajectories cross the saddle TSDS and events that would be counted in variational TST at crossings of the variational TSDS. The percentages again do not represent statistics from a real reaction but are illustrative of what might be achieved. Adapted from Truhlar and Garrett [5].

constant with respect to the position of the TSDS, therefore has the effect of excluding some unproductive trajectories and of giving us a more accurate prediction of the rate constant. VTST is computationally much more complex than conventional TST, but at this time, it's the state of the art for computing elementary gas-phase rate constants *ab initio*.

While it is not formulated in this way, the VTST procedure is equivalent to computing a transmission coefficient to reduce the weight of recrossings. Given that it reduces the value of the rate constant, $\kappa_{VTST} < 1$. (In the TST literature, κ_{VTST} is often called Γ.)

Because this book is intended for a survey course, we won't go any deeper into VTST. However, you should be aware that software is available to perform these calculations, namely Polyrate [6]. Polyrate needs a PES in order to carry out a VTST calculation. An analytic approximation to the potential energy can be used, which would normally be obtained by fitting data from extensive *ab initio* calculations. Another option is to interface Polyrate to an electronic structure program to calculate only the points needed for the VTST calculation. A number of interface programs are available, notably Gaussrate, which, as you might guess, allows Gaussian to be used to dynamically compute points on the PES [7].

Setting up a Polyrate calculation requires that you create input files to describe the calculation you want to do. Explaining even a few basic options and working through an example would take quite a bit of space, and there are other topics we want to cover. Moreover, VTST calculations require some serious computational hardware to which you may not have access. I will therefore leave this topic for interested readers to explore on their own.

7.5 Tunneling corrections

Because matter has wave properties in quantum mechanics, it sometimes behaves more like sound than like baseballs. If I throw a baseball against a solid fence, it bounces back. Assuming a sufficiently solid fence, I can only get the ball to the other side by throwing it over the fence. Classical mechanics gives me no other options. The analogy to the description of reactions in transition-state theory is hopefully obvious, with TST's classical picture of systems passing over the saddle point. On the other hand, sound passes right through solid walls, albeit with some attenuation: think about a car with a loud sound system passing by your house. Even with all the windows and doors closed, you can often hear the boom–boom of the car's speakers.

So it is with quantum mechanical matter waves. Waves, whether they are sound waves or matter waves, are hard to confine and can permeate barriers. In the case of matter waves, the barriers we are talking about are potential energy barriers, and passing through a barrier, in the context of chemical reactions, corresponds to forming a product without acquiring enough energy to reach the transition state. This phenomenon is known as **tunneling**. Figure 7.7 schematically illustrates the quantum mechanics of tunneling. You may have learned in your previous courses that the square of the amplitude of a wavefunction is the probability that the particle

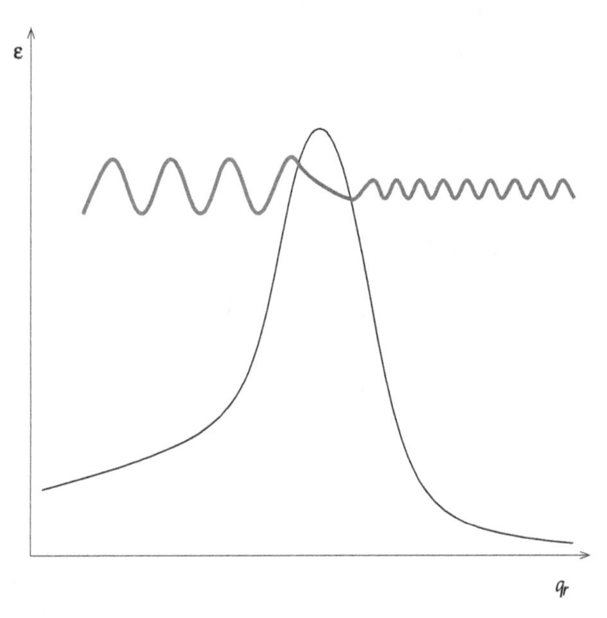

Figure 7.7. A schematic illustration of the quantum mechanics of tunneling. The effective potential is plotted vs the reaction coordinate q_r. A matter wave (the wave associated with an atom or molecule) approaches the barrier from the left. The wave is partially reflected (not shown), but some of it enters the barrier. The amplitude of the wave is attenuated as it passes through the barrier, but some of the original amplitude survives, i.e. there is leakage through the barrier.

is to be found in a certain region of space. A matter wave that enters a potential energy barrier is attenuated as it passes through. Provided the barrier is not too wide, the wave may nevertheless make it through to the other side, albeit with decreased amplitude. The ratio of the squares of the transmitted to the incident amplitudes is the probability that a wave will tunnel through the barrier and appear on the other side.

As a result of tunneling, if we only take over-the-barrier events into account when calculating a rate constant, we will be underestimating the true rate constant. But how large is this effect? The key parameters that affect tunneling are illustrated in figure 7.8. These are the reactant energy ε, the height of the barrier, here denoted by V_0, and the classical turning points q_1 and q_2. The latter are the points where a classical particle approaching the barrier from either the left or right would be reflected, since $\varepsilon = V_{\text{eff}}(q)$ at these two points. In one spatial dimension, quantum mechanics gives the tunneling probability as

$$P_{\text{tunnel}}(\varepsilon) = \exp\left(-\frac{2\sqrt{\mu_r}}{\hbar}\int_{q_1}^{q_2} \mathrm{d}q_r\sqrt{V_{\text{eff}}(q_r) - \varepsilon}\right), \tag{7.13}$$

Note the negative exponential. Accordingly, the tunneling probability decreases as any of the constituent pieces of the exponent increases. We can therefore see that the tunneling probability

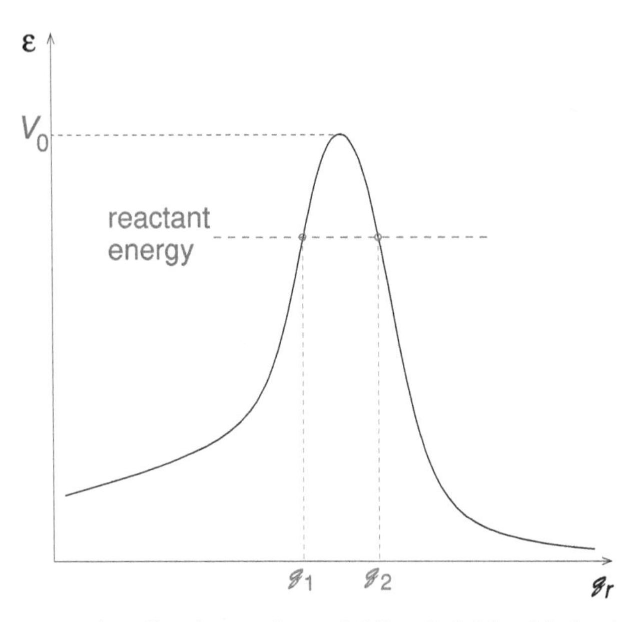

Figure 7.8. Key parameters that affect the tunneling probability: the height of the barrier, the energy of the reactants, and the width of the barrier at the reactant energy.

- decreases with increasing difference between the height of the barrier and the energy of the reactants (because of the dependence on $V_{\text{eff}}(q_r) - \varepsilon$);
- decreases with increasing barrier width (because the integral is evaluated over a wider range of coordinates which, all other things being equal, makes its value larger);
- decreases with increasing reduced mass.

Equation (7.13) gives us the tunneling probability for an event at energy ε. If we want to know how much tunneling occurs at temperature T, we have to take the Boltzmann average of this probability:

$$\kappa_{\text{tunnel}} = \int_0^\infty P_{\text{tunnel}}(\varepsilon)\frac{\exp\left(-\dfrac{\varepsilon - V_0}{k_B T}\right)}{k_B T}\mathrm{d}\varepsilon. \tag{7.14}$$

Note that the integral includes energies above V_0. Accordingly, it includes over-the-barrier events, which may seem a bit strange. In fact, equation (7.14) calculates corrections due to both tunneling for $\varepsilon < V_0$ and to nonclassical reflection for energies above V_0. The best analogy to the latter effect is perhaps the simultaneous reflection and refraction of light as it meets an interface from an optically less-dense to a denser medium. Given that it includes both tunneling and nonclassical reflection, the symbol and name of this quantity are both slightly misleading, although tunneling is usually a much larger effect than reflection.

κ_{tunnel} is applied as a multiplicative correction to the TST rate constant, i.e. it appears in the same way as a transmission coefficient[4]:

$$k = \kappa_{\text{tunnel}}\, k_{\text{TST}}.$$

Let's think about the effect that tunneling has on the temperature dependence of reactions, and in particular let's think about what happens at very low temperatures. If the temperature is sufficiently lowered, then eventually all of the molecules will be in their ground states. If there is any potential energy barrier at all between reactants and products, over-the-barrier events will cease, in accordance with the Arrhenius equation. The only possibility for reaction is then tunneling through the barrier. The tunneling probability from a single quantum state (the ground state) is temperature independent, in accordance with equation (7.13). If we now imagine studying a reaction over a wide range of temperatures, including some very low temperatures, we should get the Arrhenius behavior at high temperatures, characterized by a linear dependence of $\ln k$ on T^{-1}, but at sufficiently low temperatures where tunneling from the ground state dominates, we should see $\ln k$ become independent of T. As a result, tunneling results in curved Arrhenius plots. For reactions in which tunneling is particularly important (e.g. hydrogen atom transfers), Arrhenius plots curved due to tunneling may show up at moderate temperatures because of the different temperature dependences of tunneling and over-the-barrier events. Consider, for instance, the data of Schulz and Le Roy[5] for the $H + H_2$ reaction [8] shown in figure 7.9. The deviations from the line of best fit are not scatter. Note how close to each other the replicate points are, as well as the systematic nature of the deviations from the line. Schulz and Le Roy showed that the curvature of this plot was consistent with tunneling.

In theory, we can compute tunneling corrections directly from equations (7.13) and (7.14). This requires that we compute the potential energy along the reaction coordinate, evaluate (7.13) numerically at a series of reactant energies, then evaluate (7.14), again using a numerical method. This requires the use of numerical methods that are beyond the scope of this course, so we will focus on an approximate method instead.

Wigner developed the following formula, which is valid for barriers that are neither too high nor too wide:

$$\kappa_{\text{tunnel}} = 1 - \frac{1}{24}\left(\frac{h}{k_B T}\right)^2 V_2 - \frac{h^2}{96\mu_r k_B T}\frac{V_4}{V_2} \tag{7.15}$$

[4] Corrections can be cumulated, so if we apply both VTST and tunneling corrections, we have $k = \kappa_{\text{VTST}}\kappa_{\text{tunnel}}\, k_{\text{TST}}$.

[5] D. J. Le Roy was a great builder in Canadian science as well as a first-rate chemical kineticist. He chaired the Department of Chemistry at the University of Toronto from 1960 to 1969, a period of tremendous growth and maturation. Le Roy also served as vice-president (scientific) at the National Research Council of Canada (1969–74) and as a member of the Science Council of Canada (1974–84), an advisory body that generated reports on scientific topics of importance to public policy.

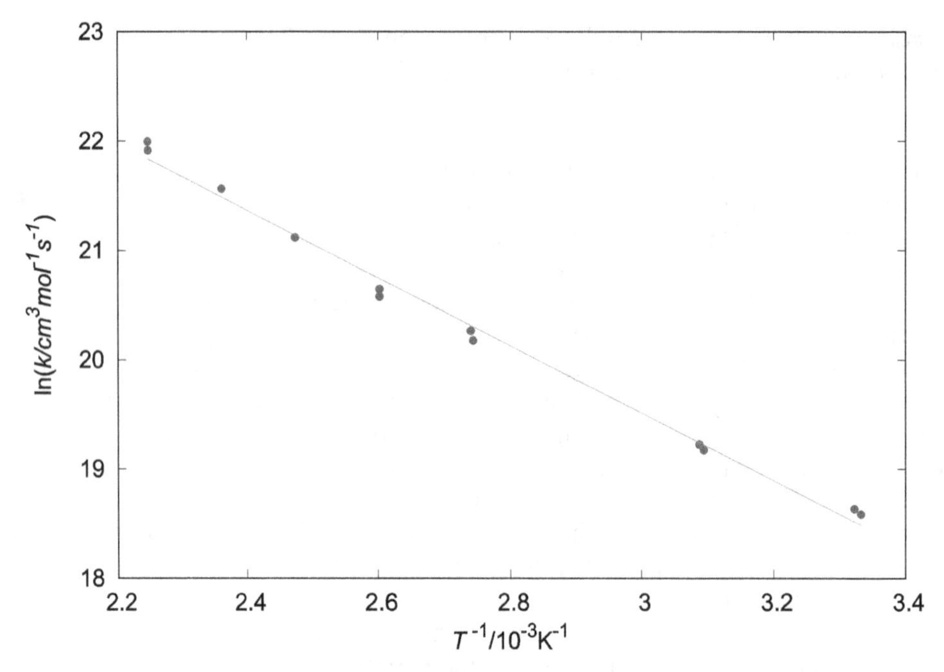

Figure 7.9. An Arrhenius plot of the data of Schulz and Le Roy for the $H + H_2$ reaction [8].

where

$$V_i = \left. \frac{d^i V_{\text{eff}}(q_r)}{dq_r^i} \right|_{q_{r(\text{TS})}}.$$

Usually, the coordinate system is such that $q_{r(\text{TS})} = 0$. Note that $V_2 < 0$, since there is a maximum along the reaction coordinate at this point. The second term in this equation is therefore positive. We typically find that V_4 is positive, so the last term is also positive. Thus, $\kappa_{\text{tunnel}} > 1$, as expected.

In order to calculate the partial derivatives V_2 and V_4, we need to calculate the potential energy along the reaction coordinate. The reaction coordinate is not uniquely defined, however. We need to pick a definition and use it. The result will depend somewhat on the definition of the reaction coordinate, but equation (7.15) is only an approximate relationship anyway. The most widely used definition of the reaction coordinate is Fukui's intrinsic reaction coordinate (IRC). This is the path of steepest descent to either side of the transition state connecting the reactant to the product. Given that the IRC is the steepest descent path, we need to start an IRC calculation from the transition state. Note also that the IRC is normally computed in mass-weighted coordinates, but to evaluate the derivatives appearing in equation (7.15), it is more convenient to use ordinary Cartesian coordinates.

Gaussian has the ability to calculate an IRC, returning the energy along this path. With a handful of points on either side of the transition state, we can calculate the derivatives V_2 and V_4 required to evaluate Wigner's tunneling correction. The catch is that we need to know how to evaluate derivatives numerically. A further issue is that Gaussian does not calculate the IRC at equally spaced points along the reaction coordinate. This leaves us relatively few practical choices for evaluating the derivatives. Fortunately, one of the methods we can use also happens to be simple to implement. We can estimate the derivative of a function $f(x)$ whose values $\{f_i\}$ are available at a set of coordinates $\{x_i\}$ by central differences:

$$\left.\frac{\mathrm{d}f(x)}{\mathrm{d}x}\right|_{x=x_i} \approx \frac{f_{i+1} - f_{i-1}}{x_{i+1} - x_{i-1}}$$

Technically, the central difference formula should only be applied when the points at which the function is sampled are evenly spaced. We're relying here on a couple of things: First of all, the spacing between adjacent pairs of points computed by Gaussian along an IRC is not wildly different from one pair to the next. Secondly, the mean value theorem of calculus tells us that there is a point between x_{i-1} and x_{i+1} where the derivative is exactly that given by the central difference formula. Without additional information, and given that x_i is near the middle of the interval between x_{i-1} and x_{i+1}, the central difference formula is as good an estimate of the derivative at x_i as anything else we might come up with. If the spacing between points was much less uniform, we would have to use a more sophisticated method to estimate the derivative, such as a method based on splines, which are piecewise polynomial functions.

Example 7.2. *Consider the following table of values of* $\sin x$:

i	1	2	3
x_i	0.9	1.0	1.1
f_i	0.7833	0.8415	0.8912

By the central difference formula, we have

$$\left.\frac{\mathrm{d}f(x)}{\mathrm{d}x}\right|_{x=1} \approx \frac{0.8912 - 0.7833}{1.1 - 0.9} = 0.5395.$$

Of course, $\mathrm{d}\sin x/\mathrm{d}x = \cos x$, *and* $\cos 1 = 0.5403$, *which is in good agreement (0.15% error) with the result of the central difference calculation.*

If we want a second derivative, all we have to do is to use the central difference formula again to calculate the derivative of the derivative. To get higher derivatives, we just keep iterating the central difference formula. Figure 7.10 shows how the calculation can be organized to calculate even derivatives at a particular point. To get the nth derivative, we need $n + 1$ points. Thus, five points of the function will let us calculate four first derivatives, three second derivatives, including the second

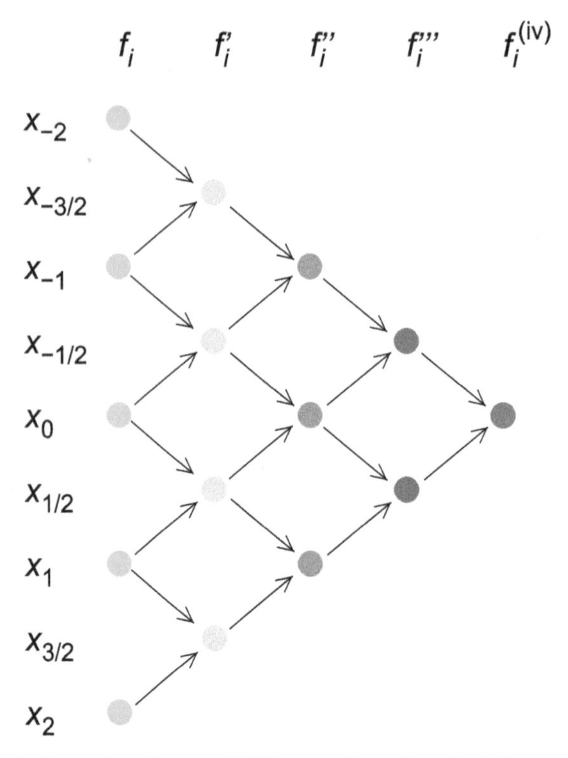

Figure 7.10. A schematic illustration of a central difference scheme for calculating even derivatives at a point x_0. We start with five function values, from which we calculate first derivatives at intermediate mesh points by central differences, then second derivatives at three mesh points, and so on.

derivative at the desired point, labeled x_0 in the figure, two third derivatives, and the desired fourth derivative.

The overall calculation procedure is as follows:

1. Find the transition state using Gaussian. Look in the `.log` file to find the reduced mass corresponding to the negative frequency, which is the frequency associated with the reactive coordinate. We will need this later.
2. Open the `.chk` file for the transition state in GaussView.
3. Open the `Gaussian Calculation Setup` dialog.
4. In the `Job Type` tab, choose an `IRC` calculation, then enter the following:

   Follow IRC: `Both directions`

   Check the `Compute more points` box and choose $N = 2$. In this case, we are using this option to *reduce* the number of points calculated, which will speed up the calculation. We need a total of five points, including the transition state itself. This option will calculate two points on either side of the transition state, plus the transition state, which is exactly what we need.

   Force Constants: `Calculate Always`.

   If we were calculating more points along the IRC, we would probably choose `Calculate Once`, and then check the `Recalculate Force`

`Constants Every nth Point` box and choose $n = 3$. The force constants are the Hooke's law constants of the vibrational modes. Calculating these takes time, but they are geometry dependent, so if we don't recalculate them from time to time, we won't get a very accurate IRC. For a small number of points, it is advantageous to just recalculate the force constants at each step. For a larger number of points (e.g. the default of ten), `Calculate once` and recalculating every few steps represents a compromise between accuracy and speed. For more speed at the cost of some accuracy, we could increase n to five in these cases.

5. Because we need to calculate spatial derivatives along the IRC, it is much simpler to use Cartesian coordinates. GaussView does not provide this option as a simple checkbox, so we need to add it manually. To do this, click on the `Edit` button at the bottom of the screen. In the editing session that opens up, you should see `irc=(calcfc,recalc=3)`. Change it to

```
irc=(calcfc,recalc=3,Cartesian).
```

6. Start the job. Go get a drink. This will take several minutes.
7. Once the job is done, open the `.log` file for your IRC calculation. Click on `Results` →`IRC/Path`. Save the data from this calculation to a text file. (The interface here is exactly the same as for coordinate scans.)
8. Import the data into a spreadsheet. Calculate the required derivatives by central differences.
9. The distance along the reaction coordinate is measured in bohr and the energy in hartree, so after computing the derivatives, you will have to convert them to SI units before applying equation (7.15). The relevant conversion factors are
   from bohr to m: $\times 5.291\,772 \times 10^{-11}$,
   from hartree to joules: $\times 4.359\,745 \times 10^{-18}$.

In section 7.2, we calculated the rate constant for the $H + H_2$ reaction at 1000 K. Let's now calculate the tunneling correction using Wigner's formula. I carried out these calculations using the ωB97X-D density functional with the aug-cc-pVTZ basis set. (Note that this is a smaller basis set than the one used for the transition state calculation. This is a deliberate choice to speed up the calculation.) The reduced mass associated with the reaction coordinate is 1.0078amu. I set up the calculation as described above, obtaining the characterization of the PES along the reaction coordinate shown in figure 7.11.

The easiest tool to use for these calculations is a spreadsheet. After importing the data into a spreadsheet, I find it convenient to insert blank rows between consecutive rows of data. This allows me to organize the calculation as illustrated in figure 7.10. In order to make this work, I need the values of the coordinates at the halfway

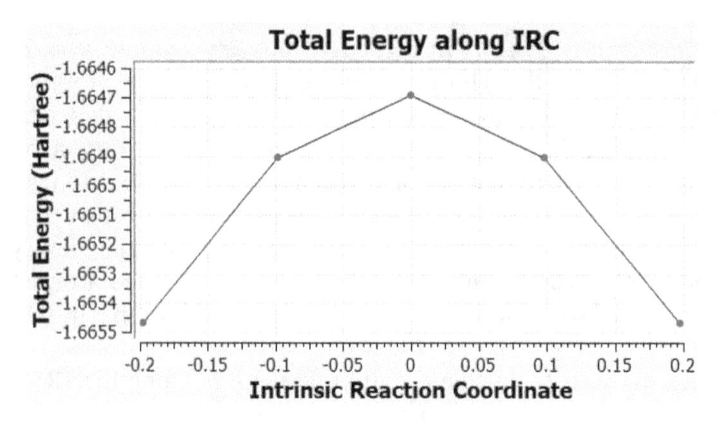

Figure 7.11. The energy along the Cartesian IRC near the transition state for the H + H_2 reaction calculated in Gaussian using the ωB97X-D density functional with the aug-cc-pVTZ basis set.

Table 7.3. The calculation of derivatives by central differences for the H + H_2 reaction. Blue coordinates are the midpoints between points calculated by Gaussian along the IRC. The data at the saddle point is highlighted in red.

q_r	V	V'	V''	V'''	$V^{(iv)}$
-0.1977	-1.665469				
-0.1480		0.0057			
-0.0983	-1.664905		-0.0354		
-0.0492		0.0022		-0.0914	
0.0000	-1.664691		-0.0444		1.8758
0.0491		-0.0022		0.0930	
0.0982	-1.664906		-0.0353		
0.1479		-0.0057			
0.1975	-1.665469				

points. These are obtained by averaging adjacent values of q_r. My data and calculated derivatives are shown in table 7.3.

The second derivative is $V_2 = -0.0444$ hartree bohr^{-2} and the fourth derivative is $V_4 = 1.8758$ hartree bohr^{-4}. Applying the conversion factors, we get

$$V_2 = \frac{(-0.0444 \text{ hartree bohr}^{-2})(4.359\,745 \times 10^{-18} \text{ J hartree}^{-1})}{(5.291\,772 \times 10^{-11} \text{ m bohr}^{-1})^2}$$
$$= -69.13 \text{ Jm}^{-2}$$

and

$$V_4 = \frac{(1.8758 \text{ hartree bohr}^{-4})(4.359\,745 \times 10^{-18} \text{ Jhartree}^{-1})}{(5.291\,772 \times 10^{-11} \text{ mbohr}^{-1})^4}$$
$$= 1.04 \times 10^{24} \text{ Jm}^{-4}.$$

In addition,

$$\mu_r = 1.0078 \text{ amu} \equiv 1.6735 \times 10^{-27} \text{ kg}.$$

We can now calculate the tunneling correction at 1000 K:

$$\begin{aligned}
\kappa_{\text{tunnel}} &= 1 - \frac{1}{24}\left(\frac{h}{k_B T}\right)^2 V_2 - \frac{h^2}{96\mu_r k_B T}\frac{V_4}{V_2} \\
&= 1 - \frac{1}{24}\left(\frac{6.626\,070 \times 10^{-34} \text{ J Hz}^{-1}}{(1.380\,649 \times 10^{-23} \text{ J K}^{-1})(1000 \text{ K})}\right)^2 (-69.13 \text{ J m}^{-2}) \\
&\quad - \frac{(6.626\,070 \times 10^{-34} \text{ JHz}^{-1})^2}{96(1.6735 \times 10^{-27} \text{ kg})(1.380\,649 \times 10^{-23} \text{ JK}^{-1})(1000 \text{ K})} \\
&\quad \times \frac{1.04 \times 10^{24} \text{ J m}^{-4}}{-69.13 \text{ J m}^{-2}} \\
&= 1 + 6.634 \times 10^{-27} + 2.99 = 3.99.
\end{aligned}$$

According to this calculation, tunneling should increase the rate of the reaction relative to our original calculation by a factor of about four. It is hard to say whether this is too large or too small because the TST calculation of section 7.2 almost certainly overestimated the over-the-barrier contribution to the rate constants, as discussed in sections 7.3 and 7.4. Moreover, the tunneling calculation itself has some potential issues. First, the calculation is highly sensitive to the fourth derivative of the potential. Getting this right requires a very accurate calculation of the potential near the saddle point. In other words, you need to do a really excellent calculation (large basis set, best available density functional, etc.) to get an accurate value for V_4. Additionally, V_4 is obtained by repeated differences, and this results in a progressive loss of precision. This is a difficult problem to avoid, but to do better would require a substantial diversion into numerical analysis.

There are two effects of quantum mechanical origin that we have not addressed in this calculation, but we should address them if we are looking to calculate tunneling corrections accurately. The first is that we're not using quite the right potential if we just use the PES, because we're ignoring the zero-point energies of the vibrational modes. As the nuclei undergo their relative motions along the reaction path, the electrons readjust, and since they provide the 'spring force' of the chemical bonds, this affects all of the vibrational frequencies. The associated zero-point energies therefore change along the reaction path. As a result, we can't count on the minimum energy of the system paralleling the PES. The resulting potential is somewhat deformed relative to the PES, an effect that really should be taken into account given the sensitivity of tunneling to the energy. This is tedious to do by hand, but of course one could write a script to extract the zero-point energies from the Gaussian output and to add them to the calculated potential energy for each point on the computed path. You would have to think hard about the format of Gaussian output files to do this, so we will leave this idea aside.

Moreover, we calculated the tunneling correction based on a one-dimensional model of tunneling, but tunneling trajectories are not confined to traveling along the reaction coordinate. They can 'cut corners' on the PES. Figure 7.12 illustrates a possible tunneling path that cuts a corner on the H_3 PES.

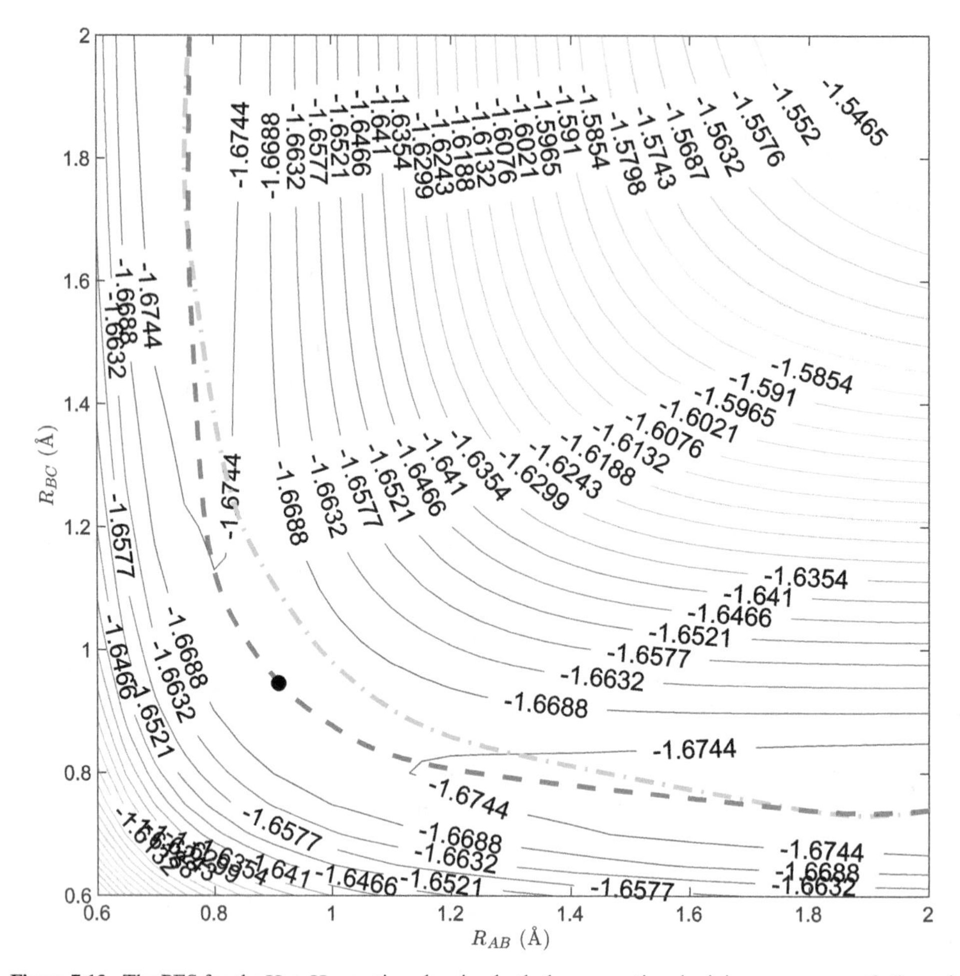

Figure 7.12. The PES for the H + H_2 reaction showing both the conventional minimum energy path through the saddle point (red, dashed) and a possible tunneling path that takes a shortcut under the PES (green, dot-dashed).

It is worth thinking a little bit about why a reacting system might take a shortcut like the one illustrated in figure 7.12. After all, the potential energy is higher along the illustrated tunneling path than along the minimum energy path, so shouldn't that make the tunneling probability smaller? It does, but it also shortens the path, which increases the tunneling probability. The highest-probability tunneling paths therefore involve a compromise between increased energy and a shorter tunneling distance.

Despite all of the caveats enumerated above, it seems likely that we have calculated a tunneling correction of the correct order of magnitude given the size of the discrepancy between the original TST calculation and the experimental rate constant. What this calculation tells us, then, is that tunneling is significant in the H + H_2 reaction. That, in itself, is useful information.

If we want to carry out a state-of-the-art calculation of tunneling corrections, Polyrate [6] again comes to the rescue. It has routines that go beyond the simple one-dimensional tunneling approximation used here. It considers the full shape of the PES, including zero-point energy corrections. The existence of a high-quality program in this area is part of the reason that we're not pursuing the corrections described above. The details have already been worked out for us.

Returning for a moment to kinetic isotope effects, we recall that changing one isotope for another doesn't change the PES. For the H_2 + D reaction, the only difference in the above calculation would be the reduced mass of the reactive mode, which is 1.0869amu. We therefore have

	$\frac{k_{TST}}{10^{-12}\,\text{cm}^3\text{molecule}^{-1}\text{s}^{-1}}$	κ_{tunnel}	$\frac{k}{10^{-12}\,\text{cm}^3\text{molecule}^{-1}\text{s}^{-1}}$
$H_2 + H$	1.11	3.99	4.43
$H_2 + D$	1.16	3.77	4.37

The predicted kinetic isotope effect for this reaction is not particularly large, according to our calculation: $k_H/k_D = 1.01$. However, we have to keep in mind that our calculations are a bit crude. We used ordinary TST (and not the more accurate variational theory), along with the one-dimensional Wigner expression for the tunneling coefficient. The true size of the kinetic isotope effect for this reaction is closer to 1.2. Again, if we want to do better, we probably need to use advanced software like Polyrate. But simple methods like the ones demonstrated in this chapter can still give us useful qualitative information about chemical reactions.

Further reading

There are many excellent reviews of modern transition-state theory. Although it's a little old, the following is an excellent starting point for someone who wants to gain a deeper appreciation of the theory:

- Truhlar D G and Garrett B C 1980 Variational transition-state theory *Acc. Chem. Res.* 13 440–8 https://doi.org/10.1021/ar50156a002

If you're interested in tunneling, the following paper is both an excellent review paper on variational TST and provides a great deal of detail on how the tunneling corrections are handled:

- Fernandez-Ramos A, Ellingson B A, Garrett B C and Truhlar D G 2007 Variational transition state theory with multidimensional tunneling *Reviews in Computational Chemistry*, vol 23, pp 125–232 (Hoboken, NJ: Wiley) https://doi.org/10.1002/9780470116449.ch3

Throughout this chapter, we followed a thermodynamic approach to transition-state theory. It is also possible to develop TST in phase space. Garrett has written an interesting historical account of this version of the theory, and Mahan has written a particularly clear introduction to it:

- Mahan B H 1974 Activated complex theory of bimolecular reactions *J. Chem. Ed.* 51 709–11 https://doi.org/10.1021/ed051p709

• Garrett B C 2000 Perspective on 'The Transition State Method': Wigner E (1938) Trans. Faraday Soc. 34 29–41. *Theor. Chem. Acc.* 103 200–4 https://doi.org/10.1007/s002149900046

Exercises

7.1 The elementary reaction $H + CO_2 \rightarrow OH + CO$ is important in combustion chemistry. How many vibrational modes would be included in the partition function $\hat{q}^{\ddagger}$? The transition state of this reaction is bent.

7.2 The elementary reaction $CO + O_2 \rightarrow CO_2 + O$ has an experimentally determined pre-exponential factor of 3.5×10^9 L mol^{-1}s^{-1} over the temperature range 2400–3000 K.

(a) Arrhenius theory treats the pre-exponential factor as a constant. Is this reasonable? Discuss this issue briefly using the theories presented in this book until now. Do *any* of them lead to a temperature-independent pre-exponential factor?

(b) The hard-sphere radii of O_2 and CO are, respectively, 1.8 and 1.9 Å. Estimate the pre-exponential factor of this reaction using simple collision theory. (Pick a reasonable temperature given the data provided.) How does your calculated rate constant compare with the experimental value? What does this tell us about the reaction?

7.3 In this question, you will study the kinetics of the conformational change

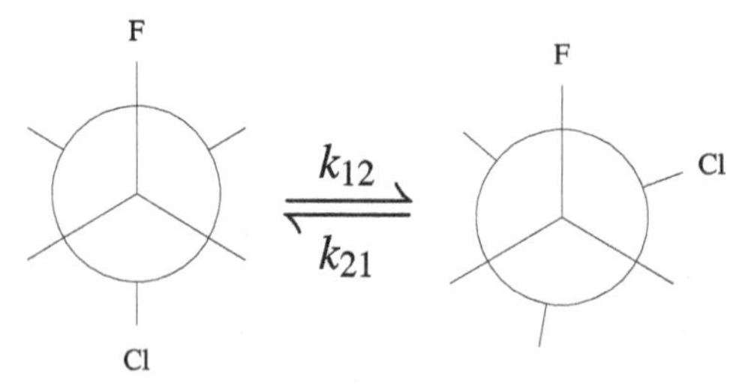

Note that we will deal with the rate constant for transition from the lowest-energy conformer to *one* of the secondary minima.

(a) Carry out a geometric optimization for
- the lowest-energy conformer,
- the secondary minimum, and
- the transition state separating them.

Report the following parameters for each of these structures in a table:
- the energy,
- the dihedral angle,
- the F–C–C bond angle,
- the Cl–C–C bond angle, and
- the carbon–carbon bond length.

(b) Calculate the rate constant k_{12} at 25 °C using transition-state theory.

(c) Calculate the reverse rate constant, k_{21}, at 25 °C.

(d) Calculate the equilibrium constant at 25 °C for the reaction. (This is a simple calculation if you remember some basic chemical kinetics.)
(e) Use the Wigner formula to estimate the importance of tunneling in this process.
(f) As noted earlier, the Wigner formula is often significantly in error, although it tends to give the correct order of magnitude for the tunneling correction for simple reactions. Based on your calculation, comment on whether it would be worthwhile to put more effort into calculating the tunneling correction for this process.
(g) If you wanted to lump together the two secondary minima, i.e. treat them as one species, what effect would this have on k_{12} and k_{21}? Explain briefly.

References

[1] Winzor D J and Jackson C M 2006 Interpretation of the temperature dependence of equilibrium and rate constants *J. Mol. Recognit.* **19** 389–407
[2] Roussel M R 2012 *A Life Scientist's Guide to Physical Chemistry* (Cambridge: Cambridge University Press)
[3] Larin I K, Spasskii A I, Trofimova E M and Proncheva N G 2018 Measurement of the rate constant of a reaction of chlorine atoms with CH_3Br in a temperature range of 298–358 K using the resonance fluorescence of chlorine atoms *Kinet. Catal.* **59** 11–6
[4] Zhang X, Hrovat D A and Border W T 2010 Calculations predict that carbon tunneling allows the degenerate Cope rearrangement of semibullvalene to occur rapidly at cryogenic temperatures *Org. Lett.* **12** 2798–801
[5] Truhlar D G and Garrett B C 1980 Variational transition-state theory *Acc. Chem. Res.* **13** 440–8
[6] Zheng J *et al* 2018 *Polyrate 17-C* (Minneapolis, MN: University of Minnesota), https://comp.chem.umn.edu/polyrate
[7] Zheng J, Bao J L, Zhang S, Corchado J C, Meana-Pañeda R, Chuang Y-Y, Coitiño E L, Ellingson B A and Truhlar D G 2017 *Gaussrate 17-B* (Minneapolis, MN: University of Minnesota), https://comp.chem.umn.edu/gaussrate
[8] Schulz W R and Le Roy D J 1965 Kinetics of the reaction $H + p\text{-}H_2 = o\text{-}H_2 + H$ *J. Chem. Phys.* **42** 3869–73

IOP Publishing

Foundations of Chemical Kinetics
A hands-on approach
Marc R Roussel

Chapter 8

Gas-phase unimolecular reactions

Gas-phase unimolecular reactions have played a pivotal role in theoretical chemical kinetics because they are, on the surface at least, so simple. And yet, it is surprisingly difficult to reconcile theory and experiment, even for these (apparently) simplest of chemical reactions. Understanding gas-phase unimolecular reactions in quantitative detail stands as one of the great theoretical achievements in 20th century chemical kinetics.

8.1 The Lindemann mechanism

There are two main categories of gas-phase unimolecular reactions, namely isomerizations such as

$$:C\equiv N-CH_3 \longrightarrow :N\equiv C-CH_3$$

and decompositions such as

$$CH_3CH_2CH_2CH_2{-}O{-}C(=CH_2){-}CH_3 \longrightarrow CH_3{-}C(=O){-}CH_3 + CH_2{=}CH{-}CH_2{-}CH_3$$

Gas-phase unimolecular reactions are often studied in a 'bath gas,' i.e. an inert gas whose pressure can be varied from experiment to experiment. It is also possible to leave out the bath gas. The reactant and product(s) then act as the bath gas. It was discovered early on that these reactions displayed different kinetics depending on the pressure regime. At low pressures, the observed rate law is $v = k_{\text{low}}[A][M]$, where A is the reactant and M is the bath gas (possibly A and the reaction products). On the other hand, at high pressures, a simple first-order rate law is observed: $v = k_{\text{high}}[A]$. The explanation for this behavior was provided by Lindemann in 1922 in a brief comment on a paper by Perrin [1]: a stable molecule does not just suddenly 'decide' to react. It has to become activated somehow. Lindemann's idea was that this

doi:10.1088/978-0-7503-5321-2ch8

activation process occurred as a result of collisions. Expressed as a mechanism (which Lindemann didn't do explicitly), this idea can be written

$$A + M \underset{k_{-1}}{\overset{k_1}{\rightleftharpoons}} A^* + M, \tag{8.1a}$$

$$A^* \xrightarrow{k_2} P, \tag{8.1b}$$

where M is, again, a bath-gas molecule (possibly a reactant or product molecule), A^* is an activated reactant molecule, and P represents the product(s). To get a rate law, we apply the steady-state approximation on the basis that A^*, being highly reactive, should be removed roughly as fast as it can be made after the initial transient rise in the concentration of this species. According to this mechanism, the rate of reaction is $v = k_2[A^*]$. The steady-state approximation for A^* reads

$$\frac{d[A^*]}{dt} = k_1[A][M] - k_{-1}[A^*][M] - k_2[A^*] \approx 0. \tag{8.2}$$

$$\therefore [A^*] \approx \frac{k_1[A][M]}{k_{-1}[M] + k_2}. \tag{8.3}$$

$$\therefore v = k_2[A^*] \approx \frac{k_1 k_2[M]}{k_{-1}[M] + k_2}[A]. \tag{8.4}$$

In the low-pressure limit (small [M]), this equation reduces to

$$v_{\text{low}} = k_1[M][A].$$

This makes sense since, at sufficiently low pressures, activating collisions should become rate limiting. At high pressures, on the other hand, we get

$$v_{\text{high}} = k_2 \frac{k_1}{k_{-1}}[A].$$

Note that k_1/k_{-1} is the equilibrium constant for collisional activation/deactivation, i.e. reaction (8.1a). At high pressures, we would expect this reaction to be in equilibrium because both directions become fast processes. The concentration of the bath gas drops out of an equilibrium treatment of the collisional activation step, leaving us with an overall rate that only depends on [A].

Assuming a constant pressure of bath gas in a given experiment, equation (8.4) describes a first-order process with rate constant

$$k_L = \frac{k_1 k_2[M]}{k_{-1}[M] + k_2} \tag{8.5}$$

(named k_L for Lindemann). We define

$$k_\infty = k_1 k_2 / k_{-1},$$

which is the high-pressure limit of k_L. Upon dividing the top and bottom of equation (8.5) by k_{-1}, we get

$$k_L = \frac{k_\infty[M]}{[M] + k_\infty/k_1}.$$

If we take the reciprocal of this equation, we have

$$\frac{1}{k_L} = \frac{1}{k_\infty} + \frac{1}{k_1}\frac{1}{[M]}.$$

A plot of k_L^{-1} vs $[M]^{-1}$ should therefore give a straight line[1]. Many experimental data sets do give a straight line when plotted in this manner. Figure 8.1 shows a typical example, in this case for the isomerization of *cis*-2-butene to *trans*-2-butene. The reactant and product act as the bath gas in this case. Note that because this is an isomerization, the total number of molecules is conserved during this reaction, so $[M] = [A] + [A^*] + [P]$ is constant. A nice straight line is obtained in the double-reciprocal plot.

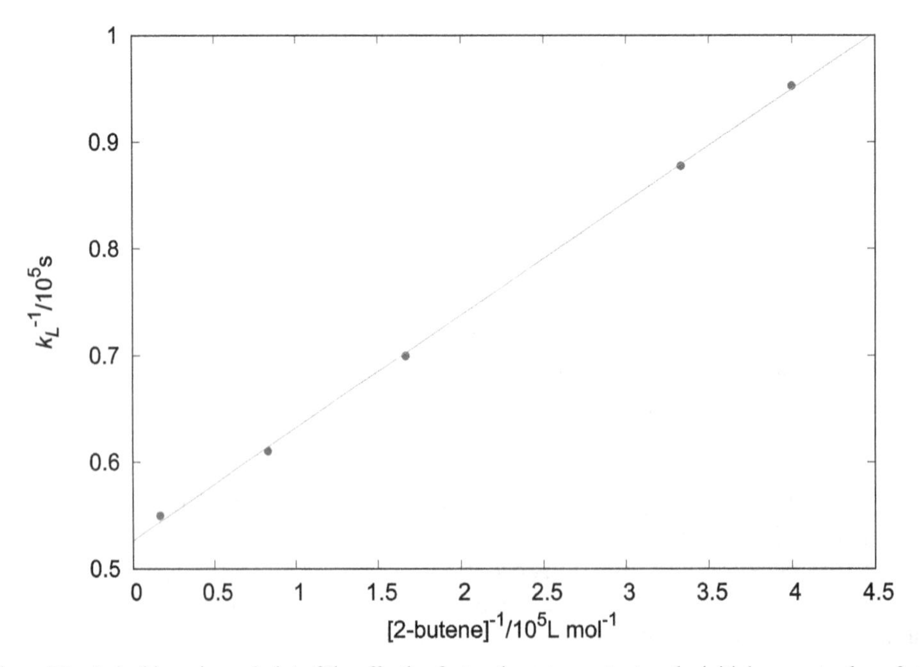

Figure 8.1. A double-reciprocal plot of the effective first-order rate constant vs the initial concentration of *cis*-2-butene for the isomerization of this molecule to its *trans* isomer at 740 K. Data from Pilling and Seakins [2].

[1] Yes, this is a double-reciprocal plot, about which I have had bad things to say elsewhere (see [3], section 15.2.1). I'm using one here because they are the typical graphs seen in the literature on the Lindemann mechanism and because we're only going to use these plots for visualization purposes. No point in being too picky just to point out (Spoiler alert!) some of the problems with the agreement between the mechanism and experiments.

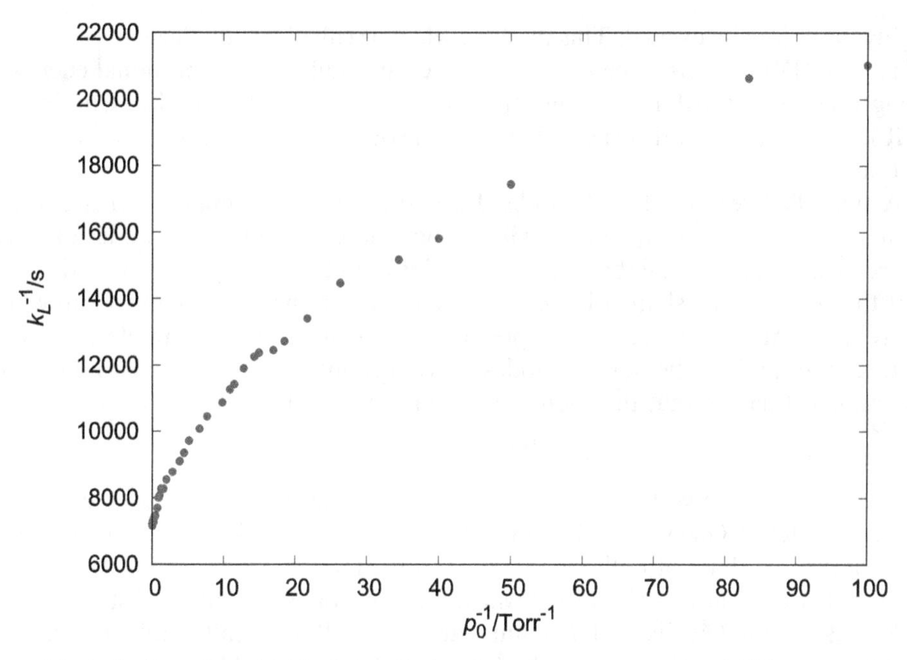

Figure 8.2. A double-reciprocal plot of the effective first-order rate constant vs initial pressure for the isomerization of 3-methylcyclobutene to *trans*-penta-1,3-diene [4].

Figure 8.2 shows some data taken over a large range of initial pressures for the isomerization of 3-methylcyclobutene to *trans*-penta-1,3-diene. The plot is clearly not linear over the whole range of initial pressures. While the Lindemann mechanism explains some features of the kinetics of unimolecular reactions, it's clearly not the final word.

And there is another problem. If the step $A + M \rightarrow A^* + M$ really just involves energy transfer during a collision, then the pre-exponential factor of k_1 should not exceed the collision-limited value. However, consider the isomerization of cyclopropane to propene, again with cyclopropane itself as the bath gas. The hard-sphere radius of cyclopropane is about 2.2 Å. This gives a collisional cross-section of 6.1×10^{-19} m^2 for collisions between two cyclopropane molecules. At a typical experimental temperature of 800 K, the mean relative speed of two cyclopropane molecules is 897 m s^{-1}. The collision-limited pre-exponential factor then works out to 1.6×10^{11} L mol^{-1}s^{-1}. However, the experimental value of the pre-exponential factor is much larger: 9×10^{18} L mol^{-1}s^{-1}. There is clearly something important that is being missed by Lindemann's theory.

8.2 Some useful mathematics

8.2.1 A combinatorics problem

In the harmonic oscillator approximation, energy is trapped in whatever normal mode holds it. However, because real molecules have anharmonic potentials, energy can be transferred from one mode to another, particularly when the molecule is highly vibrationally excited. This process, which is called intramolecular vibrational relaxation (IVR), tends to be fast, so that we can think of the vibrational energy as being randomly distributed among the molecule's vibrational modes. Considering IVR leads to statistical theories of kinetics, one of which will be discussed in the next section.

A simplified version of IVR can be described as follows: suppose that we have s normal modes sharing j quanta of vibrational energy. The normal modes of a real molecule are not identical, but we can get a lot of mileage out of the approximation that they can at least share a basic unit of energy currency. We want to know how many different ways there are to spread the j quanta over the s modes. A useful picture is to think of the normal modes as 'rooms' into which we have to distribute the quanta. Imagine that the rooms are arranged linearly:

|••| |•|•••|...|

In this diagram, the vertical lines are walls separating the rooms, and the dots are quanta of energy. One way to determine the number of possible ways of distributing the quanta is to think of both the quanta and the internal walls as movable objects. Noting that there are $s - 1$ internal walls, the question is how many distinguishable orderings of $s - 1$ walls and j quanta are there? If the walls and quanta were distinguishable, then there would be $(j + s - 1)!$ orderings. However, the j quanta are identical, so we have to divide this number by the number of indistinguishable permutations of the quanta, which is $j!$. Similarly, we divide by $(s - 1)!$ to account for the fact that the walls are indistinguishable. The number of arrangements of j quanta in s normal modes is therefore

$$W = \frac{(j + s - 1)!}{j!(s - 1)!}. \tag{8.6}$$

8.2.2 Stirling's approximation

Factorials grow quickly, so calculating W from equation (8.6) can be difficult even for values of j and s that are only moderately large. In many applications, we end up with logarithms of factorials. Stirling's approximation allows us to compute these logarithms without taking a factorial:

$$\ln N! \approx N \ln N - N.$$

This approximation works remarkably well even when N isn't particularly large. For example, $\ln 10! = 15.10$, and Stirling's approximation for this quantity is 13.03, an error of only 14%. For $N = 20$, the error is less than 6%.

8.2.3 The density of states for s harmonic oscillators

In section 2.4, we derived the following expression for the density of states of a harmonic oscillator:

$$g(\varepsilon) = (\hbar\omega_0)^{-1}.$$

This is the number of states per unit energy. Because the density of states is constant, the total number of states with energies between zero and ε is (roughly)

$$G(\varepsilon) = \varepsilon(\hbar\omega_0)^{-1}.$$

This approximation is accurate if there are a large number of energy levels between zero and ε, in other words if the levels are closely spaced compared to ε. In these cases, we can think of the vibrational levels as forming a continuum.

In section 2.2, we found that the partition function can be interpreted (with some caveats) as counting the number of states that have energies of less than k_BT. Therefore,

$$q_{\text{vib}} \approx G(k_BT) = \frac{k_BT}{\hbar\omega_0}.$$

(This approximation can also be derived from either equation (6.11) or (6.12) by assuming that $\hbar\omega_0/k_BT$ is small. The trick is to Taylor expand the numerator and denominator independently, stopping in both cases at the first nonzero term.)

We now ask: what is the density of states for s distinguishable, independent harmonic oscillators with natural frequencies $\omega_0^{(i)}$? The partition function for all of the oscillators taken together should be the product of the partition functions of the individual oscillators. To the same approximation as above, we have

$$q_s \approx \prod_{i=1}^{s} \frac{k_BT}{\hbar\omega_0^{(i)}}.$$

However, if the energy levels form a continuum, we can use equation (2.4) to relate the partition function to the density of states. Equating the two expressions for the partition function, we get

$$q_s = \int_0^\infty g_s(\varepsilon) \exp\left(-\frac{\varepsilon}{k_BT}\right) \mathrm{d}\varepsilon \approx \prod_{i=1}^{s} \frac{k_BT}{\hbar\omega_0^{(i)}},$$

where $g_s(\varepsilon)$ is the density of states for a system of s harmonic oscillators. The problem is now to find the density of states that makes these two expressions equal. This problem can be solved using a mathematical operation called an inverse Laplace transform. In order to avoid a lengthy mathematical diversion, we will just state the result here:

$$g_s(\varepsilon) = \frac{\varepsilon^{s-1}}{(s-1)! \prod_{i=1}^{s} \hbar\omega_0^{(i)}}$$

We now ask a further question: what is the probability that a molecule has a vibrational energy ε_{vib} between ε and $\varepsilon + \mathrm{d}\varepsilon$? This is just given by the Boltzmann distribution:

$$P(\varepsilon \leqslant \varepsilon_{\text{vib}} \leqslant \varepsilon + \mathrm{d}\varepsilon) = \frac{g_s(\varepsilon)}{q_s} \exp\left(-\frac{\varepsilon}{k_BT}\right) d\varepsilon.$$

We have both $g_s(\varepsilon)$ and q_s, so we could substitute both of these quantities into this equation and get the required probability. However, it will turn out to be a useful approximation to assume that the molecule's vibrational modes have a common frequency ω_0. We then have

$$q_s \approx \left(\frac{k_B T}{\hbar\omega_0}\right)^s$$

and

$$g_s(\varepsilon) \approx \frac{\varepsilon^{s-1}}{(s-1)!(\hbar\omega_0)^s}.$$

The probability that a molecule has a vibrational energy between ε and $\varepsilon + d\varepsilon$ then reduces to

$$P(\varepsilon \leqslant \varepsilon_{\text{vib}} \leqslant \varepsilon + d\varepsilon) = \frac{\varepsilon^{s-1}}{(k_B T)^s (s-1)!} \exp\left(-\frac{\varepsilon}{k_B T}\right) d\varepsilon. \tag{8.7}$$

8.3 RRK theory

In order for a molecule to undergo a particular reaction, it's not sufficient for the molecule to have enough total energy. The energy has to be stored in the appropriate normal mode, specifically the reactive mode. Implicit in the Lindemann mechanism is the idea that a collision either puts energy in that one specific mode, or it doesn't. There is no allowance made for the internal redistribution of energy. In a sufficiently complex molecule, the probability of a collision putting enough energy into one specific normal mode out of the $3N - 6$ normal modes must clearly be small, and yet there is no particular trend toward lower unimolecular rate constants with increasing molecular size. The internal redistribution of energy must therefore be important.

In the late 1920s, Rice, Ramsperger, and (independently) Kassel proposed two similar theories that extended the Lindemann treatment to include the internal redistribution of energy. While the basic ideas of these two theories were very similar, there were some differences, and it is really Kassel's version that has survived. Nevertheless, the theory is generally known as RRK theory, in honor of these three scientists.

The starting point of RRK theory is a slight elaboration of the Lindemann mechanism:

$$A + X \underset{k_{-1}}{\overset{k_1}{\rightleftharpoons}} A^* + X,$$

$$A^* \xrightarrow{k_{2K}} A^\ddagger \xrightarrow{K^\ddagger} P.$$

The new element introduced by RRK theory is a rate process for converting an energized molecule A^* into the transition state $A^‡$. A^* is a molecule that has enough energy to reach the transition state but not necessarily in the correct normal mode. The energy needed for the reaction to occur may need to flow from several normal modes into the reactive normal mode in order for the transition state to be reached. This process, intramolecular vibrational relaxation or IVR, is treated as a simple first-order reaction with a rate constant of k_{2K} (K for Kassel).

As in transition-state theory, $k^‡$ represents the very rapid process associated with crossing from reactants to products once sufficient energy is stored in the reactive mode. In fact, $A^‡$ should be the fastest decaying species in this mechanism, since it isn't even a stable molecule. Accordingly, we can apply the steady-state approximation for $[A^‡]$. Typically, we would solve for the concentration of the intermediate, but this time, we will solve for k_{2K}, for reasons that will become clear later.

$$\frac{d[A^‡]}{dt} = k_{2K}[A^*] - k^‡[A^‡] \approx 0$$

$$\therefore k_{2K} = k^‡\frac{[A^‡]}{[A^*]}$$

The ratio $p^‡ = [A^‡]/[A^*]$ can be interpreted as the probability that an energized molecule has sufficient energy in the reactive mode to form products. The last equation can be written in terms of this probability:

$$k_{2K} = k^‡ p^‡.$$

If we now apply the steady-state approximation in the normal way, we get

$$[A^‡] = \frac{k_{2K}}{k^‡}[A^*].$$

Since $v = k^‡[A^‡]$, we get $v = k_{2K}[A^*]$. In the Lindemann mechanism, we had $v = k_2[A^*]$. Comparing the two, we see that k_2 from the Lindemann mechanism corresponds to k_{2K} from the Kassel mechanism.

For simplicity, we assume that all normal modes of the reactant have the same energy. This is the idea of an energy 'currency' discussed briefly in section 8.2.1. Suppose that A^* has energy $\varepsilon^* = j\hbar\omega_0$. There is, of course, a distribution of energies, but we will worry about that later. RRK theory assumes that energy moves rapidly between vibrational modes. The degeneracy of vibrational energy level ε^* is therefore just the number of different ways of storing j quanta of vibrational energy in s vibrational modes:

$$G^* = \frac{(j + s - 1)!}{j!(s - 1)!}. \tag{8.8}$$

Now suppose that we need at least m quanta in the reactive mode in order for the reaction to occur, with $\varepsilon^‡ = m\hbar\omega_0$. The degeneracy of the set of molecules that has at least m quanta in the reactive mode is the number of ways of storing the 'extra' $j - m$ quanta in the s modes. This is

$$G^{\ddagger} = \frac{(j - m + s - 1)!}{(j - m)!(s - 1)!}. \tag{8.9}$$

Note that redistributing the $j - m$ excess quanta over all of the vibrational modes allows for some of them to end up in the reactive mode, i.e. for the reactive mode to contain more than m quanta.

Because all of the states we are considering have the same energy, each of the different ways of arranging the j quanta is equally probable. The probability $p^{\ddagger}$ that a molecule with energy ε^* has at least energy $\varepsilon^{\ddagger}$ in the reactive mode is therefore

$$p^{\ddagger} = \frac{G^{\ddagger}}{G^*} = \frac{j!(j - m + s - 1)!}{(j + s - 1)!(j - m)!}. \tag{8.10}$$

The number of quanta required to reach the transition state is large, and it is likely that many more quanta of vibrational energy than the minimum must be stored in the molecule in order for there to be a reasonable probability that the reactive mode contains m of them. Each of the terms in $p^{\ddagger}$ is therefore a factorial of a large number, to which we can apply Stirling's approximation:

$$\begin{aligned}
\ln p^{\ddagger} &= \ln j! + \ln(j - m + s - 1)! - \ln(j + s - 1)! - \ln(j - m)! \\
&\approx [j\ln j - j] + [(j - m + s - 1)\ln(j - m + s - 1) - (j - m + s - 1)] \\
&\qquad - [(j + s - 1)\ln(j + s - 1) - (j + s - 1)] \\
&\qquad - [(j - m)\ln(j - m) - (j - m)] \\
&= j\ln j + (j - m + s - 1)\ln(j - m + s - 1) - (j + s - 1)\ln(j + s - 1) \\
&\qquad - (j - m)\ln(j - m).
\end{aligned}$$

Again, the number of quanta stored in an activated molecule is large. In particular, it should be much larger than s. If we take $s - 1 = x$ to be a small quantity, then both the first and last term of this equation are of the form[2] $f(x) = (a + x)\ln(a + x)$. The first two terms of the Taylor series of this function are $f(x) \approx a\ln a + x(\ln a + 1)$. If we apply this approximation to the first and last terms of $\ln p^{\ddagger}$ and cancel terms, we get

$$\begin{aligned}
\ln p^{\ddagger} &\approx j\ln j + (j - m)\ln(j - m) + (s - 1)[\ln(j - m) + 1] \\
&\qquad - [j\ln j + (s - 1)(\ln j + 1)] - (j - m)\ln(j - m) \\
&= (s - 1)[\ln(j - m) - \ln j] \\
&= (s - 1)\ln\left(\frac{j - m}{j}\right) = \ln\left(\frac{j - m}{j}\right)^{s-1}, \\
\therefore p^{\ddagger} &\approx \left(\frac{j - m}{j}\right)^{s-1}.
\end{aligned}$$

[2] In a molecule with half a dozen atoms, s would be 12, which might not seem very small. In this case, we mean that $s - 1$ is small compared to j and to $j - m$. Technically, we should rewrite $f(x)$ as $a(1 + x/a)\ln[a(1 + x/a)]$ and take x/a as the small quantity. The result is the same.

Because of the assumption that all the normal modes have the same frequency, and given the relationship between j and the total vibrational energy ε^* and between m and $\varepsilon^\ddagger$, we can also write

$$p^\ddagger = \left(\frac{\varepsilon^* - \varepsilon^\ddagger}{\varepsilon^*}\right)^{s-1}. \tag{8.11}$$

And since $k_{2K} = k^\ddagger p^\ddagger$,

$$k_{2K} = k^\ddagger \left(\frac{\varepsilon^* - \varepsilon^\ddagger}{\varepsilon^*}\right)^{s-1}.$$

As for $k^\ddagger$, a very similar argument is used as in one version of the transition-state theory derivation. We can think of the reactive mode as a vibrational mode right up to the point at which it has stored enough energy to go over the barrier. This mode therefore has a characteristic frequency of $\nu^\ddagger$. The rate at which the barrier is crossed once enough energy has been acquired should therefore be this frequency, i.e. $k^\ddagger = \nu^\ddagger$. An attentive reader will note a small difference, notably the factor of two that appears in a similar argument in TST but not here. The difference is related to the different statistical ideas used in the two theories. In transition-state theory, we limit the transition-state complexes counted to those traveling from reactants to products when they reach the saddle point, so we have to similarly consider just the velocity in this direction. In RRK theory, if we consider a bond rupture as an example, a molecule might be in the compressional part of its vibrational cycle (in a classical picture) when IVR moves sufficient energy into the bond to break it. It would then have to complete this part of the cycle before the bond started to lengthen, eventually breaking. Because IVR might result in the bond acquiring energy $\varepsilon^\ddagger$ at any point in its vibrational cycle, we have to consider that the molecule might undergo most of a vibrational cycle before the bond breaks. Making this substitution now, we have

$$k_{2K} = \nu^\ddagger \left(\frac{\varepsilon^* - \varepsilon^\ddagger}{\varepsilon^*}\right)^{s-1}. \tag{8.12}$$

This equation gives us the rate constant $k_2 = k_{2K}$ as a function of the total energy of the molecule ε^*. Of course, different amounts of energy are gained in different collisions, so not every molecule has the same energy after collisional activation.

Going back to the Kassel mechanism, if we assume that the first step is in quasi-equilibrium, we have

$$\frac{[A^*]}{[A]} = \frac{k_1}{k_{-1}}.$$

So far, we have been treating A^* as an ensemble of molecules of fixed energy. However, the intention of the mechanism is for A^* to represent molecules with a range of different energies greater than $\varepsilon^\ddagger$. As an intermediate step from fixed energy to any sufficient amount of energy, let's think of A^* as representing molecules with energies between ε and $\varepsilon + d\varepsilon$ for some $\varepsilon > \varepsilon^\ddagger$. Since equation (8.7) gives the probability that s oscillators have a total energy between ε and $\varepsilon + \mathrm{d}\varepsilon$, the fraction of energized molecules with energy in this range should be

$$\frac{[\mathrm{A}^*]}{[\mathrm{A}]} = \frac{k_1}{k_{-1}} = \frac{\varepsilon^{s-1}}{(k_\mathrm{B}T)^s(s-1)!}\exp\left(-\frac{\varepsilon}{k_\mathrm{B}T}\right)\mathrm{d}\varepsilon. \tag{8.13}$$

The Lindemann rate constant (equation (8.5)) can be rewritten

$$k_\mathrm{L} = \frac{(k_1/k_{-1})k_2[\mathrm{M}]}{[\mathrm{M}] + k_2/k_{-1}}.$$

We have k_1/k_{-1} from equation (8.13) and $k_2 = k_{2\mathrm{K}}$ from equation (8.12). Substituting these into k_L, we get

$$\mathrm{d}k_\mathrm{RRK} = \frac{\nu^\ddagger[\mathrm{M}]\left(\frac{\varepsilon - \varepsilon^\ddagger}{\varepsilon}\right)^{s-1}\frac{\varepsilon^{s-1}}{(k_\mathrm{B}T)^s(s-1)!}\exp\left(-\frac{\varepsilon}{k_\mathrm{B}T}\right)}{[\mathrm{M}] + \frac{\nu^\ddagger}{k_{-1}}\left(\frac{\varepsilon - \varepsilon^\ddagger}{\varepsilon}\right)^{s-1}}\mathrm{d}\varepsilon \tag{8.14}$$

where we write $\mathrm{d}k_\mathrm{RRK}$, since this represents the rate constant only for reactants with energies between ε and $\varepsilon + \mathrm{d}\varepsilon$. To get the rate constant at temperature T, we just have to integrate this equation. After a little simplification, we get

$$k_\mathrm{RRK} = \int_{\varepsilon^\ddagger}^{\infty}\frac{\frac{\nu^\ddagger[\mathrm{M}]}{k_\mathrm{B}T(s-1)!}\left(\frac{\varepsilon - \varepsilon^\ddagger}{k_\mathrm{B}T}\right)^{s-1}\exp\left(-\frac{\varepsilon}{k_\mathrm{B}T}\right)}{[\mathrm{M}] + \frac{\nu^\ddagger}{k_{-1}}\left(\frac{\varepsilon - \varepsilon^\ddagger}{\varepsilon}\right)^{s-1}}\mathrm{d}\varepsilon. \tag{8.15}$$

This integral can't be evaluated analytically. It can be evaluated numerically for given values of the parameters ($\nu^\ddagger$, s, $\varepsilon^\ddagger$, k_{-1}). For k_{-1}, we usually make the **strong collision assumption**, which is that k_{-1} is strictly collision limited, i.e. that every collision of an A^* de-energizes it. Reasonable agreement with experiment is obtained if we take s to be about half of the total number of vibrational modes of the molecule. The justification for this adjustment is that only some modes couple efficiently to the reactive mode in a typical molecule. Note, however, the word 'about' in the above passage. This is a completely ad hoc adjustment, and there is no good way to tell ahead of time exactly how many modes we should include in the calculation. In fact, in order to get good agreement with experiment, it is necessary to allow s to vary with temperature.

There is another problem related to the frequency $\nu^\ddagger$. Consider the high-pressure limit of equation (8.15):

$$\begin{aligned} k_\mathrm{RRK} &\approx \int_{\varepsilon^\ddagger}^{\infty}\frac{\nu^\ddagger}{k_\mathrm{B}T(s-1)!}\left(\frac{\varepsilon - \varepsilon^\ddagger}{k_\mathrm{B}T}\right)^{s-1}\exp\left(-\frac{\varepsilon}{k_\mathrm{B}T}\right)d\varepsilon \\ &= \nu^\ddagger\exp\left(-\frac{\varepsilon^\ddagger}{k_\mathrm{B}T}\right). \end{aligned}$$

Since vibrational frequencies are never much larger than 10^{14} s^{-1}, the high-pressure pre-exponential factor should never be much larger than this either, but in practice it is often found to be much larger; values of up to 10^{17} s^{-1} are not unusual. RRK theory is a significant improvement on earlier theories of gas-phase unimolecular reactions, but it is clearly not the last word.

8.4 RRKM theory

RRKM theory was developed by Canadian Nobel Prize winner R A Marcus based on RRK theory. RRKM theory merges transition-state theory with RRK theory: as in RRK theory, equation (8.14) is used to obtain an equation for the rate constant, and k_1/k_{-1} is replaced by a probability distribution for the number of molecules with energy between ε and $d\varepsilon$. However, the correct probability distribution is used instead of equation (8.13), considering both the fact that the vibrational modes have different frequencies and that the rotational modes are coupled to the vibrational modes. Another difference is that transition-state theory is used to obtain an expression for $k_2(\varepsilon^*)$ instead of the statistical argument used in RRK theory. RRKM theory tends to give very accurate results if enough effort is put into the calculations. It predicts both ordinary rate constants and microcanonical rate constants (i.e. the dependence of the rate constant on the energy of the reactant).

The central assumption of RRK theory, which is that energy flows freely between at least some of the vibrational modes, is retained in RRKM theory. Non-RRKM behavior is typically due to a violation of this assumption. On occasion, there is an inefficient transfer of energy between some vibrational modes and others. Consequently, energy captured in one set of modes takes a long time to escape from them and participate in IVR. This results in time-dependent rate constants. An example of a type of molecule in which this might happen is an organometallic complex with two ligands attached to a heavy transition-metal atom:

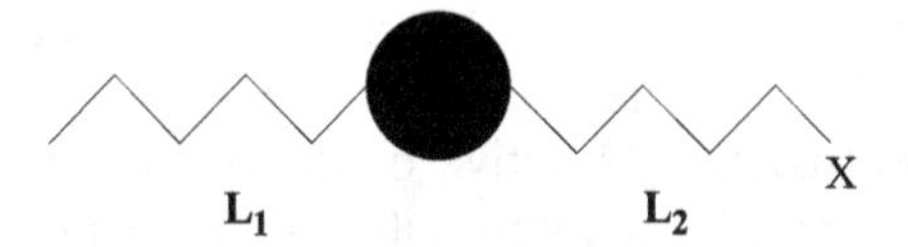

The large circle represents the transition-metal atom, and the two ligands are represented by zigzags. Note that the two ligands can be much more different than implied by this sketch. Under certain conditions, the heavy atom can act as an insulator, slowing the transfer of energy from one ligand to the other. The efficiency of energy transfer from ligand $\mathbf{L}_1$ to $\mathbf{L}_2$ depends on a host of factors: the vibrational spectra of the two ligands, the amount of energy acquired, the mass of the transition-metal atom, and the anharmonicity of the vibrational modes. For example, if the reaction under study involves breaking the $\mathbf{L}_2$ to X bond, any energy gained in $\mathbf{L}_1$ might equilibrate only slowly with $\mathbf{L}_2$. For the first short period after $\mathbf{L}_1$ has become energized, the rate of reaction might remain small, but as energy leaked over to $\mathbf{L}_2$, the rate of reaction would increase. In RRK-style theories such as RRKM, this would manifest itself in a time-dependent value of s, the number of modes involved in IVR.

Exercise

8.1 There are many variations on the Lindemann mechanism. One variation involves the formation of an exciplex (excited complex), which then goes on to react as follows:

$$A + A \underset{k_{-1}}{\overset{k_1}{\rightleftharpoons}} A_2^*$$

$$A_2^* \xrightarrow{k_2} A + \text{products}$$

(a) Derive a rate law for this mechanism. Is this mechanism consistent with experimental observations of the kinetics of gas-phase unimolecular reactions? Explain briefly.

(b) For reactions in which this mechanism is a possibility, it should compete with the ordinary Lindemann mechanism in which there is collisional activation but the molecules do not stick together. It should be possible to treat the second step of the exciplex mechanism ($A_2^* \to A + \text{products}$) using RRK theory. Which rate constant would you expect to be larger, the one for the exciplex mechanism, or the rate constant for the normal Lindemann pathway, $A^* \to$products? What does this imply in terms of the likelihood that the exciplex pathway will operate for typical reactions? Explain briefly.

8.2 During the derivation of the RRK equation, we calculated $\ln p^{\ddagger}$. Show that this term can be interpreted as the change in entropy during IVR.

8.3 Suppose that we have a molecule for which, in an RRK model, $s = 22$, which might be appropriate, for example, for the decomposition of 2,2-dimethylpropane to a *t*-butyl radical and a methyl radical: $(CH_3)_4C \to (CH_3)_3C + CH_3$.

(a) Suppose that the 22 active oscillators share 1000 quanta of vibrational energy. How many different ways are there of storing those quanta in the oscillators?

(b) Suppose that reaching the transition state requires that at least 100 quanta are stored in the reactive mode. How many different configurations of the 1000 quanta include at least 100 of them in the reactive mode?

(c) Cutting a terminal methyl group off from a hydrocarbon tends to have an energetic cost of about $350\,\text{kJ mol}^{-1}$, regardless of other details of the structure. There is an effect of the hydrocarbon's structure on the frequency associated with the dissociative mode, but this is small if the hydrocarbon is much heavier than a methyl group. Comparing the decomposition of 2,2-dimethylpropane ($s = 22$) to the decomposition of propane ($CH_3CH_2CH_3 \to CH_3CH_3 + CH_3$, $s = 13$), what energy would a 2,2-dimethylpropane molecule have to reach before it had the same probability of reaction as a molecule of propane with a total energy equivalent to 500 kJ mol^{-1}?

(d) Why would we expect the mass of the radical created by removing a methyl group to have only a small effect on the reactive mode frequency?

References

[1] Lindemann F A 1922 Discussion on 'The radiation theory of chemical action *Trans. Faraday Soc.* **17** 598–9

[2] Pilling M J and Seakins P W 1997 *Reaction Kinetics* (Oxford: Oxford Univ. Press) pp 143

[3] Roussel M R 2012 *A Life Scientist's Guide to Physical Chemistry* (Cambridge: Cambridge University Press)

[4] Frey H M and Marshall D C 1965 Thermal unimolecular isomerization of cyclobutenes. Part 5. The isomerization of 3-methylcyclobutene at low pressures *Trans. Faraday Soc.* **61** 1715–21

IOP Publishing

Foundations of Chemical Kinetics
A hands-on approach
Marc R Roussel

Chapter 9

The master equation

Gas-phase unimolecular reactions are initiated by collisions in which molecules gain vibrational energy. Intramolecular vibrational relaxation (IVR) then redistributes that energy. Both of these processes can be conceptualized as involving 'jumps' between quantum states, in one case, jumps up and down the energy ladder, and in the other, jumps between equal energy states. These jumps have their own dynamics, which we ignored in the Rice–Ramsperger–Kassel (RRK) theory. In this chapter, we will study a theoretical framework allowing us to model these jump processes, known as the master equation.

9.1 The derivation of the master equation

The idea behind the master equation is quite general and applies to many situations in the sciences, social sciences, and other fields where we can talk about systems (or components of systems) making transitions between states at random times. Accordingly, this derivation will be quite general. If it helps, think about the situations mentioned above, which involve a molecule jumping from one quantum state to another.

Suppose that we have a set of states among which a system executes random jumps at random times. In some cases, a system will be able to jump to any other state; in others, there will be rules about which jumps are allowed given the current state of the system. The jumps are instantaneous, i.e. the system doesn't spend any time between states. It's in one state or another at any given time.

Let $P_s(t)$ be the probability that a system is in a state s at time t. The states are members of a set $\mathcal{S}$ of allowed states. For each pair of states r and s, there is a **transition rate** w_{rs} such that, if the system is in state r at time t, the probability that the system jumps to state s during the subsequent time interval dt is $w_{rs}\, dt$. We want to develop an equation for the time evolution of the probabilities $P_s(t)$. Given the vector of probabilities $\mathbf{P}(t) = [P_1(t), P_2(t), \ldots]$, we can calculate the probability of being in any given state s a short time later as follows:

$$P_s(t + dt) = P_s(t) + \sum_{r \neq s} P_r(t) w_{rs} dt - \sum_{r \neq s} P_s(t) w_{sr} dt. \tag{9.1}$$

The two sums represent the change in $P_s(t)$ in a time interval dt due to transitions into state s from any other state (the first sum) and out of state s into any of the other states (the second sum). Taking the first sum as an example, the probability of jumping into state s from state r is the product of the probability that the system was in state r to start with, $P_r(t)$, times the probability that the system jumps to state s in time dt, $w_{rs}dt$.[1] In the event that the jumping rules make a certain jump impossible, the corresponding value of w_{rs} is simply zero.

We now rearrange equation (9.1) as follows:

$$\frac{P_s(t + dt) - P_s(t)}{dt} = \sum_{r \neq s} w_{rs} P_r(t) - \sum_{r \neq s} w_{sr} P_s(t).$$

In the limit as $dt \to 0$, the left-hand side becomes a derivative:

$$\frac{dP_s}{dt} = \sum_{r \neq s} w_{rs} P_r - \sum_{r \neq s} w_{sr} P_s. \tag{9.2}$$

This is the **master equation**. Note that I dropped the argument (t), since all terms are evaluated instantaneously. It is important to understand that the master equation is not a single equation. It is a set of coupled equations, one for each state of the system. The number of states, and thus the number of coupled differential equations, can be finite or countably infinite[2].

The master equation is an extremely general equation. We classify processes described by master equations based on the form taken by the transition rates w_{rs}. In general, the values of the transition rates can be time dependent and can depend on the history of the system. For example, w_{rs} could depend on how long the system has been in state r, which would be a form of history dependence. When discussing energy transfer, whether inter- or intramolecular, we only need to consider the simplest case, namely one in which the w_{rs} values are constant. If the transition rates are independent of the history of the system, we say that the system has the **Markov property**. A system with time-independent transition rates is said to be **homogeneous**. If a system's time evolution involves randomness, we describe this evolution as a **stochastic process**, 'stochastic' just being a fancy word for random. There are versions of Markov processes in which the state space is continuous. When the state space is discrete, as assumed in our derivation of the master equation, we call the resulting Markov process a **Markov chain**. The cases we will be studying in this chapter are therefore homogeneous Markov chains. Finally, a homogeneous

[1] For those of you who are statistics mavens, note that $w_{rs}dt$ is in fact a conditional jump probability.

[2] A countable infinity is one is in which each state can be numbered by an integer. They key is to create an ordering of the states that includes every possible state. Once we have this ordering, any particular state can be numbered by its position in the sequence. The classic example of this procedure is the proof that the set of rational numbers is countably infinite.

Markov chain is a type of **Poisson process**, which is to say that the waiting time until the next event (in our case, the next jump) is exponentially distributed and independent of the time origin. The independence from the time origin (the moment we designate as $t = 0$) means that the expected waiting time at any given moment is independent of how long we have been waiting so far.

Before we press on, it is useful to think about what the probabilities in the master equation mean. If we have a gas containing a large number of molecules, P_s might represent the probability that a randomly selected molecule is in quantum state s. For a large number of molecules, the number of molecules in state s would be $N_s = N_{\text{total}}P_s$. If we multiply both sides of the master equation by N_{total}, we then get an evolution equation for the number of molecules in each state:

$$\frac{dN_s}{dt} = \sum_{r \neq s} w_{rs}N_r - \sum_{r \neq s} w_{sr}N_s.$$

To emphasize: this transformation from P_s to N_s only works if we have a sufficient number of molecules that a single jump from one state to another, which changes the populations of the end and beginning states by ± 1, results in an insignificant change in each of these populations. This makes N_s an effectively continuous variable and allows us to write a differential equation for this population. Alternatively, you could think of equation (9.2) as describing the evolution of probabilities in an ensemble of identically prepared systems.

9.2 The master equation and the equilibrium distribution

Consider the following minor rewrite of the master equation, combining the two sums in equation (9.2):

$$\frac{dP_s}{dt} = \sum_{r \neq s}(w_{rs}P_r - w_{sr}P_s).$$

Now suppose that we want to find the equilibrium point of this system of equations. Equilibrium is found by setting the rate of change to zero for all of the probabilities. Thus,

$$\sum_{r \neq s}(w_{rs}P_r^{(\text{eq})} - w_{sr}P_s^{(\text{eq})}) = 0 \qquad \forall s, \tag{9.3}$$

where we use the superscript (eq) to denote equilibrium probabilities. In statistical theory, we would call a probability distribution that doesn't change over time a **stationary probability distribution**.

By construction, if we start with positive probabilities, the master equation will keep all probabilities positive, since the only negative terms in equation (9.3) are proportional to P_s. Thus, as $P_s \to 0$, $dP_s/dt \geqslant 0$. Moreover, since each term $w_{rs}P_r$ in dP_s/dt is necessarily balanced by a corresponding term $-w_{rs}P_r$ in dP_r/dt,

$$\sum\nolimits_{s \in \mathcal{S}} \frac{dP_s}{dt} = 0.$$

Since

$$\sum_{s \in \mathcal{S}} \frac{\mathrm{d}P_s}{\mathrm{d}t} = \frac{\mathrm{d}}{\mathrm{d}t} \sum_{s \in \mathcal{S}} P_s,$$

we must conclude that the sum of the probabilities is constant. Of course, we expect this, since we should have

$$\sum_{s \in \mathcal{S}} P_s = 1. \tag{9.4}$$

The system has to be in *some* state at any given time, so the sum of the probabilities over all possible states has to be exactly one.

With one additional condition, we can prove that there is a unique stationary distribution. An **ergodic Markov chain** is one in which any state can be reached from any other state in a finite number of steps. If a system does not satisfy the ergodic condition, then it may contain disconnected 'islands' of states, or 'sinks' from which the system cannot return, and the stationary distribution may then depend on the initial condition.

The solutions of a system of coupled linear differential equations either tend toward an equilibrium point[3], or they run away to infinity. The probabilities can't exceed one, or equation (9.4) would mean that at least one probability would have to be negative, which we showed wasn't possible. Accordingly, all of the probabilities are bounded between zero and one, so we can exclude solutions that blow up. If we now assume ergodicity, this ensures that the equilibrium point is unique, since the rates in and out of any given state have to balance at equilibrium, and the connectivity of an ergodic system requires this balance to be globally consistent. This only leaves the possibility of a unique, attractive equilibrium point.

We know that there is an equilibrium point, i.e. the stationary probability distribution, and we know it's unique. Now suppose that we have a microscopically reversible system, i.e. one for which (w_{rs}, w_{sr}) are either both zero or both positive for all pairs of states (r, s). An obvious solution of equations (9.3) is then obtained by setting each individual term in the summation equal to zero:

$$w_{rs} P_r^{(eq)} - w_{sr} P_s^{(eq)} = 0.$$

This is called the **detailed balance condition**. Since the stationary distribution is unique, it must be identical to the one described by the detailed balance condition. In the case of a master equation describing an ensemble of molecules, we know that the equilibrium probabilities have to obey a Boltzmann distribution. Therefore, rearranging the detailed balance condition and invoking the Boltzmann distribution, we have

$$\frac{P_r^{(eq)}}{P_s^{(eq)}} = \frac{w_{sr}}{w_{rs}} = \exp\left(-\frac{\varepsilon_r - \varepsilon_s}{k_B T}\right), \tag{9.5}$$

[3] As noted above for a non-ergodic system, the equilibrium point of a linear system may not be unique.

where r and s label individual quantum states of the system. In the case where r and s label energy levels instead of states, then we have to take the degeneracies into account:

$$\frac{w_{sr}}{w_{rs}} = \frac{g_r}{g_s}\exp\left(-\frac{\varepsilon_r - \varepsilon_s}{k_B T}\right).$$

Detailed balance fixes the ratios of the transition rates between two states. It does not, however, fix the values of these transition rates, which generally depend on a number of factors. For example, if we were to study a Markov chain for the energization and de-energization of molecules by collisions, the transition rates would depend on, among other things, the collision rate, which in turn depends on the concentrations of collision partners and on the temperature.

For harmonic oscillators and assuming weak coupling between translational and vibrational motions, it is possible to show that only vibrational transitions for which $\Delta v = \pm 1$ are likely. This will, of course, not be exactly true for real molecules, whose vibrational modes are never perfectly harmonic, but it remains a useful approximation in many cases. This assumption, namely that only transitions between adjacent vibrational levels are possible, is called the **Landau–Teller approximation**.

Suppose that states r and s have different energies. Equation (9.5) then implies that 'downward' transitions have larger transition rates than 'upward' transitions. This is what maintains a Boltzmann distribution at equilibrium. Moreover, since higher vibrational states usually have smaller energy differences from their neighboring states than is the case near the bottom of the potential well, the bias towards returning to lower energies becomes less pronounced once a molecule becomes highly excited. To the extent that the Landau–Teller approximation is valid, something close to a random walk among highly excited states results, which enhances the probability that a molecule will reach the transition-state energy once it becomes highly energized.

On the other hand, to model intramolecular vibrational relaxation, we would consider a set of states of the same energy. Then $w_{rs} = w_{sr}$.

Example 9.1. *As an example of a master equation and of its solution, we consider a simple two-level system with the master equation*

$$\frac{dP_1}{dt} = w_{21}P_2 - w_{12}P_1,$$
$$\frac{dP_2}{dt} = w_{12}P_1 - w_{21}P_2.$$

It is easy to see that $P_1 + P_2$ must be a constant, and given that they are probabilities that the system is either in state one or in state two, we must have $P_1 + P_2 = 1$. We can use this probability conservation equation to eliminate P_2*, leaving us with the single differential equation*

$$\frac{\mathrm{d}P_1}{\mathrm{d}t} = w_{21}(1 - P_1) - w_{12}P_1.$$

This equation can be solved by separation of variables:

$$\frac{\mathrm{d}P_1}{w_{21} - P_1(w_{21} + w_{12})} = \mathrm{d}t.$$

$$\therefore \int_{P_1(0)}^{P_1(t)} \frac{\mathrm{d}P_1}{w_{21} - P_1(w_{21} + w_{12})} = \int_0^t \mathrm{d}t' = t.$$

$$\therefore t = -\frac{1}{w_{21} + w_{12}} \ln\left[w_{21} - P_1(w_{21} + w_{12})\right]_{P_1(0)}^{P_1(t)}.$$

$$\therefore -t(w_{21} + w_{12}) = \ln\left(\frac{w_{21} - P_1(t)\,(w_{21} + w_{12})}{w_{21} - P_1(0)\,(w_{21} + w_{12})}\right).$$

$$\begin{aligned}\therefore P_1(t) &= \frac{w_{21}}{w_{21} + w_{12}}\left(1 - \mathrm{e}^{-t(w_{21}+w_{12})}\right) + P_1(0)\,\mathrm{e}^{-t(w_{21}+w_{12})}\\ &= \frac{1 - \mathrm{e}^{-t(w_{21}+w_{12})}}{1 + w_{12}/w_{21}} + P_1(0)\,\mathrm{e}^{-t(w_{21}+w_{12})}.\end{aligned}$$

Using the Boltzmann detailed balance condition and setting the ground-state energy to zero, we now get

$$\begin{aligned}P_1(t) &= \frac{1 - \mathrm{e}^{-t(w_{21}+w_{12})}}{1 + \frac{g_2}{g_1}\exp\left(-\frac{\varepsilon_2}{k_\mathrm{B}T}\right)} + P_1(0)\,\mathrm{e}^{-t(w_{21}+w_{12})}\\ &= \frac{g_1\left(1 - \mathrm{e}^{-t(w_{21}+w_{12})}\right)}{g_1 + g_2\exp\left(-\frac{\varepsilon_2}{k_\mathrm{B}T}\right)} + P_1(0)\,\mathrm{e}^{-t(w_{21}+w_{12})}\\ &= \frac{g_1}{q}\left(1 - \mathrm{e}^{-t(w_{21}+w_{12})}\right) + P_1(0)\,\mathrm{e}^{-t(w_{21}+w_{12})},\end{aligned}$$

where q is the partition function for a two-level system.

Note that as $t \to \infty$, $P_1 \to g_1/q$, the equilibrium Boltzmann probability for the ground state. The final equilibrium state does not depend on the initial condition, since this system is fully reversible. The relaxation time (the time required for the distance to equilibrium to be reduced by a factor of e^{-1}) is $(w_{21} + w_{12})^{-1}$, so it is directly related to the transition rates between the two levels.

There are special cases of master equations that can be solved exactly, one of which is the two-level system of the previous example. In general, though, solving the master equation poses a number of problems, most of them related to the sizes of typical master equations. It is not uncommon for molecules to have hundreds or thousands of states, and so the master equation is often a set of hundreds or thousands of coupled differential equations. Since it is a linear differential equation, it is in principle possible to solve the master equation directly by finding the eigenvalues of its coefficient matrix. However, it can be very difficult to solve large eigenvalue problems accurately. Having said that, if we are only interested in what happens as the system approaches equilibrium, it is possible to only compute the leading eigenvalues (the least negative eigenvalues), but then we have the issue of determining an initial condition for this truncated solution. Similarly, solving very large systems of differential equations numerically can run into issues that degrade the accuracy of the solution or make the solution process intolerably long.

A completely different approach is to simulate the underlying stochastic process, which we will turn to in section 9.4.

9.3 IVR in the RRK theory: a master equation approach

The RRK theory included the critical step

$$A^* \rightarrow {}^{k_{2K}}A^\ddagger \tag{9.6}$$

This reaction was intended to describe vibrational energy redistribution leading to the accumulation of sufficient energy in the reactive coordinate for a reaction to occur. The underlying picture was that the molecule would randomly wander among equal energy states until it reached a state with, again, sufficient energy in the reactive coordinate, taken to be the transition state $A^\ddagger$. The irreversibility of this reaction implies that once a molecule has reached the appropriate energy distribution to react, it is much more likely to do so than to have its energy redistributed to a nonreactive configuration. We can incorporate all of these assumptions of the RRK theory into a master equation treatment and see what predictions this theory makes.

In the IVR model corresponding to the RRK picture, $A^\ddagger$ is the set of states for which the number of quanta stored in the reaction coordinate m satisfies $m \geqslant m^\ddagger$, where $m^\ddagger$ is the minimum vibrational level that must be reached for the molecule to react. We will call the set of these states the reactive set, denoted by $\mathcal{R}$. The irreversible formation of the transition state makes $\mathcal{R}$ an **absorbing set**, i.e. a set of states from which the molecule cannot return. The states that participate in IVR that are not in the reactive set are denoted by $\mathcal{N}$.

Given that the states that participate in IVR all have the same energy, we have

$$\frac{w_{\rm sr}}{w_{\rm rs}} = \exp\left(-\frac{E_{\rm r} - E_{\rm s}}{k_{\rm B}T}\right) = 1.$$

This says that the transition rates between any given pair of states are equal. For simplicity, we assume not only pairwise equality but equal transition rates among all states participating in IVR. We therefore set $w_{\rm sr} = w_{\rm rs} = w$ for all (r, s) except for states in the reactive set. For these states, the assumptions of the RRK theory imply that $w_{\mathcal{RN}} = 0$. These states clearly violate the assumption of a Boltzmann distribution. In fact, reactions do perturb Boltzmann distributions for states near the top of the barrier, since molecules that cross the barrier to become products disturb the equilibrium statistics of the Boltzmann distribution. The RRK assumption simply represents a particularly extreme form of perturbation.

In our development of RRK theory, we saw that we could count the number of states corresponding to A^* and to $A^\ddagger$, at least if we assumed that the normal modes of the molecule all had the same frequency. These counts were given by equations (8.8) and (8.9), respectively. The number of states in the nonreactive set was then given by $G_\mathcal{N} = G^* - G^\ddagger$.

We can now write down the master equation for IVR under the assumptions of the RRK theory:

$$\frac{dP_n}{dt} = w \sum_{n' \in \mathcal{N}} (P_{n'} - P_n) - G^\ddagger w P_n \qquad \forall n \in \mathcal{N}, \tag{9.7a}$$

$$\frac{dP_r}{dt} = w \sum_{n' \in \mathcal{N}} P_{n'} \qquad \forall r \in \mathcal{R}. \tag{9.7b}$$

The last term in dP_n/dt is $-\sum_{n' \in \mathcal{R}} w P_n$, representing the transfer from state n to states in the reactive set. Note that the summand doesn't depend on n' and that there are $G^\ddagger$ such terms Also note that, while the sum in equation (9.7) should exclude $n \to n$ jumps, since $P_{n'} - P_n = 0$ when $n' = n$, it is harmless to extend the sum to all $n' \in \mathcal{N}$.

We now define $P_\mathcal{N}$ and $P_\mathcal{R}$ which are, respectively, the probabilities that the system is in the nonreactive or reactive set:

$$P_\mathcal{N} = \sum_{n \in \mathcal{N}} P_n,$$
$$P_\mathcal{R} = \sum_{r \in \mathcal{R}} P_r.$$

Using these definitions, equations (9.7a) become

$$\frac{dP_n}{dt} = w P_\mathcal{N} - w G_\mathcal{N} P_n - w G^\ddagger P_n \qquad \forall n \in \mathcal{N},$$
$$\frac{dP_r}{dt} = w P_\mathcal{N} \qquad \forall r \in \mathcal{R}.$$

We now sum these equations over $n \in \mathcal{N}$ and $r \in \mathcal{R}$, respectively:

$$\begin{aligned}\sum\nolimits_{n \in \mathcal{N}} \frac{dP_n}{dt} &= \frac{d}{dt} \sum_{n \in \mathcal{N}} P_n \\ = \frac{dP_\mathcal{N}}{dt} &= \sum_{n \in \mathcal{N}} w P_\mathcal{N} - w G_\mathcal{N} \sum_{n \in \mathcal{N}} P_n - w G^\ddagger \sum_{n \in \mathcal{N}} P_n \\ &= G_\mathcal{N} w P_\mathcal{N} - w G_\mathcal{N} P_\mathcal{N} - w G^\ddagger P_\mathcal{N} \\ &= -w G^\ddagger P_\mathcal{N}.\end{aligned}$$

Similarly,

$$\frac{\mathrm{d}P_{\mathcal{R}}}{\mathrm{d}t} = wG^{\ddagger}P_{\mathcal{N}}.$$

After all of that work, we end up with a very simple first-order differential equation for $P_{\mathcal{N}}$, with the solution

$$P_{\mathcal{N}}(t) = P_{\mathcal{N}}(0)\mathrm{e}^{-wG^{\ddagger}t}. \tag{9.8}$$

The decay in $P_{\mathcal{N}}$ is due to transfer to the set $\mathcal{R}$, i.e. to reaction (9.6). It follows that

$$k_{2\mathrm{K}} = wG^{\ddagger} \tag{9.9}$$

since this constant appears in equation (9.8) where the rate constant of a first-order rate process does.

We note in passing that choosing the value of $P_{\mathcal{N}}(0)$ is a more complicated issue than it might appear at first glance. Assuming that all states of energy ε^* are equally likely, a fraction $G^{\ddagger}/G^*$ of the molecules will land immediately in the reactive set and react immediately on energization. This creates a discontinuity in $P_{\mathcal{N}}$ at time zero which must be handled carefully in treatments that go beyond the one presented here.

We now want to compare our new expression for $k_{2\mathrm{K}}$, equation (9.9), to the original RRK expression, equation (8.12). From equation (8.10), we can write $G^{\ddagger} = p^{\ddagger}G^*$. Substituting this and using equation (8.11), equation (9.9) becomes

$$k_{2\mathrm{K}} = wG^*\left(\frac{\varepsilon^* - \varepsilon^{\ddagger}}{\varepsilon^*}\right)^{s-1}. \tag{9.10}$$

The difference between equations (8.12) and (9.10) is the factor of $\nu^{\ddagger}$ in the former vs wG^* in the latter. Since IVR is an extremely fast process, wG^* can be much, much larger than $\nu^{\ddagger}$. This is one way to understand the extremely large pre-exponential factors observed in some unimolecular reactions: they are not controlled by motion across the transition state but by the rate of IVR.

9.4 The kinetic Monte Carlo algorithm

As noted earlier, it is very difficult to solve the master equation in general. There is, fortunately, an alternative, which is to simulate the underlying Markov chain. If we start with a system in a certain state, say s_1, we need an algorithm for deciding the state the system will jump to as well as the waiting time until the jump. The probability distribution for the jump destinations will turn out to be related to the transition rates in a straightforward way. The waiting times are exponentially distributed with an exponent that also depends on the transition rates, so we will need to figure out how to generate random numbers with an exponential distribution. We turn to the second of these problems first, which will allow us to develop an important technique for stochastic simulations.

9.4.1 Generating exponentially distributed random numbers

Computer programming environments always supply a random number generator that generates uniformly distributed numbers in the interval (0, 1). The problem is that we often want to generate numbers with a different distribution. Some computer programming systems do provide facilities for generating random numbers with different distributions, and when these are available, it's often convenient to use them. However, it turns out to be extremely easy to generate numbers with an exponential distribution, so you might as well generate your own.

The problem we have is how to use uniformly distributed random numbers in the interval (0, 1) to generate any desired probability distribution. One important set of techniques, which we will not discuss here, is rejection algorithms. These are extremely powerful but are less efficient than direct methods. However, for some problems, there is really no alternative to rejection methods.

Every probability distribution has a statistic that is distributed over the interval (0, 1), which is the cumulative probability distribution. For a continuously distributed random variable x with a probability density of $p(x)$, the cumulative probability distribution is given by

$$S(x) \equiv P(x_o \leqslant x) = \int_{x_{\min}}^{x} p(x')\mathrm{d}x',$$

where x_o is a sample from the distribution, i.e. an observation. Depending on the distribution, $x_{\min}$ could be 0, $-\infty$, … For the exponential distribution in particular, we have

$$p(x) = \lambda \mathrm{e}^{-\lambda x}$$

defined on $x \in [0, \infty)$, where λ is a parameter that controls the rate of decay. Thus,

$$S(x) = \int_{0}^{x} \lambda \mathrm{e}^{-\lambda x'}\mathrm{d}x' = 1 - \mathrm{e}^{-\lambda x}.$$

Given that they are both defined on (0, 1), we can think of the computer's uniform random number as returning random values of the cumulative probability distribution. All we need to do is to invert the equation $S(x) = r$ to determine which value of x corresponds to a uniform random number r. The x values generated by this method are distributed according to the distribution $p(x)$. This works provided there is a computationally efficient way to calculate $S(x)$. It is particularly convenient if we can write down an equation for the cumulative distribution, and even better if that equation can be inverted analytically. The exponential distribution satisfies all of these criteria:

$$r = S(x) = 1 - \mathrm{e}^{-\lambda x}$$
$$\therefore x = -\ln(1 - r)/\lambda.$$

If r is uniformly distributed in (0, 1), then so is $1 - r$. Accordingly, we can save ourselves a bit of computer arithmetic by using the following:

$$x = -\ln r/\lambda. \tag{9.11}$$

(Saving unnecessary computations is important in these problems because we will often generate very large quantities of random numbers in the course of a simulation.) To emphasize: this formula lets us take a stream of uniformly distributed random numbers $r \in (0, 1)$ and transform them into a stream of exponentially distributed random numbers x.

9.4.2 Simulating single-molecule dynamics

Suppose that we are given a matrix of transition rates w_{ij} and that we want to simulate the corresponding Markov chain for a single molecule. The **kinetic Monte Carlo (KMC) algorithm** tells us how to carry out these simulations. At each step, we need to randomly generate a jump destination as well as a waiting time until this jump occurs. The waiting times are exponentially distributed, so we can generate random waiting times using equation (9.11).

Suppose that the system is in state s at time t. As long as the system remains in state s, the master equation reduces to

$$\frac{dP_s}{dt} = -\sum_{i \neq s} w_{si} P_s = -P_s \sum_{i \neq s} w_{si}.$$

Thus, the probability of being in state s decays with exponential decay constant

$$w_{\text{tot}} = \sum_{i \neq s} w_{si}. \tag{9.12}$$

This quantity replaces λ in equation (9.11).

Since w_{si} is the probability per unit time of jumping from state s to state i, the probability that the next jump will be to state i is w_{si}/w_{tot}. This gives us a conceptually simple method for choosing the jump destination. Imagine lining up the destination probabilities one after the other along the real line (figure 9.1). These probabilities

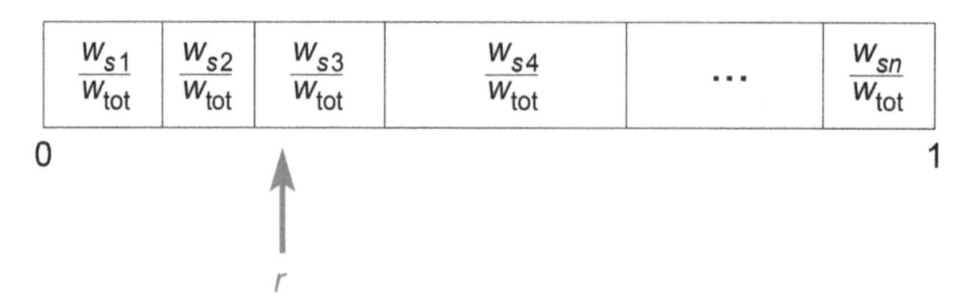

Figure 9.1. An illustration of the use of the kinetic Monte Carlo algorithm to choose the destination of a jump. A random number $r \in (0, 1)$ is used as a pointer into the corresponding interval of the real line, which we imagine as being divided up into subintervals by the destination probabilities w_{si}/w_{tot}.

occupy subintervals of (0, 1). If we generate a random number $r \in (0, 1)$, we can choose the jump according to the subinterval that r falls into.

The core of the kinetic Monte Carlo algorithm is the following sequence of steps:

1. Initialize the program:
   - Store the values of the transition rates w_{ij}.
   - Set $t = 0$.
   - Choose an initial state.
2. Repeat until the desired stopping condition is reached:
   (a) Calculate w_{tot}.
   (b) Generate two random numbers $r_{1,2} \in (0, 1)$.
   (c) Use r_1 to generate the time to the next jump:
   $$\Delta t = -\ln r_1/w_{tot}.$$
   (d) Use r_2 to pick the destination of the jump as follows: find the smallest k such that
   $$\sum_{i=1}^{k} w_{si} > r_2 w_{tot}.$$
   (e) Add Δt to the simulation time.
   (f) Change the state of the molecule according to the jump chosen.

The stopping criterion can be anything that makes sense for the problem being solved: a given number of jumps, a given amount of elapsed simulated time, a particular state reached, etc. It is also likely that you will collect some statistics as the simulation unfolds. The statistics that are useful to collect depend very much on the nature of the problem.

The steps given above simulate just one molecule's jumps. Usually, we want to simulate many molecules so we can calculate some statistics. In this case, we need to wrap much of the code above in a loop that repeats the simulation for many molecules. If you are new to computer programming however, my advice is to write simple pieces of code and to worry about repeating the computations using loops only after you are confident that your code works correctly.

Let's see how this works in practice by writing a Matlab code[4] that simulates IVR for a single molecule. Our objective is to calculate k_{2K}, the rate constant that shifts sufficient energy into the reaction coordinate through IVR for the reaction

[4] All code in this section will also run in Octave. I should caution you that, at least on my Mac, Octave runs these simulations *much* more slowly than Matlab. The final code described below ran in 11 s in Matlab but took over six minutes in Octave. If you want to save some time, you can reduce the number of realizations (which reduces the accuracy of the results) or sample energies more sparsely if you're using Octave.

to occur. For simplicity, we consider a molecule with just two vibrational modes. Some things are easier to deal with in simulations than in analytic theory, so we don't have to assume that all of the vibrational modes have the same frequency. Suppose instead that the frequencies of the two vibrational modes are related by $\omega_0^{(1)} = 2\omega_0^{(2)}$. We assume that the reaction we are modeling requires $v_2 = 10$ quanta in mode 2. Once this number of quanta accumulates in the second vibrational mode, we have reached the $A^\ddagger$ state and the simulation stops. Let's say that the time at which this occurs is t_{2K}. The rate constant is the inverse of the average time for a first-order process, i.e.

$$k_{2K} = 1/\langle t_{2K} \rangle. \tag{9.13}$$

The number of quanta of size $\hbar\omega_0^{(2)}$ stored in the two vibrational modes is

$$n = 2v_1 + v_2, \tag{9.14}$$

where, as usual, v_i is the vibrational quantum number of mode i. In this example, A^* is a molecule with $n \geqslant 10$. What we will eventually want to do is to vary n and see how the rate constant depends on this quantity, i.e. on the initial energy of A^*.

The general KMC algorithm can be simplified for the simulation of IVR. In IVR, every state has equal energy, so they are equally likely to be the destination of a jump. Thus, all we need to do is to randomly choose one of the states as the destination of any given jump. We can do this by just generating a random value of v_1, and then using equation (9.14) to calculate v_2. The possible values of v_1 are $0, 1, \ldots \lfloor n/2 \rfloor$, where $\lfloor \cdot \rfloor$ denotes the floor operation, i.e. the result of rounding a number down. We will have to be careful not to 'jump' to the same state the system is in at the start of the jump, but otherwise this is a simple exercise in picking one random state out of the set of states of equal energy. Moreover, as in section 9.3, we assume that all of the transition rates are equal. There are $\lfloor n/2 \rfloor + 1$ possible states, so given that the system is in one of these states at any given time, there are $\lfloor n/2 \rfloor$ possible destinations for a jump. The sum (9.12) is therefore very simple. It just is $w\lfloor n/2 \rfloor$.

As suggested above, the trick to writing simulation codes is to write them in small pieces, and to test each piece as we go. I'm going to start by just writing the code for one molecule undergoing IVR, and stopping as soon as that molecule reaches the $A^\ddagger$ state. Here is the code I wrote to get started:[5]

```
1  % Simulate IVR for a system with two vibrational modes related by
2  % omega1 = 2omega2, and where reaction occurs if omega2 >= nquant2_min.
3
4  % General simulation parameters
5  % Transition rate in s^{-1}. Can be adjusted to fit experimental data.
6  w = 1e15;
```

[5] Matlab code Section9_4_2_1st.m available from https://doi.org/10.1088/978-0-7503-5321-2.

```
7  % Minimum number of quanta required in mode 2 for reaction:
8  nquant2_min = 10;
9
10 % nquant2 is the number of quanta of the size of the mode 2 quanta the
11 % molecule has at t=0.
12 % Value of nquant2 used for testing:
13 nquant2 = 12
14
15 % Generate a random initial state with nquant2 quanta.
16 v1max = floor(nquant2/2);
17 v1 = randi([0,v1max]);
18 v2 = nquant2 - 2*v1
19
20 % Main simulation loop
21 t = 0;
22 wtot = w*v1max;
23 while (v2 < nquant2_min)
24     % Generate the random Delta t.
25     r = rand(1);
26     Delta_t = -log(r)/wtot;
27     % Generate the new values of v1.
28     % The while statement makes sure that there is an actual jump,
29     % i.e. that the new value of v1 is different from the current
30     % value.
31     new_v1 = randi([0,v1max]);
32     while (new_v1 == v1)
33         new_v1 = randi([0,v1max]);
34     end
35     % Update state and time.
36     v1 = new_v1;
37     v2 = nquant2 - 2*v1
38     t = t + Delta_t;
39 end
40
41 % Result of this simulation
42 disp('Final values of v1 and v2:')
43 disp([v1,v2])
44 t
```

You should save Matlab/Octave programs in files with a .m extension. If you do, then the part of the file name before the extension becomes a command you can run by typing it at the Matlab/Octave command line.

Forty-four lines of code may not seem like a short piece of code at first, but note that about half of those lines are either comments (indicated by the % character) or blank lines inserted for readability. There are also a few lines at the end to show the results, and many lines at the beginning that just set up some parameters.

First, a comment on comments: comments are good. Use many of them to explain what you are doing. You may find it difficult to remember what your own code means a few weeks later, and of course, reading someone else's code can be a real nightmare without comments[6]. Comments can be used to give the definition of a parameter (e.g. line five), or to explain how a particular piece of code works (e.g. lines 27–30), or in general anywhere that you think the intent of your code isn't obvious.

The code above was not written linearly. There were a few things I knew I would need right from the start, like a value for w, but a lot of programming involves adding detail as we go. Most of my attention was in fact focused on the main simulation loop and on what I needed to make that work. Having said that, I'm going to describe the program sequentially. The first part of the program just defines quantities we will need before we enter the main loop:

Line 6 defines the transition rate w.

Line 8 defines the minimum number of quanta that must accumulate in mode 2 in order to reach the $A^{\ddagger}$ state. It's a good idea to define such quantities as program parameters, since this lets us study the effect of this particular parameter on the results by just changing one number on one line rather than trying to hunt through the code for all of the places where this particular molecular property might appear.

Line 13 defines the initial number of quanta of size $\hbar\omega_0^{(2)}$ stored in the molecule. In later generations of the code, we will scan this parameter, but for now, we just give it an arbitrary value.

Line 16 defines the maximum number of quanta available to be stored in mode 1.

Lines 17 and 18 generate random initial values of v_1 and v_2 subject to the constraint imposed by the number of quanta available. The Matlab `randi()` function generates a single random integer, in this case between zero and `v1max`, inclusive. Note the syntax: `randi()` expects a two-element vector (delimited by the square brackets) giving the minimum and maximum integers in the range.

Line 21 sets the initial time to zero. You will note that I didn't put in a comment to explain this line. Sometimes, the intention of the programmer is reasonably obvious from the names of the variables within a certain application domain.

Line 22 calculates w_{tot}, which we need for the KMC algorithm.

You may have noticed that most of the lines end in semicolons but that some do not. A semicolon prevents Matlab from printing the result of a computation to the screen. Even a simple assignment involves the 'computation' of the right-hand side, so by default Matlab is very verbose. In this case, I am testing a new piece of code, so it's useful to see some output, e.g. the sequence of values of `v2`. While these are random values, I can look at them to make sure that the program appears to be doing what I intend it to do. In the next phase of program development, I will add semicolons to these lines, since I won't by then want to see these values, of which there will be thousands generated.

We can now look at the main loop, which runs from line 23 to line 39. The keyword `while` creates a loop whose instructions are repeated until the condition is

[6] An old programmer's joke: 'When I wrote this code, only God and I knew how it worked. Now only God knows.' (Original source unknown).

false. In this case, the program repeats the loop as long as `v2` is smaller than `nquant2_min`. The condition is tested every time the program reaches the keyword `end` at the bottom of the loop (line 39). Once this is no longer true (once there are enough quanta in mode 2 for the molecule to react), the program continues with the code following the loop. Also, if `v2` is not smaller than `nquant2_min` when the interpreter reaches line 23, the loop is never entered. This is sensible behavior in this case, since a molecule could have sufficient energy in mode 2 immediately after the collision that energized it.

Inside the loop, we need to carry out jumps according to the KMC algorithm. The Matlab function `rand()` generates matrices of random floating-point numbers distributed uniformly in (0, 1). Its arguments are the dimensions of the matrix. A single argument implies a square matrix. In particular, `rand(1)` generates a single random number. Lines 25 and 26 generate exponentially distributed random waiting times for the next jump. Line 31 generates a random value of v_1. However, we need to make sure that the new value is different from the current v_1, otherwise there wasn't a jump. The `while` loop in lines 32 to 34 ensure that this is the case by rejecting a `new_v1` that is the same as `v1` and generating a new value. Finally, lines 36 to 38 update the state of the system and time. This completes the jump to the new state.

Note that the indentation of the loops is meaningless in Matlab, unlike in Python. The `end` keyword indicates where the `while` block ends. Having said that, indentation does make code more readable and is strongly encouraged.

The last few lines of the program just output the results of the simulation. To do this, you can use the `disp()` function to output text (line 42) or values (line 43), or you can just put the name of a variable on a line by itself (line 44).

You can run your program by just typing its name in the Matlab command window, as mentioned above. The output of this program varies because it depends on some random numbers. Sometimes, the initial state corresponds to $A^{\ddagger}$ right away, and you will see something like the following output:

```
>> two_mode_IVR1
nquant2 =
    12

v2 =
    10

Final values of v1 and v2:
     1    10

t =
     0
```

(Some blank lines were deleted to save space.) In other cases, the system will have to take several IVR steps before the correct number of quanta find their way to mode 2:

```
>> two_mode_IVR1
nquant2 =
    12

v2 =
     0

v2 =
     4

v2 =
     6

v2 =
     2

v2 =
    10

Final values of v1 and v2:
     1    10

t =
   3.6261e-16
```

Because of this randomness, we will have to repeat the IVR experiment many times to collect statistics at any given energy of A*. The idea is to average the values of t over many **realizations** (statistical jargon for repetitions of a stochastic process with different random numbers) and then to calculate the rate constant using equation (9.13).

I wrote the following code for this next step in the development of our IVR code:[7]

```
1  % Simulate IVR for a system with two vibrational modes related by
2  % omega1 = 2omega2, and where reaction occurs if omega2 >= nquant2_min.
3
4  % General simulation parameters
5  % Transition rate in s^{-1}. Can be adjusted to fit experimental data.
6  w = 1e15;
7  % Minimum number of quanta required in mode 2 for reaction:
8  nquant2_min = 10;
9  nsims = 10000;  % Number of simulations at each energy
10 % Create a vector to hold the values of t from each simulation
11 % at any given value of nquant2.
12 t2K = zeros(nsims,1);
13
14 % nquant2 is the number of quanta of the size of the mode 2 quanta the
15 % molecule has at t=0.
```

[7] Matlab code Section9_4_2_2nd.m available from https://doi.org/10.1088/978-0-7503-5321-2.

```
16 % Value of nquant2 used for testing:
17 nquant2 = 13
18
19 v1max = floor(nquant2/2);
20 wtot = w*v1max;
21
22 % Generate nsims simulations.
23 for sim=1:nsims
24     % Main simulation loop
25     t = 0;
26     % Generate random values of v1 and v2 for this simulation.
27     v1 = randi([0,v1max]);
28     v2 = nquant2 - 2*v1;
29     while (v2 < nquant2_min)
30         % Generate the random Delta t.
31         r = rand(1);
32         Delta_t = -log(r)/wtot;
33         % Generate the new values of v1.
34         % The while statement makes sure that there is an actual jump,
35         % i.e. that the new value of v1 is different from the current
36         % value.
37         new_v1 = randi([0,v1max]);
38         while (new_v1 == v1)
39             new_v1 = randi([0,v1max]);
40         end
41         % Update state and time.
42         v1 = new_v1;
43         v2 = nquant2 - 2*v1;
44         t = t + Delta_t;
45     end
46
47     % The value of t is how long it took for nquant_min quanta
48     % to show up in mode 2. We store this value for later averaging.
49     t2K(sim) = t;
50 end
51
52 % Result of this simulation
53 % Plot a histogram of the t values:
54 % The following will work in either Matlab or Octave:
55 figure('DefaultAxesFontSize',18)        % Use a larger font
56 hist(t2K,20)
57 xlabel("t_{2K}/s","interpreter","tex")
58 ylabel("Count")
59 % Matlab only:
60 % For a nicer plot scaled to give probability rather than raw counts,
```

```
61  % comment out the previous three lines and uncomment the following lines:
62  %figure('DefaultAxesFontSize',18)        % Use a larger font.
63  %histogram(t2K,"Normalization","probability")
64  %xlabel("$t_{2K}/$s","interpreter","latex")
65  %ylabel("Probability")
66
67  k2K = 1.0/mean(t2K)
68  % The standard error of the mean is used to estimate the error in k2K.
69  stderr_t2K = std(t2K)/sqrt(nsims);
70  errk2K = stderr_t2K*k2K^2
```

Much of this setup is the same as before, but there are two important additions. In line 9, we establish the number of realizations to be used to compute the average. Ten thousand may seem like a large number, but because averages converge slowly, this is the correct order of magnitude to use for this kind of work. We will need somewhere to store each of the simulated IVR times. Space for these values is created in line 12. The Matlab function `zeros()` creates a matrix filled with the value zero. The two arguments of this function are the dimensions of the matrix, so in this case we are creating an array of `nsims` rows and one column, i.e. a vector. We could equally well have created a row vector with `zeros(1,nsims)`. The variable `t2K` is therefore initialized as a vector of zeros. Matlab can dynamically resize vectors or matrices, so technically it wasn't necessary to preallocate space for the results. However, memory allocation is one of the slower operations in a computer, so it often pays to preallocate space for simulation results if we know ahead of time how much space we will need.

The most visible change is likely the `for` loop that wraps around much of our previous code. A `for` loop simply repeats a set of instructions a fixed number of times. This program illustrates the simplest use of a `for` loop, which uses a counter. In Matlab, the notation `a:b` denotes a range, specifically the sequence of numbers a, $a + 1$, $a + 2$, ... up to a value not exceeding b. Here, the variable `sim` takes the values 1,2,...,`nsims`. The contents of the `for` loop are therefore repeated `nsims` times.

Each realization should have an independently generated initial state, so I was careful to move the generation of the initial values of `v1` and `v2` inside the loop. We also want to start each realization from $t = 0$. Once we reach the end of the IVR process, we need to store the time taken to put the appropriate number of quanta into mode 2. This is what line 49 does: the notation `a(i)` accesses the ith component of the vector `a`.

Once the `for` loop is complete, the vector `t2K` contains the individual values of the IVR times for each realization. I'm a fan of looking at data and not just taking it for granted that it's OK. Accordingly, I added some code to plot a histogram of the data. The `hist()` function generates a histogram. The second argument of `hist()` tells Matlab how many bins to use. (The default is ten, which didn't seem like quite enough for this data set.) I generated figure 9.2 using the Matlab-only code in lines 62 to 65.

You may note that the distribution looks exponential *except* for the first bin, which seems to have too high a probability. This is actually correct behavior for this problem, since $\frac{1}{3}$ of the time, the initial value of v_1 will be either zero or one, which

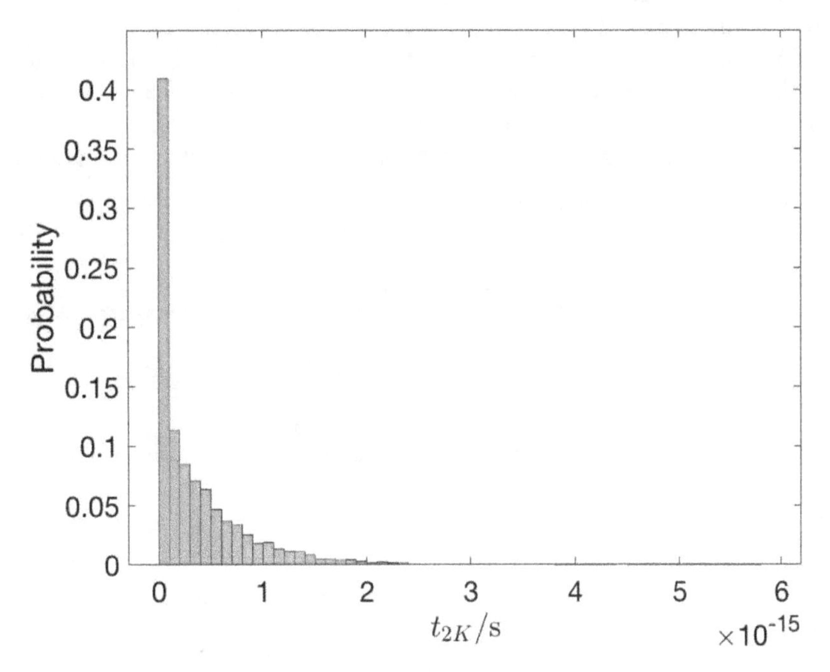

Figure 9.2. The probability distribution of the t_{2K} values generated by the second version of the kinetic Monte Carlo IVR simulation program run with $w = 10^{15}\ \mathrm{s}^{-1}$ and $n = 13$.

leaves, respectively, 13 or 11 quanta in mode 2. Accordingly, $\frac{1}{3}$ of the time, the molecule is immediately in the $A^{\ddagger}$ state, so $t = 0$.

Finally, we compute the rate constant as well as an uncertainty. We use a pair of statistical functions here: `mean()`, which computes the average of the values in a vector, and `std()`, which computes the standard deviation. The most appropriate measure of uncertainty for the average time is the standard error of the mean, which is calculated by dividing the standard deviation by the square root of the number of samples (line 69). The percentage error in k_{2K} is the same as the percentage error in the average time, which leads to line 70. If you run the program, in addition to the histogram, you should get output that looks something like this:

```
>> two_mode_IVR2
nquant2 =
    13


k2K =
   2.8229e+15


errk2K =
   3.7826e+13
```

Your results will vary slightly from these, since these are stochastic simulations. Nevertheless, we conclude that $k_{2K} = (2.82 \pm 0.04) \times 10^{15}\ \mathrm{s}^{-1}$. Because the estimate will only be within one standard error of the true mean 68% of the time, if you run the program several times, you will likely see some values that fall outside the range suggested by the standard error.

Thinking back to the problem we wanted to solve, which is to calculate k_{2K}, a quantity that depends on the energy of the energized species A*, we have one thing left to do, which is to vary the number of quanta of energy of the molecule, i.e. the parameter we have been calling nquant2. The final version of my program looks like this:[8]

```
1  % Simulate IVR for a system with two vibrational modes related by
2  % omega1 = 2omega2, and where reaction occurs if omega2 >= nquant2_min.
3
4  % General simulation parameters
5  % Transition rate in s^{-1}. Can be adjusted to fit experimental data.
6  w = 1e15;
7  % Minimum number of quanta required in mode 2 for reaction:
8  nquant2_min = 10;
9  nsims = 10000;  % Number of simulations at each energy
10 % Create a vector to hold the values of t from each simulation
11 % at any given value of nquant2.
12 t2K = zeros(nsims,1);
13 % Maximum number of omega2 quanta
14 nquant2_max = 50;
15 % nq2count is a counter for the number of nquant2 values
16 nq2count = 0;
17
18 % Loop over nquant2 values
19 for nquant2=nquant2_min:nquant2_max
20     nq2count = nq2count + 1;
21     % Generate nsims simulations.
22     for sim=1:nsims
23         % Generate a random initial state with nquant2 quanta.
24         v1max = floor(nquant2/2);
25         v1 = randi([0,v1max]);
26         v2 = nquant2 - 2*v1;
27
28         % Main simulation loop
29         t = 0;
30         wtot = w*v1max;
31         while (v2 < nquant2_min)
32             % Generate the random Delta t.
33             r = rand(1);
34             Delta_t = -log(r)/wtot;
35             % Generate the new values of v1.
36             % The while statement makes sure that there is an actual jump,
37             % i.e. that the new value of v1 is different from the current
38             % value.
39             new_v1 = v1;
40             while (new_v1 == v1)
41                 new_v1 = randi([0,v1max]);
42             end
43             % Update state and time.
```

[8] Matlab code Section9_4_2_3rd.m available from https://doi.org/10.1088/978-0-7503-5321-2.

```
44              v1 = new_v1;
45              v2 = nquant2 - 2*v1;
46              t = t + Delta_t;
47          end
48
49          % The value of t is how long it took for nquant_min quanta
50          % to show up in mode 2. We store this value for later averaging.
51          t2K(sim) = t;
52      end
53
54      % Calculate the rate constant and its error,
55      % and store the values for plotting.
56      nq2(nq2count) = nquant2;
57      k2K(nq2count) = 1.0/mean(t2K);
58      stderr_t2K = std(t2K)/sqrt(nsims);
59      errk2K(nq2count) = stderr_t2K*k2K(nq2count)^2;
60  end
61
62  figure('DefaultAxesFontSize',18)
63  errorbar(nq2,k2K,errk2K,'o')
64  xlabel('Number of quanta')
65  ylabel('$k_{2K}$/s$^{-1}$','interpreter','latex')
```

Again, much of the code is unchanged. The most significant addition is a loop over the values of `nquant2` (lines 19 to 60). This loop starts at the minimum number of required quanta and goes up to an arbitrarily selected maximum set in line 14. For each energy (each value of `nquant2`), we need to store the calculated rate constant and uncertainty. The catch is that the values of `nquant2` don't start at one; they start at `nquant2_min`, which is ten in this scenario. We could just create vectors to store the results of the calculations and ignore the first nine entries, but a nicer solution is to use a counter to keep track of where we are in some vectors that will just store the computed values. Lines 16 and 20 take care of this for us, initializing a counter called `nq2count` and then incrementing it each time through the loop over `nquant2` values. Thus, the value of `nq2count` is one on the first pass through the outer loop (when `nquant2=nquant2_min`), two on the next pass through the loop, and so on. We then use this counter to store the results in the appropriate elements of several vectors in lines 56 to 59. In particular, in line 56, we store the value of `nquant2` so that we can use it later, e.g. for plotting (line 63). (Note that we don't store `stderr_t2K`, since we only use it to calculate the uncertainty in k_{2K}.)

After all is said and done, we use the Matlab `errorbar()` function to plot the results with error bars. The first three arguments of `errorbar()` are the quantities to be plotted on the x and y axes as well as the y-value error bars. The fourth (optional) argument is the style to use for the plot. An `o` indicates that I want to use circles as the plotting symbols. A typical result of running this program is shown in figure 9.3. You should notice that essentially the same value of k_{2K} is obtained for $n = 2i$ and $2i + 1$ for integer values of i. This is because an odd quantum plays no role in this stochastic process, since the number of different states participating in IVR depends on $\lfloor n/2 \rfloor$. To save some computational time, we might have only computed k_{2K} for even values of n. To do this, we would have replaced line 19 by

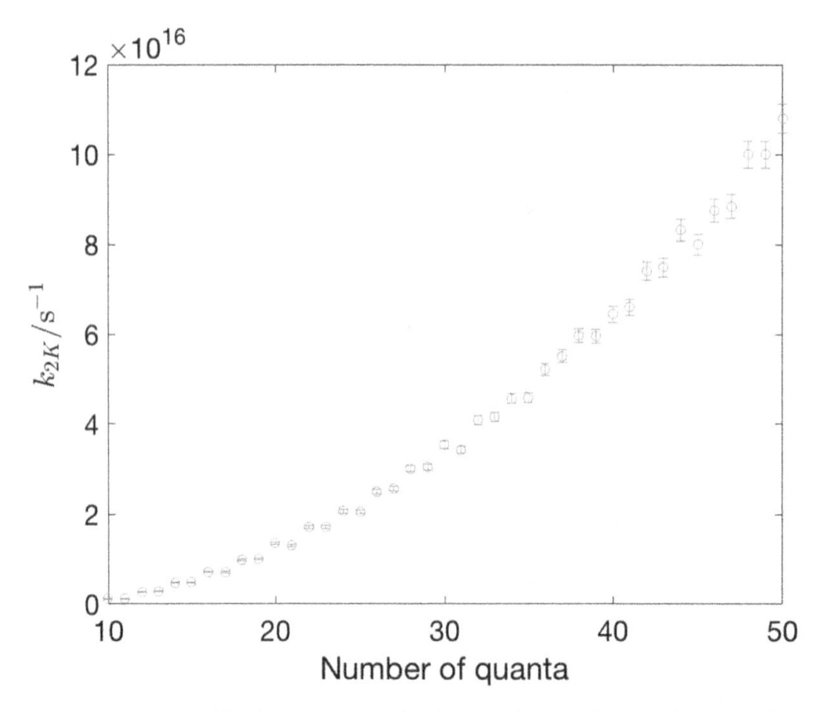

Figure 9.3. k_{2K} calculated from kinetic Monte Carlo simulations of IVR as described in the text for a molecule with two vibrational levels for which $\omega_0^{(1)} = 2\omega_0^{(2)}$; ten quanta of vibrational energy are required in mode 2 in order for the molecule to accede to the transition state. The horizontal axis represents the amount of vibrational energy stored in the molecule as the equivalent number of quanta of size $\hbar\omega_0^{(2)}$.

```
for nquant2=nquant2_min:2:nquant2_max
```

This shows a generalization of the range operator: `a:s:b` means the range from `a` to a value not exceeding `b` in steps of `s`. Thus, the line above would cause `nquant2` to step through the values `nquant2_min`, `nquant2_min+2`, etc.

9.4.2.1 An aside about the 'jump' picture of IVR

The picture of IVR presented by the KMC treatment is a classical cartoon. Quantum mechanically, at any given time, a molecule exists in a superposition of states of approximately equal energy (subject to the time–energy uncertainty principle). The time-dependent Schrödinger equation tells us how this superposition evolves over time. In the absence of external influences, the superposition should evolve towards a uniform distribution over all states of equal energy. The probability of reaction per unit time is proportional to the weight of states that carry sufficient energy in the reaction coordinate among all the states in the superposition. There are no 'jumps.' On the other hand, the master equation also describes a smooth time evolution tending towards a state where all states with the same energy have the same probability. The descriptions of IVR presented by the master equation and by the Schrödinger equation are therefore (at least) qualitatively similar, and one could hope to bring them into agreement.

For its part, the KMC algorithm simulates the same process described by the master equation. If we take a sufficient number of simulations, the time-dependent probabilities predicted by the KMC algorithm are the same as those predicted by the master equation. Thus, while the jumps simulated in the KMC algorithm have no clear quantum mechanical meaning in a more rigorous treatment of IVR, the statistics of these jumps have the right behavior. We can therefore think of the KMC algorithm as a semiclassical procedure which yields quantum-like behavior by averaging.

9.4.3 Simulating a population of molecules

In section 9.4.2, we learned how to do single-molecule kinetic Monte Carlo simulations. In order to gather statistics, we can repeat these simulations many times, as we did in our KMC treatment of IVR. This is not always the most convenient way to structure a simulation. The single-molecule approach has the advantage that it lets us visualize single-particle trajectories through the state space (by plotting the state vs t, although we did not do that), but if we really want population statistics, it would be more convenient to simulate an entire population in one go. The KMC algorithm can handle populations just as easily as it can handle single molecules, with a few very minor adjustments. We can jump (so to speak) from a single-molecule description to a population description by thinking a bit about the relationship of single-molecule transition rates to the effects of these transitions at the population level. Recall that $w_{ij}\mathrm{d}t$ is the probability that a single molecule executes a jump from molecular state i to state j in time $\mathrm{d}t$. If there are N_i molecules in state i at time t, then the probability that *one* of these molecules jumps from i to j in time $\mathrm{d}t$ is just $N_i w_{ij}\mathrm{d}t$ *provided* $\mathrm{d}t$ is sufficiently small that there is a negligible probability that two jumps occur in this period of time.

To go further, we need to establish some terminology. The **system** in this version of the kinetic Monte Carlo algorithm consists of a number of molecules distributed over the accessible molecular states. The **state of the system** is a vector of populations of molecules in each **molecular state**: $\mathbf{N} = (N_1, N_2, \ldots, N_s, \ldots)$. (Note the two different usages of the word 'state.') The **propensity** of a jump from molecular state i to molecular state j is $a_{ij} = N_i w_{ij}$. This is the probability per unit time of **one** molecule making a jump from molecular state i to state j.

The KMC algorithm for many molecules is much the same as the single-molecule algorithm, except that:

- Propensities are used to determine jump probabilities.
- Instead of updating the state of one molecule, we update the populations of the 'from' and 'to' molecular states.

Here is the algorithm in detail:

Initialization:

1. Store the w_{ij} values. In general, a matrix would be a good data structure for this purpose, although for some problems, there may be simpler solutions.

2. Store the initial populations for each molecular state in a vector $\mathbf{N}$ with components N_i.
3. Set $t = 0$.

For each simulation step:

1. Calculate the propensities $a_{ij} = w_{ij}N_i$. Again, this could be a matrix, but in special cases there may be simpler ways to store the propensities.
2. Calculate $a_{tot} = \sum_{i,j} a_{ij}$.
3. Generate two random numbers $r_{1,2} \in (0, 1)$.
4. Use r_1 to generate the time to the next jump:
$$\Delta t = -\ln r_1/a_{tot}$$
5. Use r_2 to pick the destination of the jump as follows:
   (a) Arrange the a_{ij} so that they are stored in a vector $\hat{\mathbf{a}}$ (if they aren't already).
   (b) Find the smallest k such that
$$\sum_{\ell=1}^{k} \hat{a}_\ell > r_2 a_{tot} \tag{9.15}$$
   then determine the (i, j) values corresponding to k. The details of this procedure depend on how we 'linearized' the a_{ij} values.
6. Add Δt to the simulation time.
7. Adjust the populations according to the jump:
$$N_i = N_i - 1$$
$$N_j = N_j + 1$$
8. Recalculate (at least) a_{ik} and a_{jk} $\forall k$.

As an example, let's develop a simulation program for a simple model of a unimolecular reaction, illustrated in figure 9.4. We consider an ensemble of molecules with some vibrational energy levels; for simplicity, these are taken to be equally spaced. Collisions (or possibly also the absorption and emission of infrared radiation) cause transitions between the levels. We assume that the Landau–Teller approximation holds, so that there are only transitions between adjacent levels. When the molecule reaches the top energy level, i.e. the level with vibrational quantum number v_{max}, it can react irreversibly, with a transition rate w_r. (In the figure, $v_{max} = 5$.)

We need to make some assumptions about the transition rates in order to make further progress. Let us assume that the levels are equally spaced by $h\nu$, which would be the case for a single vibrational mode. Let us further suppose that a version of the strong collision assumption holds, namely that every collision results in a change (up or down) in the vibrational energy. This means that $w_{i(i+1)} + w_{i(i-1)} = Z_A$, where Z_A is the number of collisions per unit time per reactant molecule. The work of section 3.1 implies

$$Z_A = \sigma \langle u_{rel} \rangle L[B],$$

where [B] is the concentration of bath-gas molecules. We also have the Boltzmann equilibrium condition (9.5), which is $w_{i(i+1)}/w_{(i+1)i} = e^{-\Delta\varepsilon/k_BT}$. One way to satisfy these

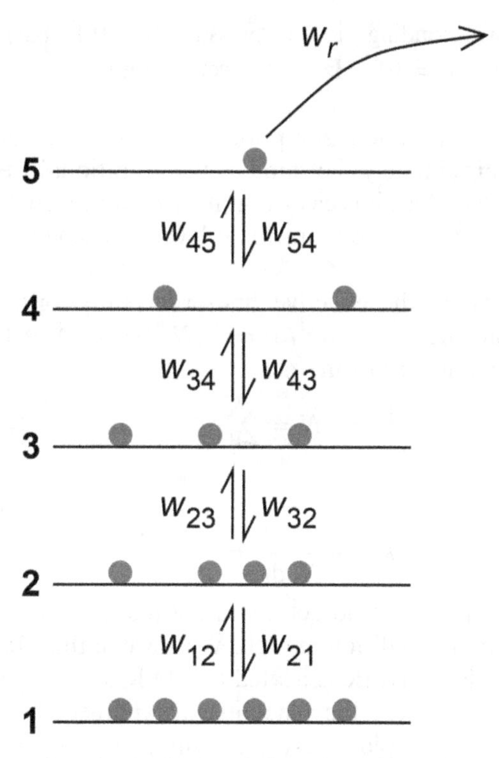

Figure 9.4. A simple picture of a unimolecular chemical reaction. The numbered lines are vibrational states. Each red ball represents one molecule. A molecule can be energized or de-energized by collisions, causing jumps between adjacent energy levels according to the Landau–Teller approximation. When a molecule reaches the top vibrational level, shown here as the fifth energy level, it can react to form products irreversibly.

equations is for all of the 'upward' transition rates to be the same, say w_{up}, and for all of the 'downward' rates to also be the same, say w_{down}. We then have

$$w_{\mathrm{up}} + w_{\mathrm{down}} = Z_{\mathrm{A}}$$

and

$$\frac{w_{\mathrm{up}}}{w_{\mathrm{down}}} = \mathrm{e}^{-\Delta\varepsilon/k_{\mathrm{B}}T}.$$

This pair of equations is easily solved. We get

$$w_{\mathrm{down}} = \frac{Z_{\mathrm{A}}}{1 + \mathrm{e}^{-\Delta\varepsilon/k_{\mathrm{B}}T}}$$

and

$$w_{\mathrm{up}} = \frac{Z_{\mathrm{A}}\mathrm{e}^{-\Delta\varepsilon/k_{\mathrm{B}}T}}{1 + \mathrm{e}^{-\Delta\varepsilon/k_{\mathrm{B}}T}}.$$

Given the values of the temperature, cross-section, masses of the reactant and bath gas molecules, the concentration of the bath gas, and the vibrational spacing, we can calculate both of these transition rates. For the sake of argument, taking $T = 600$ K, $R_{\mathrm{AB}} = 4$ Å,

$[B] = 2 \text{ mol m}^{-3}$ (corresponding to a pressure of 0.1 bar), $M_A = 64 \text{ g mol}^{-1}$, $M_B = 40 \text{ g mol}^{-1}$ and $\nu = 10^{13}$ Hz, we get roughly $w_{down} = 3 \times 10^8 \text{ s}^{-1}$ and $w_{up} = 1 \times 10^8 \text{ s}^{-1}$.

As for w_r, we make the same assumption as in transition-state or RRK theory, which is that it is approximately the same as the vibrational frequency, thus about 10^{13} s^{-1}. Given this value of w_r, it is clear that moving up through the ladder of states is rate determining. Once the top vibrational level is reached, reaction occurs with nearly 100% efficiency.

We want to determine the effective first-order rate constant. Since this is a unimolecular reaction, the rate is $dN/dt = -kN$, where N is the total number of reactant molecules remaining at time t:

$$N = \sum_{i=1}^{v_{max}} N_i.$$

Accordingly,

$$k = -\frac{1}{N}\frac{dN}{dt} = -\frac{d\ln N}{dt}.$$

If we just plot $\ln N$ vs t, we should get a straight line of slope $-k$.

We will need an initial condition as well. We assume that the reactants start out cold and that the reaction mixture is heated to 600 K at $t = 0$. (Obviously, in a real experiment, we could not heat the reactants instantaneously, although there are methods that achieve very rapid temperature jumps [1], section 16.3.1.) Thus, the initial populations are $N_1 = N_{init}$, $N_2 = N_3 = \cdots = 0$.

We are now ready to write a program. In some ways, a population simulation is simpler than the IVR problem we tackled previously because we don't need to embed loops inside each other, so I will just present my final program here:[9]

```
1  % Program to simulate a simple model of a unimolecular reaction with vmax
2  % equally spaced energy levels, and reaction from the last level.
3
4  % Clear all variables and arrays.
5  clear
6
7  % Number of vibrational levels:
8  vmax = 10
9
10 % Transition rates in s^{-1}
11 wdown = 3e8
12 wup = 1e8
13 wr = 1e13
14
15 % Number of molecules and initial populations:
16 t = 0;
17 Ntot = 1000
18 N = zeros(vmax,1);
19 N(1) = Ntot;
20
```

[9] Matlab code Section9_4_3.m available from https://doi.org/10.1088/978-0-7503-5321-2.

```
21 % Preallocate a propensity array.
22 % Organization of the propensities:
23 %   [a12 a23 a34 ... a(vmax-1)vmax a21 a32 a43 ... a(vmax)(vmax-1) areact]
24 % There are 2(vmax-1) + 1 propensities.
25 % Propensities for i=1 to vmax-1 are the propensities for an upward jump
26 %   from level i to level i+1.
27 % Propensities for i=vmax to 2(vmax-1) are the propensities for a downward
28 %   jump from level i+2-vmax to level i+1-vmax.
29 % Propensity 2(vmax-1) + 1 = 2vmax - 1 is the propensity of reaction.
30 a = zeros(2*vmax-1,1);
31
32 % A counter for the number of simulation steps
33 nsteps = 0;
34
35 % Main simulation loop: continue until we run out of molecules.
36 while (Ntot > 0)
37     % Calculate propensities.
38     a(1:vmax-1) = wup*N(1:vmax-1);
39     a(vmax:2*(vmax-1)) = wdown*N(2:vmax);
40     a(2*vmax-1) = wr*N(vmax);
41
42     % KMC selection of time and next jump.
43     a_sums = cumsum(a);
44     atot = a_sums(end);
45     r = rand(2,1);
46     Delta_t = -log(r(1))/atot;
47     k = find(a_sums > r(2)*atot,1,'first');
48
49     % What reaction did we pick? Change the populations accordingly.
50     if (k < vmax)
51         % This is an upward jump.
52         start_state = k;
53         N(start_state) = N(start_state) - 1;
54         N(start_state+1) = N(start_state+1) + 1;
55     elseif (k < 2*vmax-1)
56         % Downward jump.
57         start_state = k + 2 - vmax;
58         N(start_state) = N(start_state) - 1;
59         N(start_state-1) = N(start_state-1) + 1;
60     else    % Reaction
61         N(vmax) = N(vmax) - 1;
62         Ntot = Ntot - 1;
63     end
64
65     % Increment t, then store the current total number of molecules.
```

```
66      t = t + Delta_t;
67      nsteps = nsteps + 1;
68      tval(nsteps) = t;
69      Ntotval(nsteps) = Ntot;
70  end
71
72  % Simulation done. Plot ln(Ntot) vs t.
73  % The last value of Ntotval will always be zero so we can't take its log.
74  % Get rid of this point.
75  nsteps = nsteps - 1;
76  tval = tval(1:nsteps);
77  Ntotval = Ntotval(1:nsteps);
78  figure('DefaultAxesFontSize',18)
79  plot(tval,log(Ntotval))
80  xlabel('t/s')
81  ylabel('ln(N_{total})','interpreter','tex')
82  hold    % This will allow us to add a line to the plot.
83
84  % Fit the data to get a rate constant.
85  % line_coeffs(1) will be the slope and line_coeffs(2) the intercept.
86  line_coeffs = polyfit(tval,log(Ntotval),1);
87  % The first time point is at t=0. The last time point is stored in t.
88  t_endpoints = [0,t];
89  % Calculate the corresponding N values from the equation of the line:
90  N_endpoints = line_coeffs(1)*t_endpoints + line_coeffs(2);
91  plot(t_endpoints,N_endpoints)
92  hold off
93  % The rate constant is -(slope).
94  k = -line_coeffs(1)
```

The program starts with a precaution, which we might have used in our previous programs: the keyword `clear` on line five clears all variables in the Matlab workspace. Especially when we are developing a program, and will therefore be running it repeatedly to test it, it is often a good idea to start a program with `clear` so that variable values from previous runs don't interfere with program execution.

Lines 21 to 34 describe the layout of the propensity array in detail. This extended comment is useful for at least two reasons: first, it forced me to think clearly about how the propensities would be organized in the propensity vector, and second, this will of course be useful to anyone who wants to use this code, especially if they want to modify it to solve a related problem. It is often the case that programs are written by modifying existing code. Carefully constructed comments become incredibly important in such cases.

The main loop starts at line 36. It may be useful to compare the contents of this loop to the description of the algorithm as you work through it.

Lines 38 to 40 calculate the propensities using Matlab's ability to carry out vector operations. Note the values of N_i used in the calculations of the propensities. Upward jumps start from levels 1 to $v_{\max} - 1$. Downward jumps start from levels 2 to $v_{\max}$. In addition, compare these calculations to the description of the propensity array.

Lines 43 to 47 carry out the key steps of the KMC algorithm. The Matlab `cumsum()` function calculates the partial sums of its argument. Thus, `cumsum(a)` returns a vector containing the values

$$\left[a_1, a_1 + a_2, \ldots, \sum_{i=1}^{k} a_i, \ldots \sum_{\forall i} a_i\right].$$

Note that the last component of this vector is the quantity called a_{tot} in the KMC algorithm. We take advantage of this in line 44 to avoid a separate calculation of a_{tot}. Note the use of the keyword `end` to indicate the last element in a vector.

The selection of the next jump occurs in line 47. This uses a very powerful Matlab function called find. The first argument of find is a condition that should be satisfied by the sought-after elements of an array, in this case the jump selection condition (9.15). The second argument is the number of elements desired and the third argument controls the direction of the search. In this case, we want one element (second argument), and the first one that is found (third argument). The return value of this function is the position (or, in general, the positions) of the element(s) satisfying the condition. In other languages, this Matlab one-liner would be replaced by a `while` loop that checked the partial sums one by one.

Once we have determined which jump happens, we need to figure out what process this corresponds to: an upward or downward jump, or reaction, as well as the state prior to the jump. We then need to adjust the populations accordingly. This is what happens in lines 50 to 63. Here we use a conditional, an `if` control structure, to choose the appropriate case so that we can do the bookkeeping for the populations. You should again compare the conditions in this block of code to the layout of the propensity array. The first condition (line 50) checks whether the selected jump is an upward jump. The `start_state` is the state from which the jump is taken. Both the populations of this state and of the state to which a molecule is jumping, the next-highest state, are updated. If the condition in line 50 was satisfied, after executing the code following this condition, the program jumps to the matching `end` statement. The other conditions in this block are then not tested.

If `k` was larger than or equal to `vmax`, then the first condition fails and the program checks the second condition. If this condition is satisfied then we have a downward jump, with appropriate adjustments made to the populations. You should spend a bit of time with pen and paper to verify the logic of line 57, which computes the state from which the jump is initiated. To do this, you will again need to think about the layout of the propensity array. If it helps, pick a specific value for `vmax`.

If neither the `if` nor the `elseif` condition is satisfied, then the only remaining possibility is that the 'jump' selected by the KMC algorithm is in fact the reaction of a molecule in the top energy level. In this case, we need to adjust `Ntot`, which we use to keep track of how many reactant molecules we have left.

Once the populations have been adjusted based on the selected jump, we can increment the time, and store both the time and the total number of reactant molecules remaining for later analysis.

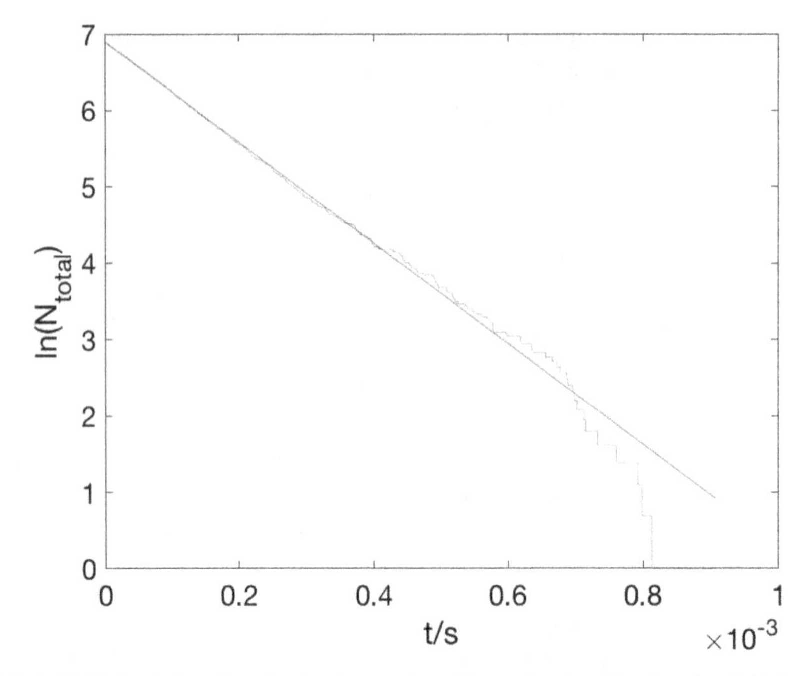

Figure 9.5. A KMC simulation of a unimolecular reaction with ten vibrational levels and an initial population of 1000 molecules. The other parameters are given in the text.

The rest of the program plots the data, fits a line to it, plots that line on the same graph as the data, and returns the rate constant. The `hold` keyword (line 82) tells Matlab that we will want to add to the graph we have created. Line 86 fits a line to the ln N vs t data we have collected. The `polyfit()` function takes x and y values as arguments along with the order of the polynomial to fit to the data. It returns a vector of coefficients, starting with the highest power, so in this case, `line_coeffs(1)` will be the slope of the line, and `line_coeffs(2)` will be the intercept. We only need two points on a line to plot it. Lines 88 and 90 calculate two points on the line of best fit. Once this line has been plotted `hold off` tells Matlab that our graph is complete.

I ran this program with an initial population of 1000 molecules. The graph produced by Matlab is shown in figure 9.5. Towards the end of the simulation, when there are few molecules left, the results start to deviate from the line. However, because there are so many more points collected when there is a large number of molecules, this final phase has a negligible effect on the fit. The value of the rate constant computed by Matlab from this simulation was 6.6×10^3 s^{-1}. Running the program again gives us an idea of the error bars for this value, which are of the order of 0.1×10^3 s^{-1}.

9.5 Theories or models?

So far, we have discussed a number of theories of chemical kinetics, each with its strengths and weaknesses. Transition-state theory is easy to apply but involves some dubious assumptions that can partly be fixed by 'bolt-on' modifications (e.g. the variational formulation, tunneling corrections). The Lindemann mechanism gives at least a qualitative explanation of the pressure dependence of gas-phase unimolecular reactions but fails some quantitative tests. RRK theory explains some of the failures of Lindemann theory but has its own quantitative problems. The Rice–Ramsperger–Kassel–Marcus (RRKM) theory does better still but runs into tricky problems in reactions where a bottleneck impedes energy flow. And we can use a master equation treatment to understand why the rate constant k_2 from the Lindemann mechanism is sometimes surprisingly large. Some of these theories build on each other, but often they are based on completely different principles. So which theory is 'right'? And how do we choose which one to use?

The desire to find one true theory of chemical kinetics is completely understandable. However, the theories covered in this book are not quite like the great theories of physics, where each theory neatly contains the previous theory as a special limit[10]. Instead, we need to think of these theories, despite the use of the latter term, as **models**. Each of these models provides a 'picture' (typically of a mathematical sort) that helps us to understand what is happening during a chemical reaction. Of course, they also allow us to calculate rate constants, or at least to understand the factors that affect the rate constant, and each of them, with sufficient effort, gives excellent results for some reactions, even if it fails for others. We dare to hope, however, that the failures are instructive and that they point to further refinements (e.g. variational transition-state theory to address the recrossing problem) or to new models (e.g. RRK theory to address some of the failings of the Lindemann theory). The answer to the question of which of these theories is 'right' is therefore that all of them are right in the cases that are appropriate to them, and all of them are wrong if we expect any one of them to answer all questions.

To put it another way: the purpose of much of the material presented in this book is not to develop one definitive theory that allows us to calculate rate constants for any conceivable reaction but to equip us with a suite of models that provide insights into the factors that determine the rates of chemical reactions: from TST, we learn about the importance of the region near the saddle point separating reactants from products in determining the rate of reaction; from the Lindemann mechanism, we learn about the importance of collisional energy transfer in the gas phase; from RRK(M) theories, we learn about the key role played by intramolecular energy transfer; and so on. This perspective on models of kinetic events will become especially important when we turn to reactions in solution in the next part of this book, given that the complexity of the problem increases greatly once we include solvent molecules. There, theories will provide tools that help us to *understand* effects that we may not be able to *quantitatively* predict from first principles.

If I may be permitted one personal anecdote, during my comprehensive examination as a doctoral student at the University of Toronto, Professor Paul Brumer asked me to select and briefly discuss some of the great problems of contemporary chemical physics. (For context, this conversation occurred in the early 1990s.) One of the problems I mentioned was computing rate constants in solution

[10] Even in physics, the situation isn't as tidy as we would like, given that gravity and the other fundamental forces don't play nicely with each other.

from first principles. I believe I said that 'not much progress' had been made on this problem. One of my other examiners, Professor Ray Kapral, had a visible reaction to this statement, one of those 'care to rephrase that?' looks. Professor Kapral had done (and continues to do) a lot of exciting work on understanding the effects of coupling of reactants to solvents on rates of reaction, and I should have known better than to say such a thing. I explained that (at the time) it wasn't possible to calculate a rate constant for a reaction in solution fully from first principles to experimental accuracy. That explanation got me out of immediate trouble. Professor Kapral was incredibly generous about my faux pas and took this opportunity to lead me through an interesting discussion on the purpose of chemical theory and of the role of models in understanding chemical behavior, which the text of this section largely reflects. What could have been a comprehensive exam disaster therefore became a learning opportunity for me, and one for which I continue to be grateful.

Further reading

The theory of Markov processes is useful in a variety of applications. The following books provide clear introductions to the subject:

- Bharucha-Reid A T 1960 *Elements of the Theory of Markov Processes and Their Applications* (New York: McGraw-Hill)
- Gillespie D T 1992 *Markov Processes: An Introduction for Physical Scientists* (Boston, MA: Academic)

If you want to go a little deeper into the theory and applications of stochastic processes, Gardiner's book could be your next step.

- Gardiner C W 1985 *Handbook of Stochastic Methods* 2nd edn (Berlin: Springer)

Exercises

9.1 In this question, you will analyze a master equation for a very simple model of a chemical reaction. Suppose that we have n equally spaced non-degenerate energy levels with spacing ΔE between adjacent levels. Molecules make transitions between adjacent levels only (the Landau–Teller approximation) with transition rates w_{ij}. Once the molecule reaches the nth level, it can irreversibly form a product with transition rate w_p.

(a) Let $P_i(t)$ be the probability that a molecule is in state i at time t. Write down the master equation for this model.

Note: you will need to treat the lowest energy level and the nth energy level as special cases.

(b) Suppose that for all upward transitions, $w_{ij} = w_+$ is a constant. What is w_-, the transition rate for the downward transitions? The 'expected' answer requires an assumption. What is that assumption?

(c) In the rest of this question, fix $n = 5$, take $w_+ = 10^7\ \mathrm{s^{-1}}$, $w_p = 10^6\ \mathrm{s^{-1}}$, $\Delta E = 10^{-21}$ J and $T = 700$ K. Calculate w_-.

(d) The solution of the master equation with initial conditions $P_1(0) = 1$, $P_2(0) = 0$, $P_3(0) = 0$, $P_4(0) = 0$, and $P_5(0) = 0$ is

$$P_1(t) = 0.036e^{\lambda_1 t} + 0.129e^{\lambda_2 t} + 0.242e^{\lambda_3 t} + 0.271e^{\lambda_4 t} + 0.323e^{\lambda_5 t},$$
$$P_2(t) = -0.091e^{\lambda_1 t} - 0.206e^{\lambda_2 t} - 0.106e^{\lambda_3 t} + 0.241e^{\lambda_4 t} + 0.162e^{\lambda_5 t},$$
$$P_3(t) = 0.107e^{\lambda_1 t} + 0.008e^{\lambda_2 t} - 0.278e^{\lambda_3 t} + 0.210e^{\lambda_4 t} - 0.048e^{\lambda_5 t},$$
$$P_4(t) = -0.084e^{\lambda_1 t} + 0.181e^{\lambda_2 t} - 0.060e^{\lambda_3 t} + 0.180e^{\lambda_4 t} - 0.218e^{\lambda_5 t},$$
$$P_5(t) = 0.032e^{\lambda_1 t} - 0.116e^{\lambda_2 t} + 0.217e^{\lambda_3 t} + 0.151e^{\lambda_4 t} - 0.284e^{\lambda_5 t},$$

with

$$\lambda_1 = -3.817 \times 10^7 \text{ s}^{-1},$$
$$\lambda_2 = -2.775 \times 10^7 \text{ s}^{-1},$$
$$\lambda_3 = -1.487 \times 10^7 \text{ s}^{-1},$$
$$\lambda_4 = -1.433 \times 10^5 \text{ s}^{-1},$$
$$\lambda_5 = -4.430 \times 10^6 \text{ s}^{-1}.$$

(i) Verify that this solution satisfies the initial condition, taking into account the finite precision of the data.

(ii) Verify that the solution satisfies the equation for dP_1/dt. (You could of course also verify the other equations.)

(iii) Obtain the probability distribution for the time of reaction.

(iv) Calculate the rate constant for this reaction.

9.2 We can use the theory of stochastic kinetics to describe reactions at the single-molecule level. Suppose that we have a molecule with two states. An example is 5,10,15,20-tetraphenylporphyrin, which can exist in one of two protonation states:

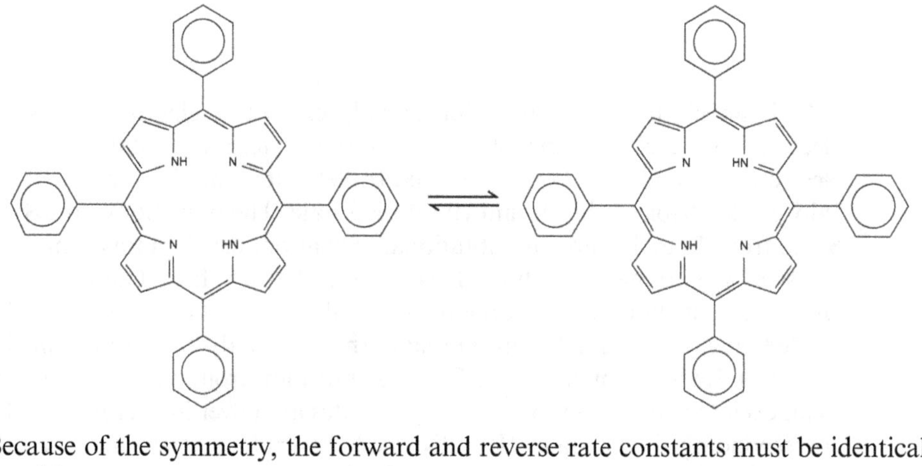

Because of the symmetry, the forward and reverse rate constants must be identical. Let this common rate constant be k.

(a) Let P_{I} be the probability of being in the first of the two protonation states, and P_{II} be the probability of being in the second. Write down the master equation for the two states of a single molecule of this substance.
(b) What is the stationary probability distribution?
Note: it may not be necessary to do a calculation.
(c) Sketch $P_{\mathrm{I}}(t)$ assuming $P_{\mathrm{I}}(0) = 1$. Label your graph as thoroughly as possible.
(d) Assume that we are able to detect which isomer is present at any given time.
What data would you have to collect to verify the curve you sketched in part (c)?

9.3 In this question, you will develop and briefly study a model of fluorescence and phosphorescence kinetics. The model is represented by the following diagram:

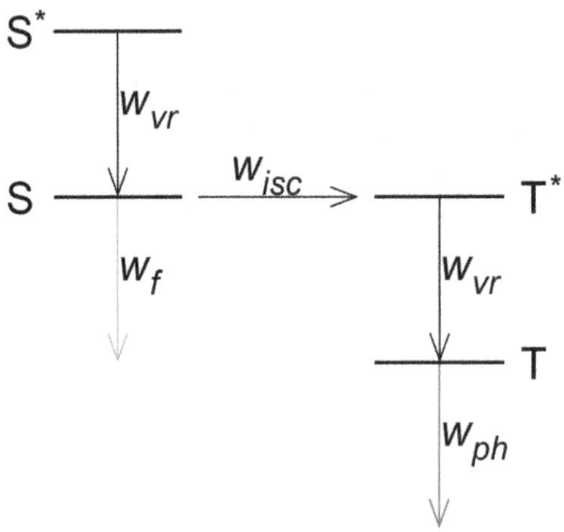

S* is an electronically and vibrationally excited singlet state, S is an electronically excited singlet state in its ground vibrational state, T* is an electronically and vibrationally excited triplet state, and T is an electronically excited triplet in its ground vibrational state. The transitions from S* to S and from T* to T represent vibrational relaxation, which in this context is a loss of vibrational excitation. This tends to be very fast. The transition with rate w_f is fluorescence, a relatively rapid process of emission from the excited singlet to a singlet ground state (not explicitly represented in the diagram). Intersystem crossing (ISC) is a transition from an excited singlet to an excited triplet state. Because triplet states are lower in energy than the corresponding singlet states (Hund's rule), the triplet state formed by ISC will typically be vibrationally excited. For simplicity, we are assuming that vibrational relaxation occurs at the same rate for the singlet and triplet.

Once the triplet has lost its vibrational excitation, it can undergo the slow emission process of phosphorescence to return to the singlet ground state.

For this assignment, assume the following transition rates:

$$w_{vr} = 10^{11}\ s^{-1}$$
$$w_{f} = 10^{8}\ s^{-1}$$
$$w_{isc} = 10^{8}\ s^{-1}$$
$$w_{ph} = 10^{-2}\ s^{-1}$$

Write a population kinetic Monte Carlo simulation of this system assuming that all molecules start in the S^* state. Store the time at which each fluorescence or phosphorescence event occurs. (These should be stored in two separate vectors, a vector of fluorescence event times and a vector of phosphorescence event times.) The output of your program should consist of two separate histograms, one of the fluorescence event times and one of the phosphorescence event times. You should simulate at least 1000 molecules to get reasonable statistics. As with all other things statistical, more is better.

The fluorescence (phosphorescence) decay time is the time it takes for the fluorescence (phosphorescence) to fall to $1/e$ of its original value. For exponential decay, which you will observe here, the decay time is equal to the average of the event times. Calculate and report the fluorescence and phosphorescence decay times.

How are the decay times related to the rate constants? Provide a brief explanation and/or a few equations to support your answer.

Reference

[1] Roussel M R 2012 *A Life Scientist's Guide to Physical Chemistry* (Cambridge: Cambridge University Press)

IOP Publishing

Chapter 10

The chemical master equation

The analysis of chemical kinetics experiments is almost entirely based on differential equations for the concentrations. We can write differential equations because a mole is a very large number; therefore, a typical experimental system contains an awful lot of molecules. The change in the concentrations when one reactive event occurs is accordingly negligible, such that concentrations behave like continuous variables. But what if we look at small systems containing few molecules? There must be a point at which it is no longer reasonable to treat the concentrations as continuous. We would then need a different kind of theory, one that takes the discrete nature of matter into account. We could try to develop a theory based on tracking individual molecules, their quantum states, orientations, and so on. Molecular dynamics simulations do this, always with some simplifications, and, in a way, the master-equation treatments of molecular energetics of the previous chapter were building towards a theory of reaction kinetics incorporating many details of the molecular state. However, approaches of this nature become increasingly cumbersome as the number of molecules grows. Moreover, this may be more detail than we need. In some studies, we may only need a mesoscopic[1] counterpart of mass-action kinetics, which is to say a theory that describes how the number of molecules in a small system evolves with time. In contrast with macroscopic systems, for systems with few molecules, each reaction creates a perceptible step change in the concentrations of reactants and products. Moreover, all the sources of randomness previously discussed—random collisions, intermolecular energy exchange, intramolecular vibrational relaxation, … make the timing of each reactive event unpredictable. Accordingly, we need a statistical theory. We will, in fact, end up with a type of master equation and a corresponding simulation algorithm, so that we will be able to talk about the statistical properties of a kinetic process but not about the exact

[1] A mesoscopic description falls between a microscopic description, typically understood to be one that includes many molecular details, and a macroscopic description.

doi:10.1088/978-0-7503-5321-2ch10

sequence of reactive events. Theories of kinetics built on statistical ideas are called **stochastic**, the latter being a synonym of random.

The stochastic theory that will be presented in this chapter applies equally well to gas-phase reactions as to solution-phase reactions. It is developed here because it follows naturally from our treatment of master equations in the previous chapter. But this theory will be particularly useful in biochemistry, where some key molecules are present in extremely small numbers in a solution environment. For example, think about a human gene present in a cell in two copies. Suppose that when a repressor protein is bound to the gene, transcription can't be initiated. We surely can't treat the number of active (unbound) copies of the gene as a continuous variable, since it can only take the values zero, one, or two, depending on whether both copies are bound by the repressor, one copy is bound, or neither, respectively. This is an extreme, of course, but both biochemistry and nanotechnology often involve systems with just a few dozen molecules of a particular type.

10.1 The chemical master equation

In conventional kinetics, the state of a system at a fixed temperature is specified by its concentrations, which could be arranged in a vector. Similarly, we will consider a vector of the populations of each type of molecule in the system. Thus, let $\mathbf{N}(t) = (N_1(t), N_2(t), \ldots, N_n(t))$ be the composition vector of the system, where N_i is the number of molecules of type i and n is the number of distinct chemical species. We limit ourselves to isothermal and isochoric (constant volume) systems such that $\mathbf{N}(t)$ is a complete specification of the state of the system. We take no account of internal states of the molecules[2]. Each N_i is a simple count of how many molecules of type i we have at time t. We use $\mathcal{N}$ to denote the space of all possible vectors $\mathbf{N}$ for a given system. This space may be infinite if we are dealing with an open system, or it may be finite (although possibly very large) for a closed system. The key quantity in the theory to be developed here will be $\mathbf{P}(\mathcal{N}, t)$, the probability distribution over the space $\mathcal{N}$. The components of this vector are the probabilities of each possible state[3]. Note that as the reaction proceeds, this probability distribution changes, hence the dependence on t.

---

[2] It would be possible to account for internal molecular states by letting N_i represent the number of molecules of a certain type in a particular quantum state or in a particular conformation. The theory developed here could be applied to a model with this level of detail provided we were content to treat transitions between states as simple reactions, e.g. as first-order reactions, or in cases involving collisions, as second-order processes. Needless to say, the resulting models would tend to be complex, but there are situations in which such a detailed model allows us to study the effects of intramolecular processes on the behavior of a mechanism without resorting to detailed mechanical descriptions.

[3] Given that the only possible values of the components of the population vector $\mathbf{N}$ are nonnegative integers, the states are denumerable, which is to say that they can be numbered. This is true even if there is an infinite number of states: it is possible to build a list of all possibilities that creates a unique ordering, so any given population vector can be assigned an index denoting its place in the list. If you're curious how this is done, look up a proof that the rational numbers are denumerable. The same trick can be used for the $\mathbf{N}$ vectors. This is important because it means that, even in a system with an infinite number of states, we can create an ordering such that each component of the probability vector can be assigned a specific position within the vector.

We will consider well-mixed systems such that the probability of finding a particular molecule inside a subvolume ΔV is $\Delta V / V$.[4] If this assumption holds, then we can treat reactions as random events obeying a probability law that we will develop below. Physically, is this reasonable? Consider the two most common cases:

- First-order reactions occur when some essentially random condition is met within a molecule (e.g. intramolecular vibrational relaxation (IVR) puts enough energy into a reactive mode). Mixing issues are not directly relevant to these processes.
- In a well-mixed system, the collisions necessary for a second-order reaction to occur are random events.

The latter idea, which we first ran into in section 2.1, is worthy of further discussion. It is based on Boltzmann's assumption of molecular chaos, also known as the Stosszahlansatz, which, as noted earlier, was put on a firmer footing in the 20th century with the development of ideas in nonlinear dynamics. Boltzmann's insight was that collisions cause a rapid loss of memory of a trajectory's history because of all of the details that affect their outcomes. In other words, knowing where the molecules in a system are at a particular moment in time as well as their momenta and the details of their current states doesn't give us enough information to figure out where they came from. Poincaré would later emphasize the converse predictive process: we can't predict what will happen to the molecules after a few collisions. This property is known as **sensitive dependence on initial conditions** and it is fundamental to the technical definition of chaos. If we can't predict in detail what molecules are going to do, then the best we can do is to develop statistical theories. To spin the matter in a more positive way, because molecular collisions repeatedly reapportion the molecules' energies and momenta, particle trajectories can be treated as essentially random. Unless there are other physicochemical interactions that create long-lasting correlations between reactive events, chemical reactions can therefore be treated as Markov (memoryless) processes provided we do not inquire about events on timescales of the order of the mean collision time. This in turn means that we can write a master equation for $\mathbf{P}(\mathcal{N}, t)$.

While there will be some similarities with the theory presented in the previous chapter, there will be significant differences too. The 'transitions' to be modeled by the chemical master equation are changes in the composition of the system caused by chemical reaction. For example, an occurrence of a chemical reaction $A \rightarrow B$ changes the state of the system, assuming the system contains just these two chemical species, from (N_A, N_B) to $(N_A - 1, N_B + 1)$. In this theory, the transition rates are called

[4] This is not a trivial property. There are many situations in which we can't count on a system being well mixed in this sense. For example, in chain reactions, there is often a reaction front, which has mostly reacted material behind it and mostly unreacted material ahead of it. Intermolecular forces can also create relatively large, transient inhomogeneities in some cases. (Fog would be a familiar example.) The assumption that the system is well mixed is, however, the same one necessary for the validity of the law of mass action, absent diffusion or advection terms. This assumption therefore seems like a good place to start for a theory of stochastic kinetics. Interestingly, many theories of aggregation are based on stochastic models, so understanding this theory properly will also allow you to transition into studying systems with inhomogeneities if you so choose.

reaction propensities. These propensities depend on the state of the system at time t, for the same reasons that concentrations appear in the law of mass action: the more reactant molecules we have in the system, the more likely we are to have reactions involving those molecules. We therefore use $a_r(\mathbf{N})$ to denote the propensities, where r labels a particular reaction and $\mathbf{N}$ denotes a particular composition vector.

For each reaction, we can define a **stoichiometric vector** ν_r that gives the change in the numbers of each species as a result of reaction r. Returning to the $A \rightarrow B$ reaction used as an example above, $\nu_r = (-1, 1)$. Let's use subscripts i and f to denote the initial and final states corresponding to reaction r. We then have $\mathbf{N}_f = \mathbf{N}_i + \nu_r$, or, putting it the other way, $\mathbf{N}_i = \mathbf{N}_f - \nu_r$. Armed with this notation, we can now write the **chemical master equation** (CME):

$$\frac{dP(\mathbf{N}, t)}{dt} = \sum_{r \in \mathcal{R}} a_r(\mathbf{N} - \nu_r)P(\mathbf{N} - \nu_r, t) - \sum_{r \in \mathcal{R}} a_r(\mathbf{N})P(\mathbf{N}, t) \qquad (10.1)$$

where $P(\mathbf{N}, t)$ is the probability that the composition of the system is $\mathbf{N}$ at time t, and $\mathcal{R}$ is the set of reactions that can occur in the system. There is one such equation for every possible composition $\mathbf{N}$ of the system. Each term in the first sum on the right-hand side gives the probability per unit time that reaction r occurs, taking into account both the propensity of the reaction and the probability that the system was in state $\mathbf{N} - \nu_r$ prior to the reaction, thus changing the composition to $\mathbf{N}$. Each of these terms depends on the probability that the system was in the appropriate initial state before the reaction and on the propensity to react if the system was in this initial state. The second sum, on the other hand, gives the probability per unit time that the system transitions out of state $\mathbf{N}$ due to each of the reactions that can occur.

I will briefly note that we did not derive this equation because this derivation would have been essentially the same as that of the general master equation in section 9.1. The only difference is that the transition probabilities are called propensities, and these propensities depend on $\mathbf{N}$. But how do the propensities depend on $\mathbf{N}$? This is our next item for discussion.

10.2 The reaction propensities

The propensity $a_r(\mathbf{N})$ is the probability per unit time that reaction r occurs, *given that* the composition of the system is $\mathbf{N}$. Let's start by looking at the most common types of reaction.

- **First-order reactions:** $X_i \rightarrow$ products. Suppose that the probability per unit time that any given molecule of X_i reacts is κ_r. If Δt is sufficiently small, then the probability that a molecule of X_i reacts, given that there are N_i molecules of this type, is $\kappa_r N_i \Delta t$. It is important to pick a small Δt because we want to exclude the possibility that two reactions occur in this small interval of time, which would change the propensities. Since the propensity is the probability of reaction per unit time given that the system is in the specified state,

$$a_r = \kappa_r N_i. \qquad (10.2)$$

Note that the propensity only depends on the population of the reactant of an elementary reaction, which is why I didn't show any products for the reaction: the products don't matter.

- **Heteromolecular second-order reactions:** $X_i + X_j \rightarrow$ products, with $j \neq i$

  We start by figuring out the probability that a *particular* pair of molecules of types i and j meet and react in time Δt. From collision theory, the reaction volume explored per unit time in the gas phase is $v_{ij}\sigma_{ij}$.[5] This represents the fraction $(v_{ij}\sigma_{ij}/V)\Delta t$ of the total volume V in time Δt. The latter is thus the probability of a collision between any particular pair of molecules in time Δt. If a fraction η_r of the collisions lead to reaction, then the reaction probability in time Δt for a particular pair of molecules of types i and j is $(\eta_r v_{ij}\sigma_{ij}/V)\Delta t = \kappa_r \Delta t$. There are $N_i N_j$ pairs of molecules, so if we take a sufficiently small Δt, the probability that one reaction of type r occurs per unit time is

$$a_r = \kappa_r N_i N_i.$$

- **Homomolecular second-order reactions:** $X_i + X_i \rightarrow$ products. The argument is exactly as above, first considering collisions between two selected molecules of type i and then multiplying by the number of pairs of molecules one can select from the N_i molecules available. The collision rate is, however, given by $\frac{1}{2}v_{ii}\sigma_{ii}$, leading to $\kappa_r = \frac{1}{2}\eta_r v_{ii}\sigma_{ii}/V$. From your basic statistics, you may recall that the number of ways of choosing two molecules from a set of N_i identical molecules is $N_i(N_i - 1)/2$. The propensity is therefore

$$a_r = \kappa_r N_i(N_i - 1)/2.$$

In general, the reaction propensity can always be written as the product of a **stochastic rate constant**, denoted by κ_r, with units of inverse time,[6] and a combinatorial factor h_r which gives the number of different combinations of *reactant* molecules that could be formed prior to the reaction:

$$a_r = \kappa_r h_r(\mathbf{N}).$$

Example 10.1. *Let's work out the CME for the reaction*

$$A + B \underset{\kappa_{-1}}{\overset{\kappa_1}{\rightleftarrows}} \mathrm{C}.$$

We will assume that this is the only possible reaction in a closed system. We always need to choose an ordering for the chemical species populations in the state vector. In this case, the obvious choice is (N_A, N_B, N_C). *The forward and reverse reactions have stoichiometry vectors*

[5] We will discuss a parallel theory of diffusion-limited reactions in chapter 11. We could substitute the encounter rates from this theory for the collision rates of gas-phase theory.

[6] Many authors use c_r for the stochastic rate constant.

$$\boldsymbol{\nu}_1 = (-1, -1, 1),$$
$$\boldsymbol{\nu}_{-1} = (1, 1, -1).$$

The propensities are

$$a_1 = \kappa_1 N_A N_B,$$
$$a_{-1} = \kappa_{-1} N_C.$$

We always need an initial condition for these problems. The initial conditions can take many forms, but in this case, suppose that we start with a system of known composition, namely $N(0) = (4, 3, 0)$. *The space of all possible compositions is then*

$$\mathcal{N} = \{(4, 3, 0), (3, 2, 1), (2, 1, 2), (1, 0, 3)\}.$$

Correspondingly, the probability space for this problem is

$$\mathbf{P}(\mathcal{N}) = \{P(4, 3, 0), P(3, 2, 1), P(2, 1, 2), P(1, 0, 3)\}.$$

Moreover, the initial condition in probability space is fixed as follows by the initially known composition:

$$\mathbf{P}(\mathcal{N}, t = 0) = \{P(4, 3, 0) = 1, P(3, 2, 1) = 0, P(2, 1, 2) = 0, P(1, 0, 3) = 0\}.$$

In some problems, there is some uncertainty about the initial composition, and this would be reflected by an initial condition in which the probability is distributed in some way over several states.

We can now specialize the chemical master equation (10.1) to this system. The probability of a general state obeys the equation

$$\begin{aligned}\frac{\mathrm{d}P(N_A, N_B, N_C)}{\mathrm{d}t} &= \kappa_1(N_A + 1)(N_B + 1)P(N_A + 1, N_B + 1, N_C - 1) \\ &\quad + \kappa_{-1}(N_C + 1)P(N_A - 1, N_B - 1, N_C + 1) \\ &\quad - (\kappa_1 N_A N_B + \kappa_{-1} N_C)P(N_A, N_B, N_C).\end{aligned} \tag{10.3}$$

It is worth spending a minute to think about the terms in this equation. The first term corresponds to the forward reaction, which uses up one molecule each of A and B to produce a molecule of C. Therefore, the state (N_A, N_B, N_C) *gains probability due to this reaction through transitions from the initial state* $(N_A + 1, N_B + 1, N_C - 1)$. *The first term therefore expresses the transfer of probability from the latter state to* (N_A, N_B, N_C), *which is why both the propensity and the probability are evaluated at* $(N_A + 1, N_B + 1, N_C - 1)$. *You can analyze the second term similarly. The term* $-\kappa_1 N_A N_B P(N_A, N_B, N_C)$ *expresses the loss of probability from state* (N_A, N_B, N_C) *due to the forward reaction taking into account the probability of the system being in this state. And again, you should be able to articulate the meaning of the* $-\kappa_{-1} N_C P(N_A, N_B, N_C)$ *term in the equation.*

There is a small problem: some of the terms in equation (10.3) don't make sense for every state. For example, if we are writing down the equation for $P(4, 3, 0)$, *the first term would be* $\kappa_1(5)(4)P(5, 4, -1)$, *but this particular system can't have* $N_A = 5$ and $N_B = 4$, *not to mention that* $N_C = -1$ *has no physical interpretation in the context of a chemical system. To address this, we always attach* **boundary conditions** *to a CME which assign a zero probability to any physically unreachable states on the boundary of the probability space. In this case, for example, we would set* $P(5, 4, -1) = 0$ *for all time. Given the appropriate boundary conditions, we can expand this CME to*

$$\frac{dP(4, 3, 0)}{dt} = \kappa_{-1}P(3, 2, 1) - 12\kappa_1 P(4, 3, 0),$$

$$\frac{dP(3, 2, 1)}{dt} = 12\kappa_1 P(4, 3, 0) + 2\kappa_{-1}P(2, 1, 2) - (6\kappa_1 + \kappa_{-1})P(3, 2, 1),$$

$$\frac{dP(2, 1, 2)}{dt} = 6\kappa_1 P(3, 2, 1) + 3\kappa_{-1}P(1, 0, 3) - (2\kappa_1 + 2\kappa_{-1})P(2, 1, 2),$$

$$\frac{dP(1, 0, 3)}{dt} = 2\kappa_1 P(2, 1, 2) - 3\kappa_{-1}P(1, 0, 3).$$

Note that

$$\frac{dP(4, 3, 0)}{dt} + \frac{dP(3, 2, 1)}{dt} + \frac{dP(2, 1, 2)}{dt} + \frac{dP(1, 0, 3)}{dt} = 0.$$

This implies that $P(4, 3, 0) + P(3, 2, 1) + P(2, 1, 2) + P(1, 0, 3)$ *is a constant, which makes sense because the probabilities should add to one. The CME therefore conserves probability, i.e. if we start with correct initial conditions for which* $\sum_{\mathbf{N}\in\mathcal{N}} P(\mathbf{N}, 0) = 1$, *then the sum of the probabilities will remain one for all time.*

10.3 The stationary distribution

In ordinary chemical kinetics, a closed chemical system has a unique equilibrium composition, usually given as a set of concentrations. Open chemical systems can have stable steady states as well, which may or may not be unique. In stochastic kinetics, we have a **stationary probability distribution** instead. This is a distribution that is time independent. The majority of realistic stochastic chemical systems have a stationary distribution that appears as the unique long-term limit of a system's evolution. At least in theory, calculating this distribution is straightforward: we just need to solve the equation $d\mathbf{P}(\mathcal{N})dt = 0$ with the added probability conservation condition that $\sum_{\mathcal{N}} P(\mathbf{N}) = 1$.

In classical kinetics, a closed chemical system obeys the principle of detailed balance: when the system reaches equilibrium, each elementary reaction is itself in equilibrium, i.e. $v_i^{(+)} = v_i^{(-)}$, where $v_i^{(+)}$ is the forward rate of elementary reaction i and $v_i^{(-)}$ is the rate of the corresponding reverse reaction. A similar property holds in stochastic kinetics, although it has to be rephrased in terms of the rates of probability transfer due to a reaction and its reverse.

Suppose that the elementary reactions are numbered 1 to ρ, with the same number used for the forward and reverse reactions. We write $a_r^{(+)}$ for the propensity of the forward reaction, and $a_r^{(-)}$ for the propensity of the corresponding reverse reaction. If the stoichiometric vector of reaction r is ν_r in the forward direction, then the stoichiometric vector of the reverse reaction is $-\nu_r$. We will use this fact to write the equations below in terms of the stoichiometric vector of the forward reaction only. If we separate the forward and reverse reactions, we can write the CME (equation (10.1)) in the form

$$\frac{dP(\mathbf{N}, t)}{dt} = \sum_{r=1}^{\rho}[a_r^{(+)}(\mathbf{N} - \nu_r)P(\mathbf{N} - \nu_r, t) - a_r^{(+)}(\mathbf{N})P(\mathbf{N}, t)] + \sum_{r=1}^{\rho}[a_r^{(-)}(\mathbf{N} + \nu_r)P(\mathbf{N} + \nu_r, t) - a_r^{(-)}(\mathbf{N})P(\mathbf{N}, t)].$$

Recall that ν_r is specifically the stoichiometric vector of the forward reaction, hence the appearance of the positive sign in $\mathbf{N} + \nu_r$ in the sum over reverse reactions (the second line of the CME above). Since both sums run over the same values of r, we can combine and rearrange them as follows:

$$\frac{dP(\mathbf{N}, t)}{dt} = \sum_{r=1}^{\rho}\Big\{[a_r^{(+)}(\mathbf{N} - \nu_r)P(\mathbf{N} - \nu_r, t) - a_r^{(-)}(\mathbf{N})P(\mathbf{N}, t)] + [a_r^{(-)}(\mathbf{N} + \nu_r)P(\mathbf{N} + \nu_r, t) - a_r^{(+)}(\mathbf{N})P(\mathbf{N}, t)]\Big\}.$$

Notice that each pair of square brackets contains a difference of rates for the forward and corresponding reverse reactions. In the stochastic context, for a stationary probability distribution obeying detailed balance, each bracketed pair of terms is zero. Mathematically, this reduces to

$$a_r^{(+)}(\mathbf{N} - \nu_r)P(\mathbf{N} - \nu_r, t) = a_r^{(-)}(\mathbf{N})P(\mathbf{N}, t)$$

for every reaction and for every possible value of $\mathbf{N}$.[7] In the case of closed chemical systems, the stochastic detailed balance property is always satisfied by the stationary distribution, just as the conventional detailed balance property holds in classical mass-action kinetics at equilibrium.

Example 10.2. *In this example, we will calculate the stationary distribution for the Michaelis–Menten mechanism*

$$\mathrm{E} + \mathrm{S} \underset{\kappa_{-1}}{\overset{\kappa_1}{\rightleftarrows}} \mathrm{C} \underset{\kappa_{-2}}{\overset{\kappa_2}{\rightleftarrows}} \mathrm{E} + \mathrm{P}$$

and verify the detailed balance condition for a system with initial conditions $\{N_E(0) = 1, N_S(0) = 3, N_C(0) = N_P(0) = 0\}$ *and stochastic rate constants* $\kappa_1 = 10$, $\kappa_{-1} = 1$, $\kappa_2 = 5$ *and* $\kappa_{-2} = 2\ \mathrm{s}^{-1}$.

[7] If you're wondering why I didn't also write $a_r^{(-)}(\mathbf{N} + \nu_r)P(\mathbf{N} + \nu_r, t) = a_r^{(+)}(\mathbf{N})P(\mathbf{N}, t)$, it's because this would generate the same set of detailed balance conditions. In other words, this equation would be redundant with the one given above.

The ordering of the variables in the population vector is arbitrary. I chose the ordering (N_E, N_S, N_C, N_P)*, but any other permutation would do just as well. The CME is*

$$\begin{aligned}\frac{dP(\mathbf{N}, t)}{dt} = {} & \kappa_1(N_E + 1)(N_S + 1)P(N_E + 1, N_S + 1, N_C - 1, N_P) \\ & + \kappa_{-1}(N_C + 1)P(N_E - 1, N_S - 1, N_C + 1, N_P) \\ & + \kappa_2(N_C + 1)P(N_E - 1, N_S, N_C + 1, N_P - 1) \\ & + \kappa_{-2}(N_E + 1)(N_P + 1)P(N_E + 1, N_S, N_C - 1, N_P + 1) \\ & - P(N_E, N_S, N_C, N_P)[\kappa_1 N_E N_S + \kappa_{-1} N_C + \kappa_2 N_C + \kappa_{-2} N_E N_P].\end{aligned}$$

We also need the boundary conditions

$$\begin{aligned}P(-1, N_S, N_C, N_P) &= P(N_E, -1, N_C, N_P) = P(N_E, N_S, -1, N_P) \\ &= P(N_E, N_S, N_C, -1) = 0.\end{aligned}$$

For the specific parameters of this example, we get

$$\begin{aligned}dP(1, 3, 0, 0)/dt &= P(0, 2, 1, 0) - 30P(1, 3, 0, 0), \\ dP(0, 2, 1, 0)/dt &= 30P(1, 3, 0, 0) + 2P(1, 2, 0, 1) - 6P(0, 2, 1, 0), \\ dP(1, 2, 0, 1)/dt &= P(0, 1, 1, 1) + 5P(0, 2, 1, 0) - 22P(1, 2, 0, 1), \\ dP(0, 1, 1, 1)/dt &= 20P(1, 2, 0, 1) + 4P(1, 1, 0, 2) - 6P(0, 1, 1, 1), \\ dP(1, 1, 0, 2)/dt &= P(0, 0, 1, 2) + 5P(0, 1, 1, 1) - 14P(1, 1, 0, 2), \\ dP(0, 0, 1, 2)/dt &= 10P(1, 1, 0, 2) + 6P(1, 0, 0, 3) - 6P(0, 0, 1, 2), \\ dP(1, 0, 0, 3)/dt &= 5P(0, 0, 1, 2) - 6P(1, 0, 0, 3).\end{aligned}$$

If you add up the right-hand sides of these rate equations, you get zero. Verifying probability conservation is one way to check that we haven't made a small error somewhere in writing down the CME.

To find the stationary distribution, we need to set the rate equations equal to zero, add the probability conservation equation, and solve. The fact that the sum of the rate equations is zero means that one of these equations is redundant. We can therefore throw one of them out before forming our system of equations. It doesn't matter which one you eliminate. I find that the simplest thing is to get rid of the last equation. The equation we need to solve is linear in the probabilities, so it can be rewritten as a matrix equation:

$$\begin{bmatrix} -30 & 1 & 0 & 0 & 0 & 0 & 0 \\ 30 & -6 & 2 & 0 & 0 & 0 & 0 \\ 0 & 5 & -22 & 1 & 0 & 0 & 0 \\ 0 & 0 & 20 & -6 & 4 & 0 & 0 \\ 0 & 0 & 0 & 5 & -14 & 1 & 0 \\ 0 & 0 & 0 & 0 & 10 & -6 & 6 \\ 1 & 1 & 1 & 1 & 1 & 1 & 1 \end{bmatrix} \begin{bmatrix} P(1, 3, 0, 0) \\ P(0, 2, 1, 0) \\ P(1, 2, 0, 1) \\ P(0, 1, 1, 1) \\ P(1, 1, 0, 2) \\ P(0, 0, 1, 2) \\ P(1, 0, 0, 3) \end{bmatrix} = \begin{bmatrix} 0 \\ 0 \\ 0 \\ 0 \\ 0 \\ 0 \\ 1 \end{bmatrix}.$$

The last row of the matrix is the probability conservation equation.

Matlab (matrix laboratory) is an ideal tool for solving this kind of equation, as its name suggests. It was originally created as a kind of matrix calculator for students studying numerical linear algebra. If we think of the equation above as **AP** = **b**, *we can enter the matrix* **A** *as follows in Matlab:*

```
>> A=[-30 1 0 0 0 0 0;
30 -6 2 0 0 0 0 ;
0 5 -22 1 0 0 0 ;
0 0 20 -6 4 0 0 ;
0 0 0 5 -14 1 0;
0 0 0 0 10 -6 6;
1 1 1 1 1 1 1]

A =
   -30     1     0     0     0     0     0
    30    -6     2     0     0     0     0
     0     5   -22     1     0     0     0
     0     0    20    -6     4     0     0
     0     0     0     5   -14     1     0
     0     0     0     0    10    -6     6
     1     1     1     1     1     1     1
```

Matrix entries in a row can be separated by spaces or commas. Rows are separated by semicolons. The simplest way to enter the vector **b** *is to create a vector of zeros and then replace the last value by one:*

```
>> b=zeros(7,1);
>> b(7)=1

b =
     0
     0
     0
     0
     0
     0
     1
```

Note that we need a column vector, so it's important to create a vector with seven rows and one column.

Now here comes the Matlab magic. To solve a system of linear equations such as our **AP** = **b**, *Matlab provides a '\' operator that gives us the solution in one step:*

```
>> P=A\b

P =
      0.0000
      0.0008
      0.0020
      0.0396
      0.0495
      0.4953
      0.4127
```

If you want to see the first component of **P**, *you can just type P(1) at the command line. Going back now to our original probability space, we have the stationary probability density*

$$P(1, 3, 0, 0) = 3 \times 10^{-5}, \quad P(1, 1, 0, 2) = 0.0495,$$
$$P(0, 2, 1, 0) = 8 \times 10^{-4}, \quad P(0, 0, 1, 2) = 0.4953,$$
$$P(1, 2, 0, 1) = 2.0 \times 10^{-3}, \quad P(1, 0, 0, 3) = 0.4127.$$
$$P(0, 1, 1, 1) = 0.0396,$$

There are two ways to interpret this distribution. One is to think of a single system that we allow to come to equilibrium, then peek into once in a while to observe the populations. If we do that, we should see the $(0, 0, 1, 2)$ *state roughly 50% of the time, the (1,0,0,3) state roughly 41% of the time, and so on. In order for this to be true, we need to space our observations by an interval of time sufficient for many reactions to have taken place, i.e. long enough for the state to become uncorrelated from the last observation. The other interpretation would involve an ensemble of identically prepared systems; at any given time, we would expect to see half of them in the* $(0, 0, 1, 2)$ *state, 41% in the (1,0,0,3) state, and so on. These two perspectives are interchangeable because of the ergodicity of the underlying Markov process.*

We can now verify whether detailed balance is obeyed for this stationary distribution. If so, we should have

$$\kappa_1 N_E N_S P(N_E, N_S, N_C, N_P) = \kappa_{-1}(N_1 + 1)P(N_E - 1, N_S - 1, N_C + 1, N_P)$$

and

$$\kappa_2 N_C P(N_E, N_S, N_C, N_P) = \kappa_{-2}(N_E + 1)(N_P + 1)P(N_E + 1, N_S, N_C - 1, N_P + 1)$$

for all possible values of (N_C, N_E, N_P, N_S). *As an example, consider the second of these detailed balance conditions, and the* $(N_C, N_E, N_P, N_S) = (0, 1, 1, 1)$ *state, for which*

$$\kappa_2 N_C P(N_E, N_S, N_C, N_P)$$
$$= 0.198 = \kappa_{-2}(N_E + 1)(N_P + 1)P(N_E + 1, N_S, N_C - 1, N_P + 1).$$

By all means check that detailed balance is observed for other states for both reactions.

10.4 The relationship between the chemical master equation and mass-action kinetics

There is a sense in which the chemical master equation is a more general theory than classical mass-action kinetics, given that it takes into account the discrete nature of matter. We have more than a century and a half of evidence that supports the validity of the law of mass action. If the chemical master equation is a correct theory, it must therefore reduce to the law of mass action in the macroscopic limit, in the same way that quantum mechanics reduces to classical mechanics in the limit of large mass. We can approach this 'chemical correspondence principle' by examining the statistical properties of elementary reactions predicted by the CME. Specifically, we need to show that the average behavior predicted by the CME is in the mass-action form when the number of molecules is macroscopically large, while the **coefficient of variation** (CV), the ratio of the standard deviation to the average, a measure of relative fluctuations, becomes vanishingly small.

10.4.1 First-order reactions

The propensity for a first-order reaction, given by equation (10.2), already looks a lot like the mass-action rate, so we can guess that the relationship between stochastic and classical mass-action kinetics will be particularly simple in this case. The master equation for the first-order reaction A → products is

$$\frac{\mathrm{d}P(N,\,t)}{\mathrm{d}t} = \kappa(N+1)P(N+1,\,t) - \kappa NP(N,\,t), \tag{10.4}$$

where κ is the reaction's stochastic rate constant and N is the number of molecules of A. (Since there is only one type of molecule, we don't need any subscripts.)

This CME describes how the *probability* of having N molecules of A evolves with time. Mass-action kinetics, on the other hand, describes how the number of molecules (usually presented as a concentration) evolves over time. The link between these two quantities will be established, in the first instance, by taking an average relative to the probability distribution $\mathbf{P}(\mathcal{N},\,t)$.

Suppose that we start with A_0 molecules of A, making A_0 the maximum number of molecules of A the system might contain at any $t > 0$. This maximum implies the boundary condition $P(A_0 + 1,\,t) = 0$.

Given the probability distribution $\mathbf{P}(\mathcal{N},\,t)$, the average number of molecules of A at time t is

$$\langle N\rangle(t) = \sum_{N=0}^{A_0} NP(N,\,t).$$

$$\therefore\ \frac{\mathrm{d}\langle N\rangle}{\mathrm{d}t} = \sum_{N=0}^{A_0} N\frac{\mathrm{d}P(N,\,t)}{\mathrm{d}t}.$$

Note that, in the CME treatment, N is an element of the state space over which the probability distribution is defined. $\mathcal{N}$ is a fixed set, so its elements do not have a time dependence. The time-dependent quantity in this equation is $P(N, t)$, which is why the time derivative only acts on the probability inside the summation. Equation (10.4) gives us $\mathrm{d}P(N, t)/\mathrm{d}t$, so we can substitute that into the rate of change of $\langle N \rangle$:

$$\begin{aligned}\frac{\mathrm{d}\langle N \rangle}{\mathrm{d}t} &= \sum_{N=0}^{A_0} N[\kappa(N+1)P(N+1, t) - \kappa N P(N, t)] \\ &= \kappa\left\{\sum_{N=0}^{A_0-1} N(N+1)P(N+1, t) - \sum_{N=0}^{A_0} N^2 P(N, t)\right\}.\end{aligned}$$

The first sum can be stopped at $N = A_0 - 1$ because of the boundary condition. We want to rewrite the first sum so that it is a sum over $P(N, t)$ rather than $P(N + 1, t)$. This allows us to cancel terms between the two sums. To replace $P(N + 1, t)$ in the first sum by $P(N, t)$, we need to shift the arguments in the summand down by one and start the summation from $N = 1$. If you have trouble seeing this, try writing down the first few terms of the summation from the last line above and the first few terms from the line below.

$$\begin{aligned}\frac{\mathrm{d}\langle N \rangle}{\mathrm{d}t} &= \kappa\left\{\sum_{N=1}^{A_0} (N-1)N P(N, t) - \sum_{N=0}^{A_0} N^2 P(N, t)\right\} \\ &= \kappa\left\{\sum_{N=0}^{A_0} N(N-1)P(N, t) - \sum_{N=0}^{A_0} N^2 P(N, t)\right\}.\end{aligned}$$

The extension of the summation range in the first term in the preceding line is allowed because $N(N - 1)P(N, t)$ is zero when $N = 0$. If we now combine the two sums, the $N^2P(N, t)$ terms cancel, leaving us with

$$\frac{\mathrm{d}\langle N \rangle}{\mathrm{d}t} = -\kappa \sum_{N=0}^{A_0} N P(N, t) = -\kappa\langle N \rangle. \tag{10.5}$$

At this point, we have made no assumptions and obtained something that looks like the law of mass action for a first-order reaction. We can conclude that the *average* number of molecules (or concentration, if we divide both sides of the equation by LV) obeys the law of mass action. But that's not enough. In well-controlled experiments carried out in laboratory-scale apparatus, the law of mass action is closely obeyed, and no detectable fluctuations fall outside the precision of the measuring instrument. We therefore also need to show that the fluctuations in N are negligible before we can conclude that the CME has the law of mass action as its macroscopic limit.

To talk about fluctuations, we would usually evaluate the standard deviation. The variance (square of the standard deviation) is defined as follows:

$$\sigma^2 = \langle N^2 \rangle - \langle N \rangle^2 = \sum_{N=0}^{A_0} N^2 P(N, t) - \langle N \rangle^2.$$

Let's work out how the variance changes over time:

$$\frac{d\sigma^2}{dt} = \sum_{N=0}^{A_0} N^2 \frac{dP(N, t)}{dt} - 2\langle N \rangle \frac{d\langle N \rangle}{dt}. \tag{10.6}$$

Using equations (10.4) and (10.5), we get

$$\begin{aligned}
\frac{d\sigma^2}{dt} &= \sum_{N=0}^{A_0} N^2[\kappa(N+1)P(N+1, t) - \kappa N P(N, t)] + 2\kappa\langle N \rangle^2 \\
&= \kappa\left\{\sum_{N=0}^{A_0-1} N^2(N+1)P(N+1, t) - \sum_{N=0}^{A_0} N^3 P(N, t) + 2\langle N \rangle^2\right\} \\
&= \kappa\left\{\sum_{N=1}^{A_0} (N-1)^2 N P(N, t) - \sum_{N=0}^{A_0} N^3 P(N, t) + 2\langle N \rangle^2\right\} \\
&= \kappa\left\{\sum_{N=0}^{A_0} N(N^2 - 2N + 1)P(N, t) - \sum_{N=0}^{A_0} N^3 P(N, t) + 2\langle N \rangle^2\right\} \\
&= \kappa\left\{\sum_{N=0}^{A_0} N(1 - 2N)P(N, t) + 2\langle N \rangle^2\right\}.
\end{aligned}$$

We now expand the sums and use the definitions of $\langle N \rangle$, $\langle N^2 \rangle$, and σ^2 to get

$$\begin{aligned}
\frac{d\sigma^2}{dt} &= \kappa\left\{\sum_{N=0}^{A_0} N P(N, t) - 2\sum_{N=0}^{A_0} N^2 P(N, t) + 2\langle N \rangle^2\right\} \\
&= \kappa\{\langle N \rangle - 2(\langle N^2 \rangle - \langle N \rangle^2)\} \\
&= \kappa\{\langle N \rangle - 2\sigma^2\}.
\end{aligned} \tag{10.7}$$

We can think of equations (10.5) and (10.7) as a system of coupled first-order ordinary differential equations. Equation (10.5) can be solved on its own, since the right-hand side only depends on $\langle N \rangle$. You should be familiar with the solution of this simple first-order equation:

$$\langle N \rangle(t) = \langle N \rangle_0 e^{-\kappa t},$$

where $\langle N \rangle_0$ is the initial value of $\langle N \rangle$. Substituting this solution into equation (10.7), we get

$$\frac{d\sigma^2}{dt} = -2\kappa\sigma^2 + \kappa\langle N \rangle_0 e^{-\kappa t}. \tag{10.8}$$

This equation can be solved by the method of undetermined coefficients. To use this method, we first need to solve the homogeneous equation

$$\frac{d\sigma_h^2}{dt} = -2\kappa\sigma_h^2.$$

The general solution of this equation is, again, the first-order decay equation:

$$\sigma_h^2 = Ae^{-2\kappa t},$$

where A is a constant. According to the method of undetermined coefficients, the solution of equation (10.8) takes the form

$$\sigma^2(t) = Ae^{-2\kappa t} + Be^{-\kappa t}.$$

The second term above is an exponential of the same form as the inhomogeneous term (the term that doesn't involve σ^2 in equation (10.8)) but with an undetermined coefficient B. If we differentiate this general solution with respect to t and set it equal to the right-hand side of equation (10.8), we get, after a bit of algebra, $B = \langle N \rangle_0$. On the other hand, if we evaluate the general solution at time zero, we get $\sigma^2(0) \equiv \sigma_0^2 = A + B$, so $A = \sigma_0^2 - \langle N \rangle_0$. The solution of the differential equation for σ^2 is therefore

$$\sigma^2(t) = e^{-2\kappa t}(\sigma_0^2 - \langle N \rangle_0) + \langle N \rangle_0 e^{-\kappa t}.$$

The first term on the right-hand side decays faster than the second term because of its larger negative exponent. Accordingly, after a little while, that first term will be negligible compared to the second. The second term is just $\langle N \rangle(t)$. After a transient, the variance will therefore tend to

$$\sigma^2 \sim \langle N \rangle.$$

The standard deviation therefore tends to

$$\sigma \sim \sqrt{\langle N \rangle}.$$

A standard deviation of $\sqrt{N}$ is a common feature of population processes, whatever their origin.

The CV therefore tends towards the value

$$\mathrm{CV} = \frac{\sigma}{\langle N \rangle} \sim \langle N \rangle^{-1/2}.$$

For large values of $\langle N \rangle$, CV $\to 0$, so fluctuations become relatively negligible. This is exactly what we needed to show. It follows that, in the limit of large $\langle N \rangle$, the dynamics are governed by the mass-action equation (10.5). If we divide both sides of this equation by LV and make the association $[\mathrm{A}] = \langle N \rangle / LV$, we get

$$\frac{d[\mathrm{A}]}{dt} = -\kappa[\mathrm{A}].$$

κ appears as the mass-action rate constant, i.e.

$$k = \kappa$$

for a first-order reaction.

10.4.2 A + B bimolecular reactions

The CME for a heteromolecular bimolecular reaction is[8]

$$\frac{dP(N_A, N_B, t)}{dt} = \kappa(N_A + 1)(N_B + 1)P(N_A + 1, N_B + 1, t) - \kappa N_A N_B P(N_A, N_B, t).$$

Because the reaction uses up one molecule of B for every molecule of A, stoichiometry imposes the constraint that $N_A - N_B = A_0 - B_0 = \Delta$ is constant, where A_0 and B_0 are the initial numbers of molecules of A and B, respectively. For simplicity, we treat the case $\Delta = 0$, i.e. the case in which the experiment is carried out using equal concentrations of A and B. Then, $N_A = N_B$, and we can rewrite the CME in terms of a single molecule number, $N \equiv N_A = N_B$:

$$\frac{dP(N, t)}{dt} = \kappa(N + 1)^2 P(N + 1, t) - \kappa N^2 P(N, t).$$

We now want to work out a differential equation for the average number of molecules. The derivation is a close parallel to the one for first-order reactions, so I will get on with the math without commenting on every step:

$$\langle N \rangle = \sum_{N=0}^{A_0} NP(N, t).$$

$$\therefore \frac{d\langle N \rangle}{dt} = \sum_{N=0}^{A_0} N \frac{dP(N, t)}{dt}$$

$$= \kappa \left\{ \sum_{N=0}^{A_0} N(N + 1)^2 P(N + 1, t) - \sum_{N=0}^{A_0} N^3 P(N, t) \right\}$$

$$= \kappa \left\{ \sum_{N=0}^{A_0} (N - 1)N^2 P(N, t) - \sum_{N=0}^{A_0} N^3 P(N, t) \right\}$$

$$= -\kappa \sum_{N=0}^{A_0} N^2 P(N, t),$$

$$\therefore \frac{d\langle N \rangle}{dt} = -\kappa \langle N^2 \rangle. \tag{10.9}$$

At this point, it is useful to think about where we are going. For the A + B reaction with equal reactant concentrations, the law of mass action gives the rate equation

$$\frac{d[A]}{dt} = -k[A][B] = -k[A]^2. \tag{10.10}$$

Equations (10.9) and (10.10) look superficially similar, but there's a small problem: $\langle N^2 \rangle \neq \langle N \rangle^2$. In order for the average to have the same rate law as predicted by the law of mass action, we need $\langle N \rangle^2$ in equation (10.9), not $\langle N^2 \rangle$. If we can show that

[8] I'm using κ generically for the stochastic rate constants. All stochastic rate constants have the same units of inverse time. Hopefully, this isn't confusing, even though they are distinct quantities.

the standard deviation is small compared to the average in the macroscopic limit, i.e. that the CV is small, then $\langle N^2 \rangle \approx \langle N \rangle^2$. And of course, finding that the standard deviation is small also proves that the fluctuations are small.

We start off in similar way to the approach used for first-order reactions by writing down the variance and working out its time derivative:

$$\sigma^2 = \langle N^2 \rangle - \langle N \rangle^2,$$

$$\therefore \frac{\mathrm{d}\sigma^2}{\mathrm{d}t} = \frac{\mathrm{d}}{\mathrm{d}t}\left[\sum_{N=0}^{A_0} N^2 P(N, t) - \langle N \rangle^2\right]$$

$$= \sum_{N=0}^{A_0} N^2 \frac{\mathrm{d}P(N, t)}{\mathrm{d}t} - \frac{\mathrm{d}}{\mathrm{d}t}\langle N \rangle^2$$

$$= \kappa\left\{\sum_{N=0}^{A_0} N^2(N+1)^2 P(N+1, t) - \sum_{N=0}^{A_0} N^4 P(N, t)\right\} - 2\langle N \rangle \frac{\mathrm{d}\langle N \rangle}{\mathrm{d}t}$$

$$= \kappa\left\{\sum_{N=0}^{A_0} (N-1)^2 N^2 P(N, t) - \sum_{N=0}^{A_0} N^4 P(N, t)\right\} + 2\kappa\langle N \rangle\langle N^2 \rangle$$

$$= \kappa\left\{\sum_{N=0}^{A_0} N^2(1 - 2N)P(N, t) + 2\langle N \rangle\langle N^2 \rangle\right\}$$

$$= \kappa\{\langle N^2 \rangle - 2\langle N^3 \rangle + 2\langle N \rangle\langle N^2 \rangle\}.$$

We now use $\langle N^2 \rangle = \sigma^2 + \langle N \rangle^2$, and we get, after some algebra,

$$\frac{\mathrm{d}\sigma^2}{\mathrm{d}t} = \kappa\{\sigma^2(1 + 2\langle N \rangle) + \langle N \rangle^2 - 2(\langle N^3 \rangle - \langle N \rangle^3)\}. \tag{10.11}$$

We introduce another quantity, the third central moment of a distribution, μ_3, defined as follows:

$$\mu_3 = \langle (N - \langle N \rangle)^3 \rangle = \langle N^3 \rangle - \langle N \rangle^3 - 3\sigma^2\langle N \rangle.$$

To get the last equality, we expand the cube, take the average, and use the definition of the variance. Isolating $(\langle N^3 \rangle - \langle N \rangle^3)$ from this expression and substituting it into equation (10.11), we get

$$\frac{\mathrm{d}\sigma^2}{\mathrm{d}t} = \kappa\{\sigma^2(1 - 4\langle N \rangle) + \langle N \rangle^2 - 2\mu_3\}.$$

We have another problem: we don't know μ_3. We are going to assume, without proof, that when $\langle N \rangle$ is large, μ_3 is small. This is equivalent to assuming that the distribution is not highly skewed. This leaves us with

$$\frac{\mathrm{d}\sigma^2}{\mathrm{d}t} = \kappa\{\sigma^2(1 - 4\langle N \rangle) + \langle N \rangle^2\}. \tag{10.12}$$

If we also rewrite the rate equation for the average in terms of the variance and $\langle N \rangle$, we get

$$\frac{\mathrm{d}\langle N \rangle}{\mathrm{d}t} = -\kappa(\sigma^2 + \langle N \rangle^2). \tag{10.13}$$

Equations (10.12) and (10.13) describe trajectories in the $(\langle N \rangle, \sigma^2)$ plane[9]. To finish the argument, we need to analyze how these trajectories behave. Trajectories cross the curve defined by $\mathrm{d}\sigma^2/\mathrm{d}t = 0$ horizontally, and because $\mathrm{d}\langle N \rangle/\mathrm{d}t < 0$, they cross it from right to left (figure 10.1). This curve, called the σ^2 nullcline, is given by

$$\sigma^2 = \frac{\langle N \rangle^2}{4\langle N \rangle - 1}.$$

At large $\langle N \rangle$, this curve is nearly linear, as shown in the figure. If the system's statistics start below this curve, the variance will rise and the average will fall, crossing the nullcline horizontally. After the nullcline is crossed, the sign of $\mathrm{d}\sigma^2/\mathrm{d}t$ changes, so the variance starts to fall.

Now let's consider what happens to trajectories as they cross the line $\sigma^2 = \langle N \rangle$. First note that throughout the region above the nullcline, $\mathrm{d}\sigma^2/\mathrm{d}t$ is negative, so from a qualitative viewpoint, the motion in the phase plane is still down and to the left. We can calculate the slope of the trajectories crossing this line as follows: first, we calculate the slope of a trajectory passing through an arbitrary point in the phase plane by

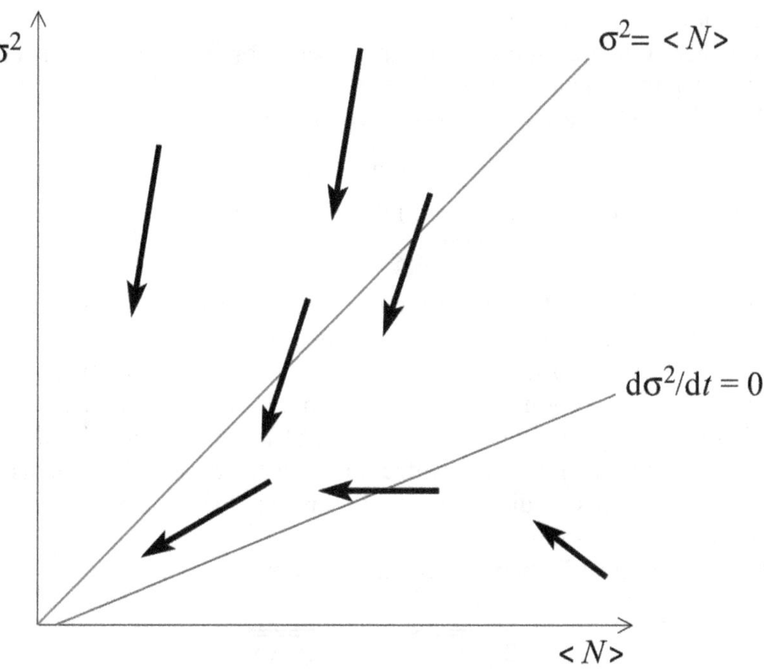

Figure 10.1. A sketch of the evolution of the statistics $\langle N \rangle$ and σ^2 in phase space. The arrows show the qualitative co-evolution of the variables in different regions of the plane. The blue curve is the σ^2 nullcline, which is crossed horizontally from right to left. The red curve is the line $\sigma^2 = \langle N \rangle$. Trajectories approaching from above enter the region under this curve. This means that trajectories are eventually trapped between these two curves.

[9] This is a phase space for this problem (given that we have neglected μ_3), which is to say that knowing the current values of these two coordinates is sufficient to trace out trajectories showing how $\langle N \rangle$ and σ^2 evolve within the space. The technique deployed in this section is known as phase-plane analysis.

$$\begin{aligned}\frac{\mathrm{d}\sigma^2}{\mathrm{d}\langle N\rangle} &= \frac{\mathrm{d}\sigma^2/\mathrm{d}t}{\mathrm{d}\langle N\rangle/\mathrm{d}t} \\ &= \frac{\kappa\{\sigma^2(1-4\langle N\rangle)+\langle N\rangle^2\}}{-\kappa(\sigma^2+\langle N\rangle^2)} \\ &= \frac{\sigma^2(4\langle N\rangle-1)-\langle N\rangle^2}{\sigma^2+\langle N\rangle^2}.\end{aligned}$$

We can now calculate the slopes of trajectories crossing the line $\sigma^2 = \langle N\rangle$:

$$\left.\frac{\mathrm{d}\sigma^2}{\mathrm{d}\langle N\rangle}\right|_{\sigma^2=\langle N\rangle} = \frac{3\langle N\rangle - 1}{\langle N\rangle + 1}.$$

We are interested in the case where $\langle N\rangle$ is large. In this limit,

$$\left.\frac{\mathrm{d}\sigma^2}{\mathrm{d}\langle N\rangle}\right|_{\sigma^2=\langle N\rangle} \to 3.$$

This is a higher slope than the slope of the line $\sigma^2 = \langle N\rangle$. This forces trajectories to enter the region under the line, as shown in figure 10.1. Figure 10.1 puts the whole picture together. Whether approaching from below or from above, trajectories are trapped between the nullcline and the line $\sigma^2 = \langle N\rangle$. Since, at large $\langle N\rangle$, the nullcline approaches $\sigma^2 = \langle N\rangle/4$, after a transient, we must have

$$\langle N\rangle/4 < \sigma^2 < \langle N\rangle.$$

The standard deviation must therefore fall in the interval

$$\sqrt{\langle N\rangle}/2 < \sigma < \sqrt{\langle N\rangle}.$$

We see again that the standard deviation must be proportional to the square root of the population. Using techniques from statistical mechanics not covered here, we can show that we expect to see root-N fluctuations around an equilibrium point. The work presented in this section, however, shows that we can expect fluctuations of this magnitude to set in *on the way* to equilibrium, at least for a bimolecular reaction. You will note that we can only guarantee this after the decay of transients, even for the very simple systems studied here, so this result does not necessarily hold for systems far from equilibrium.

Finally, the CV must lie between the limits shown below:

$$\frac{1}{2\sqrt{\langle N\rangle}} < \mathrm{CV} < \frac{1}{\sqrt{\langle N\rangle}}.$$

This proves that, assuming the distribution has a small skew, the CV must be small in the macroscopic limit.

The small CV computed above implies that $\langle N^2\rangle - \langle N\rangle^2$ must be small compared to $\langle N\rangle^2$. In other words, the error caused by replacing $\langle N^2\rangle$ by $\langle N\rangle^2$ is small. Equation (10.9) then becomes

$$\frac{\mathrm{d}\langle N\rangle}{\mathrm{d}t} \approx -\kappa\langle N\rangle^2.$$

To convert this equation to units of concentration, we make the substitution $[A] = \langle N \rangle / LV$. After a bit of rearrangement, we get

$$\frac{d[A]}{dt} = -LV\kappa[A]^2.$$

Comparing this equation to the mass-action rate equation (10.10), we conclude that

$$\kappa = \frac{k}{LV}.$$

The mass-action rate constant is independent of volume. This equation therefore implies that *the stochastic rate constant for a second-order reaction depends on the volume*. This is essentially due to the conversion from an equation for the number of molecules to one for the concentration. In general, stochastic rate constants depend on the volume, with the exception of the stochastic rate constant for a first-order reaction.

10.4.3 A + A bimolecular reactions

If we repeat the work of the last section for a reaction of the type $A + A \rightarrow$ products, we find

$$\kappa = \frac{2k}{LV}.$$

The factor of two arises from the combinatorial factor $h_r = N_A(N_A - 1)/2$ that appears in the propensity of an A + A reaction.

This section was very technical. Maybe you will find a use for the techniques used here at some point in your career, and maybe not. Rather than dwell on the technical aspects, it may be worth reflecting on what we have learned. First, the chemical master equation predicts the average behavior in accord with the law of mass action. This result can be extended to more complex systems using, as you can imagine, more sophisticated arguments than those deployed here. Second, the fluctuations along a trajectory eventually scale as the square root of the molecular population. This simple fact lets us rapidly estimate whether we expect fluctuations to be important in a chemical or biochemical system. If we are studying an RNA molecule present in cells at an average of a dozen copies, the distribution of RNA numbers across a population of cells would have a standard deviation of about 3.5, so we would expect relatively large differences to be seen between cells, which could have an effect on cellular function. On the other hand, a very abundant RNA that averages 1000 copies per cell would have a standard deviation that is only 3% of the average, which would not be expected to lead to functionally significant effects.

Before we move on, and following the discussion of section 9.5, note that this chapter has added yet another way to think about reactions and their rate constants, this time showing that a statistical model of chemical reactions reduces to mass-action kinetics in the macroscopic limit. This is a perspective that was perhaps implied by collision theory, but here we see that the ideas that led to collision theory fit within an even larger framework of stochastic kinetics, of which we have only scratched the surface. We can already see that this theory will let us talk about the size of fluctuations, an important topic for mesoscopic systems. Random fluctuations, in turn, open up a range of behaviors rarely hinted at in our beaker-scale experiments. In particular, dynamic equilibrium, which appears as a simple balance of rates in bulk kinetics, becomes a statistical phenomenon involving never-ending random wanderings in the neighborhood of the equilibrium composition. This is a much more dynamic picture of chemical dynamics near an equilibrium point than that provided by the smooth curves of chemical kinetics and hopefully one that helps you think about chemical systems and their equilibria in a different way.

10.5 The CME and the curse of dimensionality

You may have noticed that the CME examples we have looked at so far involved very few molecules. There is a reason for that. The CME requires a differential equation for each possible state of the system, i.e. for each distinct composition vector. However, the number of different composition vectors grows very rapidly with the size of the system.

Let's think about what happens when we study a set of systems with the same concentration but different volumes. As the volume increases, so does the number of molecules, since $N = LcV$. To keep things simple, suppose that we have n different isomers that are mutually interconverted by chemical reactions. If, at a certain volume, we have N molecules in total, the number of possible states of the system is

$$\frac{(N+n-1)!}{N!(n-1)!} \approx \left(\frac{eN}{n-1}\right)^{n-1}.$$

The first part of this equation was obtained by asking how many different ways we can allocate the N molecules to the n isomeric states. The second part was obtained by applying Stirling's approximation to the factorials. The number of states to be considered therefore grows as N^{n-1}, which in turn is proportional to V^{n-1}. As a rule of thumb, the exponent is roughly the number of chemical reactions in the system, which we earlier called ρ. Note that in the case of n isomers that can only be converted in sequence, $A_1 \rightleftharpoons A_2 \rightleftharpoons \cdots \rightleftharpoons A_n$, ρ is exactly $n-1$. Thus, the number of equations grows as V^{ρ}. This is very rapid growth, resulting in unmanageably large CMEs for almost any realistic system. The explosive growth in the dimension of the probability space with increasing volume leads to a **curse of dimensionality**, i.e. to intractably large systems of equations.

As an example, consider a chemical substance present in a living cell at a concentration of 10 nM that is converted between $n = 4$ different forms by reactions occurring in the cell. In a bacterial cell with a volume of about 10^{-15} L, 10 nM corresponds to about six molecules. The number of different states to be considered by a master-equation model under these conditions is 72 072, already an impressive number. We certainly won't be writing those equations down in a notebook! If we were to scale up the volume to that of a human cell (10^{-12} L), then we would have about 6000 molecules and over 2.34×10^{24} states! The CME therefore suffers very badly from the curse of dimensionality, leading to sets of equations too large to be handled directly.

10.6 The Gillespie stochastic simulation algorithm

We can get around the issues with the CME by carrying out simulations and collecting statistics. The simulation method presented here, which is due to Daniel Gillespie [1], is a variation on the kinetic Monte Carlo algorithms discussed in the last chapter. It enables simulations of realizations of a random chemical process whose probability distribution evolves according to the CME. By carrying out many simulations, we can collect statistics allowing us to compute averages, standard deviations, population distributions, or indeed any other statistical quantities that happen to catch our fancy.

As in the kinetic Monte Carlo algorithms, we will be using the transition rates, the propensities, both to generate the time to the next reaction and to determine which reaction will occur next. The main difference is that every time a reaction occurs, some of the propensities change, so we need to recalculate the propensities after every step.

Here then is the Gillespie algorithm for a single realization:

1. Initialize the program:
   (a) Set $t = 0$.
   (b) Store the population vector.
   (c) Calculate the propensities a_{r}.
2. Repeat until the desired stopping condition is reached:
   (a) Calculate $a_0 = \sum_{\mathrm{r}} a_{\mathrm{r}}$.
   (a_0 in the Gillespie algorithm corresponds to w_{tot} in the kinetic Monte Carlo algorithm.)
   (b) Generate two random numbers $r_{1,2} \in (0, 1)$.
   (c) Use r_1 to generate the time to the next reaction:

$$\Delta t = -\frac{1}{a_0} \ln r_1.$$

   (d) Use r_2 to choose the next reaction. The logic is identical to that presented in figure 9.1: choose the smallest μ for which

$$\sum_{r=1}^{\mu} a_{\mathrm{r}} > r_2 a_0.$$

   (e) Update the populations according to the stoichiometry of reaction μ.
   (f) Add Δt to t.
   (g) Recompute the propensities.

We can put a loop around the core algorithm to obtain as many realizations as we need to get good statistics.

As an example, I provide below the code for simulating the Michaelis–Menten reaction. Given your past experience with kinetic Monte Carlo codes, the comments in the code should be sufficient to let you follow the logic.[10]

[10] Matlab code Section10_6.m available from https://doi.org/10.1088/978-0-7503-5321-2.

```
1  % Carry out a Gillespie simulation of the Michaelis-Menten mechanism
2  % Reactions:
3  %    (1) E + S -> C
4  %    (2) C -> E + S
5  %    (3) C -> E + P
6
7  % Rate constants in units consistent with M, s:
8  k1 = 1e6
9  k2 = 30
10 k3 = 100
11
12 % Assume reaction in a bacterium of volume
13 V = 1e-15
14
15 % Stochastic rate constants:
16 L = 6.022e23;
17 kappa = zeros(3,1);
18 kappa(1) = k1/(L*V);
19 kappa(2) = k2;
20 kappa(3) = k3;
21
22 % Initial concentrations (M):
23 cE = 10e-9
24 cS = 50e-9
25
26 % Initial populations
27 % Note that the floor function rounds down.
28 NE0 = round(cE*L*V);
29 NS0 = round(cS*L*V);
30
31 % Repeat Nrepeat times to collect statistics.
32 Nrepeat = 1000; % Not a particularly large value as these things go.
33
34 % Collect data at intervals
35 tcollect = 10;
36
37 for rep=1:Nrepeat
38     % Initialize time and populations for each repetition
39     t = 0;
40     NE = NE0;
41     NS = NS0;
42     NC = 0;
43     NP = 0;
44
45     % Initial propensities:
```

```
46      a = zeros(3,1);
47      a(1) = kappa(1)*NE*NS;
48      a(2) = kappa(2)*NC;
49      a(3) = kappa(3)*NC;
50
51      % For each trajectory, we're going to store values of NS at a sequence
52      % of t values.
53      % Store the first point.
54      NSstore(rep,1) = NS;
55      tstore(rep,1) = t;
56      istore = 1;
57      % Time of next data collection.
58      tnext_collect = tcollect;
59
60      % For each simulation, continue until all of the substrate has been
61      % converted to product.
62      while NP < NS0
63          r = rand(2,1);
64          sum_a = cumsum(a);
65          a0 = sum_a(3);
66          deltat = -log(r(1))/a0;
67          % tnext_collect falls between the last time point and the one we
68          % are computing now, so store the state before the update.
69          if t+deltat > tnext_collect
70              istore = istore + 1;
71              NSstore(rep,istore) = NS;
72              tstore(rep,istore) = tnext_collect;
73              tnext_collect = tnext_collect + tcollect;
74          end
75          % Update the time
76          t = t + deltat;
77          % Choose the next reaction and update the populations accordingly
78          mu = find(sum_a>r(1)*a0,1,'first');
79          if mu==1
80              NE = NE - 1;
81              NS = NS - 1;
82              NC = NC + 1;
83          elseif mu==2
84              NE = NE + 1;
85              NS = NS + 1;
86              NC = NC - 1;
87          else
88              NC = NC - 1;
89              NE = NE + 1;
90              NP = NP + 1;
```

```
91            end
92            % Recalculate the propensities.
93            a(1) = kappa(1)*NE*NS;
94            a(2) = kappa(2)*NC;
95            a(3) = kappa(3)*NC;
96        end
97    end
98
99    % We need to find the longest time series since the end time is itself a
100   % random variable.
101   ilongest = find(tstore(:,end),1);
102
103   % Plot a few time series
104   % Create a figure with a large font size for the axis labels
105   figure('DefaultAxesFontSize',18);
106   for rep=1:5
107       plot(tstore(ilongest,:),NSstore(rep,:))
108       hold on
109   end
110   xlabel("t/s")
111   ylabel("N_S","interpreter","tex")
112   hold off
113
114   % Calculate and plot the average as a function of time.
115   % The second argument of the mean() function indicates the dimension along
116   % which to average values. In this case, we want to average across
117   % realizations, which was the first dimension of the NSstore array.
118   NSav = mean(NSstore,1);
119   figure('DefaultAxesFontSize',18)
120   plot(tstore(ilongest,:),NSav)
121   xlabel("t/s")
122   % \langle and \rangle give us nice-looking angle brackets
123   ylabel("\langle N_S\rangle","interpreter","tex")
124
125   % We can calculate distributions at any time point.
126   % Here, we calculate the distribution of NS at the midpoint of the simulation.
127   itime = floor(size(tstore,2)/2);
128   % Print the time corresponding to the distribution on the Matlab console.
129   tstore(1,itime)
130   figure('DefaultAxesFontSize',18)
131   % The following will work in Matlab only. Comment out this line and
132   % uncomment the following one if you're using Octave.
133   histogram(NSstore(:,itime),"Normalization","probability")
134   %hist(NSstore(:,itime),20)
135   xlabel("N_S","interpreter","tex")
```

Running this program generates figures 10.2, 10.3, and 10.4. The first shows several individual trajectories. While the overall trend is the same, they are clearly not identical. The average of all of the computed trajectories, on the other hand, is much smoother (figure 10.3). Finally, we can calculate the probability distribution as a function of time. As an illustration, figure 10.4 shows the probability distribution for the variable N_S at $t = 280$ s.

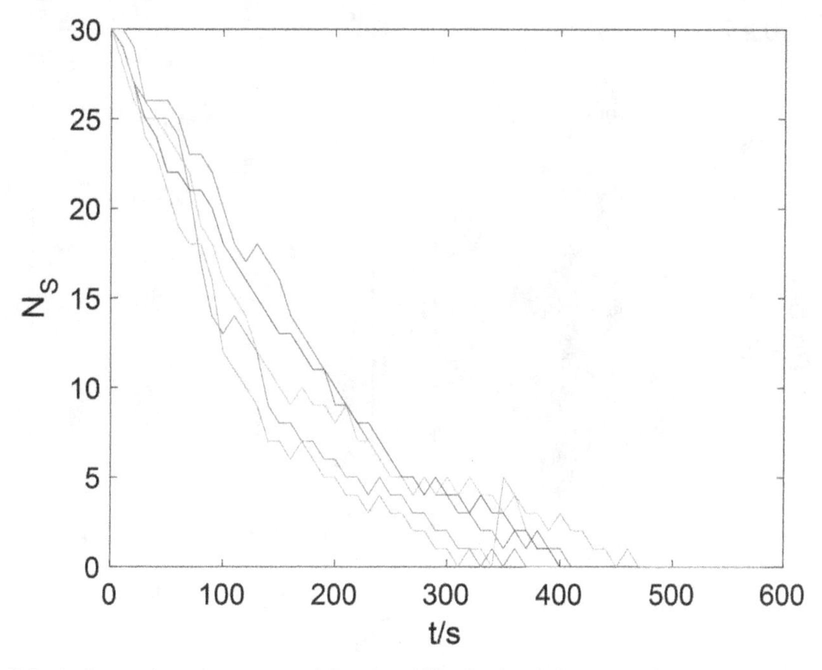

Figure 10.2. A few trajectories generated by the Gillespie simulation code for the Michaelis–Menten mechanism.

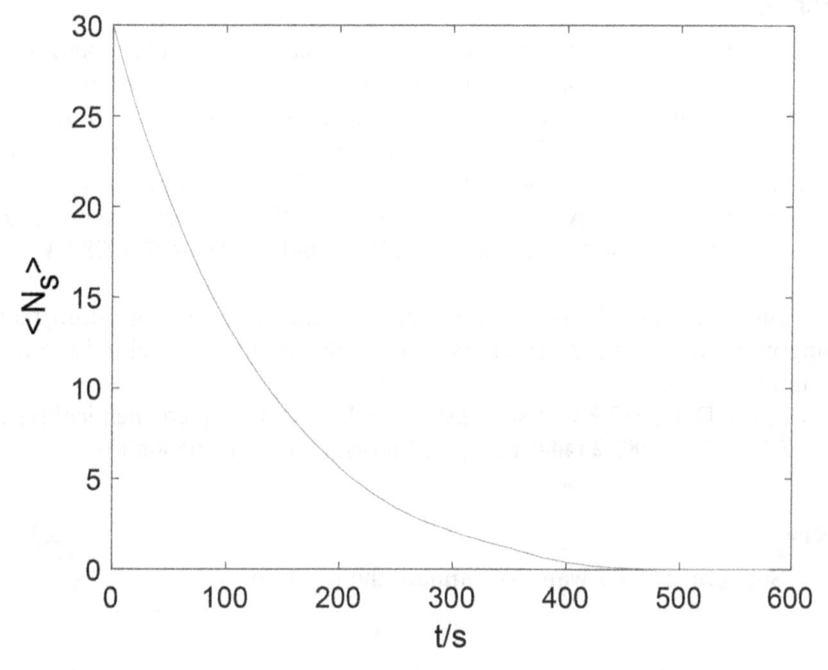

Figure 10.3. The average of the trajectories generated by the Gillespie simulation code for the Michaelis–Menten mechanism.

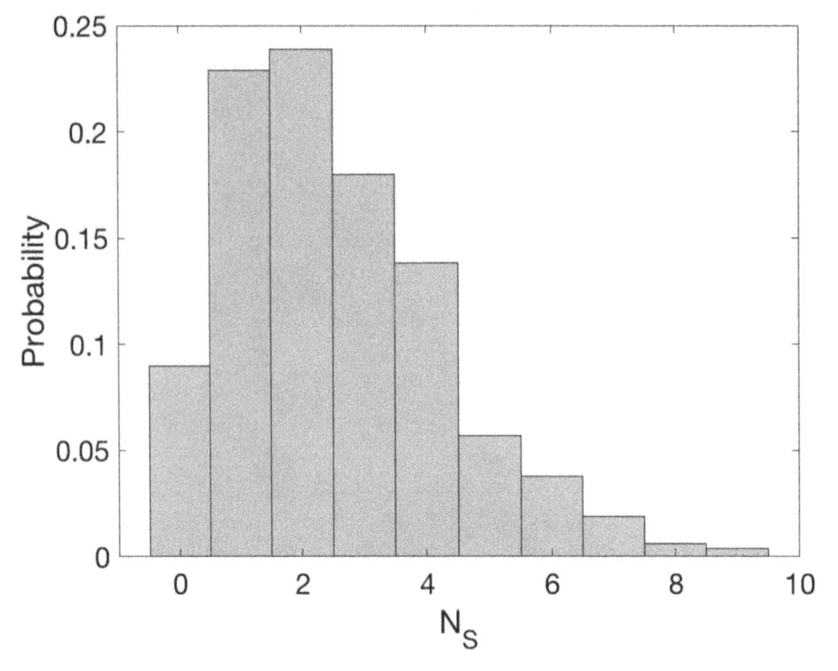

Figure 10.4. The distribution of N_S values generated by the Gillespie simulation code for the Michaelis–Menten mechanism at $t = 280$ s.

Further reading

I skipped the derivation of the chemical master equation here. This equation has a deep heritage, going back at least to Delbrück in 1940 [2], with more readily recognizable versions of the CME appearing in works by Singer [3], Bartholomay [4, 5], and McQuarrie [6] in the 1950s and 1960s. Although the paper is a little quirky, if you want to study a derivation of the CME, I recommend

- Gillespie D T 1992 A rigorous derivation of the chemical master equation *Physica A* **188 404–25** https://doi.org/10.1016/0378-4371(92)90283-V

I generally recommend that students learning the Gillespie algorithm read the following paper by Gillespie, which explains both the thinking behind the method and its implementation:

- Gillespie D T 1977 Exact stochastic simulation of coupled chemical reactions *J. Phys. Chem.* **81 2340–61** https://doi.org/10.1021/j100540a008

Exercise

10.1 Suppose that we want to simulate the reaction

$$2A \underset{\kappa_-}{\overset{\kappa_+}{\rightleftarrows}} \mathrm{B}$$

using the Gillespie algorithm.

(a) If the mass-action rate constants have the values $k_+ = 1.4 \times 10^6$ L mol^{-1}s^{-1} and $k_- = 1.0 \times 10^{-5}$ s^{-1}, and the reaction volume is 1.5 nL, what are the stochastic rate constants?

(b) Suppose that at some particular point in time, $N_A = 2304$ and $N_B = 1232$. Your computer generates the following two random numbers: 0.1637 and 0.6379. Determine which reaction occurs next according to the Gillespie algorithm and the time at which it will occur. Show all the steps of the calculation. Also determine the composition after the reaction has occurred.

10.2 Write a Gillespie simulation program for a single realization of the reversible reaction $A \underset{\kappa_{-1}}{\overset{\kappa_1}{\rightleftharpoons}} B$. Take $\kappa_1 = \kappa_{-1} = 1$. (The units of time don't matter here.) The equilibrium composition clearly has $N_A = N_B$. Starting from a series of equilibrium compositions with increasing numbers of molecules, run simulations of single realizations and plot N_B vs N_A for each simulation. I recommend that you run your simulations for 100 time units. What trend(s) do you notice as you increase the number of molecules in the system?

10.3 Write a Gillespie simulation program for the Lindemann mechanism with rate constants $k_1 = 10$ L mol^{-1}s^{-1}, $k_{-1} = 100$ L mol^{-1}s^{-1}, and $k_2 = 20$ s^{-1} in a container of volume 1 μL at an initial pressure of A of 1 mPa and a bath gas pressure of 15 kPa. Your program should have the following outputs:

(a) a few representative traces of the number of molecules of A vs t;

(b) the average N_A vs t curve;

(c) the distribution of 'extinction times,' i.e. the times at which the last molecule of A is converted to P.

References

[1] Gillespie D T 1976 A general method for numerically simulating the stochastic time evolution of coupled chemical reactions *J. Comput. Phys.* **22** 403–34

[2] Delbrück M 1940 Statistical fluctuations in autocatalytic reactions *J. Chem. Phys.* **8** 120–4

[3] Singer K 1953 Application of the theory of stochastic processes to the study of irreproducible chemical reactions and nucleation processes *J. R. Stat. Soc. Ser. B. Stat. Methodol.* **15** 92–106

[4] Bartholomay A F 1962 A stochastic approach to statistical kinetics with application to enzyme kinetics *Biochemistry* **1** 223–30

[5] Bartholomay A F 1962 Enzymatic reaction-rate theory: a stochastic approach *Ann. N. Y. Acad. Sci.* **96** 897–912

[6] McQuarrie D A 1963 Kinetics of small systems. I *J. Chem. Phys.* **38** 433–6

Part II

Solution-phase kinetics

IOP Publishing

Foundations of Chemical Kinetics
A hands-on approach
Marc R Roussel

Chapter 11

Diffusion-influenced reactions

So far, we have mostly focused on theory that is useful for thinking about gas-phase reactions. In the remainder of this book, we turn our attention to reactions in solution. The presence of a solvent has both advantages and disadvantages from the point of view of developing theories of chemical kinetics. On the one hand, we lose the neat treatments of gas-phase kinetics facilitated by the isolation of a reacting system from any other direct influences during the process of reaction. On the other hand, the constant interaction with the solvent allows for some averaging that would not have been possible for gas-phase molecules in their splendid isolation.

11.1 Some useful concepts from vector calculus

We don't need to become experts in vector calculus, but a few basic ideas and some notation will be useful.

The **gradient operator** ∇ is defined as follows:

$$\nabla = \left(\frac{\partial}{\partial x}, \frac{\partial}{\partial y}, \frac{\partial}{\partial z}\right).$$

The above notation is a shortcut for defining an operator. An operator always has to act on a function. The gradient of a function $f(x, y, z)$ is

$$\nabla f = \left(\frac{\partial f}{\partial x}, \frac{\partial f}{\partial y}, \frac{\partial f}{\partial z}\right).$$

The gradient has a simple geometric interpretation: it is a vector that points in the direction of greatest increase in f.

The **divergence** of a vector-valued function $\mathbf{u}(x, y, z)$ is

$$\nabla \cdot \mathbf{u} = \frac{\partial u_x}{\partial x} + \frac{\partial u_y}{\partial y} + \frac{\partial u_z}{\partial z}.$$

doi:10.1088/978-0-7503-5321-2ch11

A vector-valued function describes a vector field. You can think of a vector field as associating a vector with every point in space[1]. A positive value of the divergence at a point indicates that the vector field describes local expansion; a negative value is indicative of local compression.

The **Laplacian operator** is

$$\nabla \cdot \nabla = \nabla^2 = \frac{\partial^2}{\partial x^2} + \frac{\partial^2}{\partial y^2} + \frac{\partial^2}{\partial z^2}.$$

As with operators in general, the Laplacian only makes sense if we are applying it to a function, say $f(x, y, z)$:

$$\nabla \cdot \nabla f = \nabla^2 f = \frac{\partial^2 f}{\partial x^2} + \frac{\partial^2 f}{\partial y^2} + \frac{\partial^2 f}{\partial z^2}.$$

The Laplacian is the divergence of a gradient, so you can interpret it based on the meanings of those quantities: the gradient defines a vector field that locally points in the steepest uphill direction. If this vector field has a negative divergence, then these steepest uphill directions are being squeezed together. Loosely speaking, you can think of these directions of steepest ascent as being directed into a channel in this case. The Laplacian is also negative at a local maximum where all the nearby directions of steepest ascent point towards the maximum. In a fluid analogy, the maximum is a 'sink' for the gradient vector field. Conversely, at a local minimum, the Laplacian is positive, since the gradient vector field behaves as if the steepest ascent vectors originate at the minimum and expand away from it.

Equations for the gradient, divergence, and Laplacian were given above in the usual Cartesian coordinate system. These operators can, of course, be written in other coordinate systems. The formulas are generally nontrivial and should be looked up. We will be using the spherical polar coordinate system for some of our work on diffusion. In this coordinate system, which we first encountered in section 3.2, the gradient is given by

$$\nabla = \hat{\mathbf{r}}\frac{\partial}{\partial r} + \frac{\hat{\boldsymbol{\theta}}}{r}\frac{\partial}{\partial \theta} + \frac{\hat{\boldsymbol{\phi}}}{r \sin\theta}\frac{\partial}{\partial \phi}, \tag{11.1}$$

where $\hat{\mathbf{r}}$, $\hat{\boldsymbol{\theta}}$, and $\hat{\boldsymbol{\phi}}$ are unit vectors in the corresponding directions (figure 11.1). The Laplacian is

$$\nabla^2 = \frac{1}{r^2}\frac{\partial}{\partial r}\left(r^2\frac{\partial}{\partial r}\right) + \frac{1}{r^2 \sin\theta}\frac{\partial}{\partial \theta}\left(\sin\theta\frac{\partial}{\partial \theta}\right) + \frac{1}{r^2\sin^2\theta}\frac{\partial^2}{\partial \phi^2}. \tag{11.2}$$

Note that these operators are applied to a function 'from the left.' For example,

$$\nabla^2 f = \frac{1}{r^2}\frac{\partial}{\partial r}\left(r^2\frac{\partial f}{\partial r}\right) + \frac{1}{r^2 \sin\theta}\frac{\partial}{\partial \theta}\left(\sin\theta\frac{\partial f}{\partial \theta}\right) + \frac{1}{r^2\sin^2\theta}\frac{\partial^2 f}{\partial \phi^2}.$$

[1] We analyzed a vector field in the phase-plane analysis of section 10.4.2. There, we treated $(\mathrm{d}\langle N\rangle/\mathrm{d}t, \mathrm{d}\sigma^2/\mathrm{d}t)$ as a vector field describing the evolution of the statistics $\langle N\rangle$ and σ^2 through the phase plane.

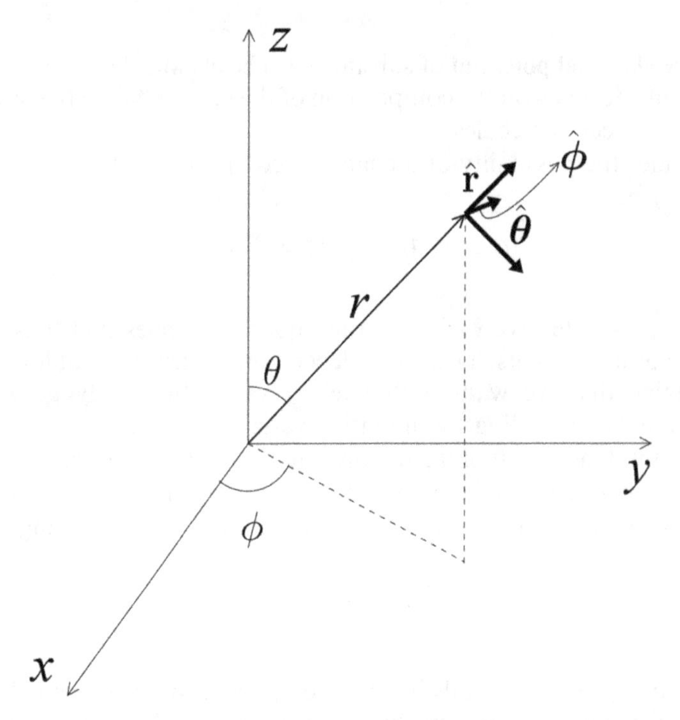

Figure 11.1. Spherical coordinate unit vectors. The unit vectors vary from point to point in space, unlike those of the Cartesian coordinate system. Here, $\hat{\mathbf{r}}$ is a unit vector in the direction of the line from the origin to the point, $\hat{\boldsymbol{\theta}}$ is a unit vector pointing in the direction of increasing θ, and similarly $\hat{\boldsymbol{\phi}}$ is a unit vector pointing in the direction of increasing ϕ.

11.2 The chemical potential

You may recall the equation

$$dG = V\,dp - S\,dT \tag{11.3}$$

from your previous study of thermodynamics. This equation tells us how the Gibbs free energy depends on pressure and temperature in a system of fixed composition. From this equation, we can 'read' the partial derivatives of G with respect to p while holding T constant and with respect to T while holding p constant:

$$\left.\frac{\partial G}{\partial p}\right|_T = V \qquad \text{and} \qquad \left.\frac{\partial G}{\partial T}\right|_p = -S.$$

Most chemically interesting systems are mixtures with n_1 moles of species 1, n_2 moles of species 2, and so on. What happens if we add some amount of species i to the system? How does that affect the free energy of the system? It should be possible to phrase this in terms of a partial derivative:

$$\mu_i = \left.\frac{\partial G}{\partial n_i}\right|_{p,T,n_{j\neq i}}.$$

μ_i is called the **chemical potential** of substance i. The chemical potential of any given species generally depends on the composition of the system due to the intermolecular forces acting between molecules.

If we consider the possibility of a changing composition, then the differential of the free energy becomes

$$dG = V\,dp - S\,dT + \sum_i \mu_i\,dn_i.$$

Since G and n_i are extensive variables, this equation implies that μ_i is an intensive variable[2]. Accordingly, μ_i itself can only depend on intensive variables (p, T, c_i, etc).

Now imagine that we want to 'make' a system by slowly growing it from nothing, all the while holding the intensive variables p and T and all the concentrations constant. One way to imagine this is to think about adding a tiny droplet at a time, each containing all of the components in the correct proportions, until we reach the final volume of the system. Since neither p nor T change during this process, we have

$$dG = \sum_i \mu_i\,dn_i,$$

with the sum being taken over all chemical components of the system. This equation describes how the total free energy changes as we add each droplet. Integrating this equation, we get

$$G = \sum_i \int_0^{n_i} \mu_i\,dn_i = \sum_i \mu_i \int_0^{n_i} dn_i = \sum_i \mu_i n_i. \tag{11.4}$$

On the left-hand side, we just get G because the free energy of the nothing we started with is zero. Also note that we were able to pull the chemical potential out of the integral because it only depends on intensive variables that are held constant during this process. Equation (11.4) gives us some insight into the meaning of the chemical potential: $\mu_i n_i$ is the part of the Gibbs energy of a system that can be attributed to species i. Thus, μ_i itself is the molar Gibbs energy of a substance in a particular solution (or, in general, in any mixture).

Given the connection of μ_i to the Gibbs free energy, it should not be very surprising that μ_i obeys the equation

$$\mu_i = \mu_i^\circ + RT\ln a_i,$$

where μ_i° is the chemical potential under standard conditions and a_i is the activity of species i: $a_i = \gamma_i c_i / c^\circ$ for a solute. In the most commonly used convention[3], $c^\circ = 1\ \text{mol L}^{-1}$.

[2] Extensive variables depend on the size of the system. Intensive variables do not.

[3] If you spend enough time with your nose in thermodynamics texts, you will sooner or later run into a number of other conventions. This one will be more than adequate for our purposes.

11.3 Diffusion

11.3.1 The transport equation

Roughly speaking, diffusion is the random motion of molecules as they bump into each other, exchanging momentum and energy with each collision. To build a more formal theory of macroscopic diffusion, imagine a small (imaginary) surface of area dA immersed in a fluid. The net number of molecules of type i passing through this surface in a specified direction per unit time divided by the area is the **flux**[4], J, of species i. If the surface is oriented perpendicularly to the x axis and we count the net number of molecules traveling to the right (i.e. right-traveling minus left-traveling), we then have J_x, the x component of the flux vector. The flux can be expressed in various units. We will assume throughout this section that its units are mol m^{-2}s^{-1}.

Now consider figure 11.2. We want to know the rate of change of the concentration inside the illustrated prism if there is a flux in the x direction of $J_x(x)$. Note that we consider the dependence of the flux on the position along the x axis. J_x is the number of moles of particles flowing through a surface perpendicular to the x axis per unit area per unit time, so the change in the number of moles of material in the prism per unit time is the difference between the rate at which molecules are entering, $AJ_x(x)$, and the rate at which they are leaving, $AJ_x(x + \Delta x)$. Thus,

$$\frac{\Delta n}{\Delta t} = A[J_x(x) - J_x(x + \Delta x)].$$

If we divide both sides by V, the volume of the prism, we get the change in concentration per unit time. Since $V = A\Delta x$, we get

$$\frac{\Delta c}{\Delta t} = \frac{[J_x(x) - J_x(x + \Delta x)]}{\Delta x}.$$

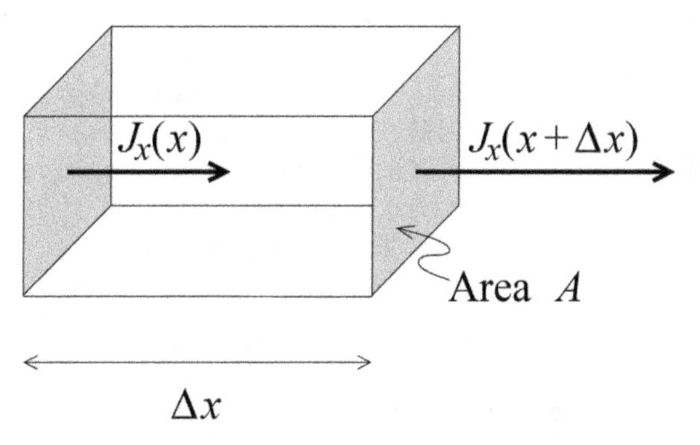

Figure 11.2. A small prism of cross-sectional area A and length Δx, showing the fluxes entering and leaving this prism, which may be different.

[4] We could write J_i, but for now there is no chance of confusion, and we will shortly need subscripts for the Cartesian directions. A little later, we will go back to using subscripts to denote chemical species in a mixture.

If we then let $\Delta x \to 0$ and $\Delta t \to 0$, we get

$$\frac{\partial c}{\partial t} = -\frac{\partial J_x}{\partial x}.$$

So far, we have only considered the flux in the x direction. Of course, there is nothing special about this Cartesian direction, and we could repeat the entire exercise with the y and z directions to obtain the effects of those fluxes on the concentration. The overall change in concentration is the sum of the effects of the fluxes in the three Cartesian directions, i.e.

$$\frac{\partial c}{\partial t} = -\frac{\partial J_x}{\partial x} - \frac{\partial J_y}{\partial y} - \frac{\partial J_z}{\partial z} = -\nabla \cdot \mathbf{J}. \tag{11.5}$$

This equation makes sense given the interpretation we previously discussed for the divergence: if $\nabla \cdot \mathbf{J} > 0$, which indicates an expanding vector field, then $c(x, y, z, t)$ locally decreases. Conversely, if $\nabla \cdot \mathbf{J} < 0$, suggesting a compression of the vector field, then the local concentration increases. Equation (11.5) is called the **transport equation**. It applies to any kind of transport: diffusion, advection[5], etc. All we need to know to apply the transport equation is how the flux varies with spatial position.

11.3.2 The 'driving force' of diffusion and Fick's second law

The term 'driving force' in the title of this section was placed in quotation marks deliberately. We are not going to be calculating a physical force. Rather, we will be calculating a virtual force, i.e. a force *implied* by the free energy change when a molecule moves through a concentration gradient. The arguments we will use are, to use a term I have used elsewhere, peculiar (see [1], section 18.2). We will at times be using equations more appropriate for macroscopic objects than for molecules, but in the end we will obtain equations that have proven useful for understanding diffusion and that are surprisingly accurate in their predictions in a variety of cases. So let's dive in.

Suppose we have an inhomogeneous system, in the sense that the composition of the system and therefore the chemical potentials of its components, vary from point to point. We will for now just consider diffusive motion along the x axis. Suppose that we take one molecule of substance i from the vicinity of position x along the x axis, and move it to position $x + \mathrm{d}x$. The corresponding change in free energy for that one molecule is

$$\mathrm{d}G = \left[\mu_i(x + \mathrm{d}x) - \mu_i(x)\right]/L.$$

Note that we had to divide by Avogadro's number, L, to convert to free energy per molecule, given that the chemical potentials are molar free energies.

Recall that $\mathrm{d}G = \mathrm{d}w_{\mathrm{rev}}$, the reversible work measured relative to the system. If $\mathrm{d}G$ is positive, an external force had to do work on the molecule to move it from x to $x + \mathrm{d}x$. The work done relative to the source of the external force is

[5] Advection means being carried along by a velocity field, such as a particle carried in a river.

$$\mathrm{d}w = -\left[\mu_i(x + \mathrm{d}x) - \mu_i(x)\right]/L.$$

From the definition of work, $dw = F_{i,x}\, dx$, so

$$F_{i,x} = -\frac{1}{L}\frac{\mu_i(x + \mathrm{d}x) - \mu_i(x)}{\mathrm{d}x}$$

or, in the limit as $\mathrm{d}x \to 0$,

$$F_{i,x} = -\frac{1}{L}\frac{\partial \mu_i}{\partial x}. \tag{11.6}$$

This is the so-called **driving force of diffusion**. Note that this force acts against the gradient, i.e. it tends to push molecules towards regions of lower chemical potential.

It is great fun to think about the effects of non-ideality on diffusion. For now, however, let's assume that the solution behaves ideally. Thus,

$$\mu_i = \mu_i^\circ + RT \ln(c_i(x)/c^\circ).$$

We can evaluate the derivative in equation (11.6) using the chain rule. The result is

$$F_{i,x} = -\frac{RT}{Lc_i}\frac{\partial c_i}{\partial x} = -\frac{k_\mathrm{B}T}{c_i}\frac{\partial c_i}{\partial x}.$$

There are, of course, similar equations for the driving force of diffusion in the y and z directions. Thus, in three dimensions,

$$\mathbf{F}_i = -\frac{k_\mathrm{B}T}{c_i}\nabla c_i.$$

Assuming ideal behavior, the force acts against the concentration gradient. The tendency is for molecules to move to areas of lower concentration from areas of higher concentration and thus for concentrations to even out across the system, barring other effects[6].

Since $F = ma$, the driving force of diffusion must be balanced by some other force, or else molecules would constantly accelerate. Molecules bump into other molecules as they move through solution, and this provides the opposing force that limits the velocities they can achieve. If molecules were macroscopic objects, we would describe the effect of these collisions as a drag force. In yet another peculiar aspect of this theory, we assume that we can treat the resistance molecules experience as they move through a fluid as a drag, using the same equations that we use to describe drag on macroscopic objects.

[6] Nonideal effects, for example, can lead to some degree of self-segregation, in which a system spontaneously organizes into regions where a particular solute (or group of solutes) has a high concentration and others where this solute is much less abundant. In systems with both chemical reactions and diffusion, we can observe Turing patterns under certain conditions [2], in which we again get sustained concentration differences within a system, often organized as stripes or spots. Paradoxically, Turing patterns are *caused* by diffusion ([3], section 13.2), even though diffusion is normally a homogenizing effect.

From the theory of hydrodynamics, we know that the drag force experienced by an object moving through a fluid at low relative speeds is given by $\mathbf{F}_i^{(d)} = -f_i\,\mathbf{u}_i$, where f_i is a frictional (drag) coefficient and $\mathbf{u}_i$ is the mean drift velocity. We say 'drift velocity' because this is an average velocity down the chemical potential (or concentration) gradient. The constant collisions with other molecules mean that this speed is anything but constant. However, assuming that there is a mean drift velocity, the drag force should be equal and opposite to the driving force of diffusion in order for a constant drift speed to be reached. Thus,

$$-f_i\,\mathbf{u}_i = \frac{k_B T}{c_i}\nabla c_i. \tag{11.7}$$

The flux is the number of moles of material per unit area per unit time. This can be separated into two factors: the concentration (moles per unit volume) multiplied by the drift velocity (distance traveled per unit time), i.e. $\mathbf{J}_i = c_i\mathbf{u}_i$. Rearranging equation (11.7), we get

$$c_i\mathbf{u}_i = -\frac{k_B T}{f_i}\nabla c_i = \mathbf{J}_i.$$

We define the **diffusion coefficient**

$$D_i = k_B T / f_i \tag{11.8}$$

so that

$$\mathbf{J}_i = -D_i \nabla c_i. \tag{11.9}$$

This equation is known as **Fick's first law**. The derivation of Fick's first law involves a number of assumptions, some obvious, some less so. Accordingly, Fick's first law has a limited range of applicability (ideally behaving solutes, gradients not too large). Moreover, note that the frictional coefficient f_i depends on both the solute and the solvent environment, so D_i does too.

We can now combine the transport equation (11.5) with Fick's first law, equation (11.9), to obtain the **diffusion equation**, also known as **Fick's second law**:

$$\frac{\partial c_i}{\partial t} = D_i \nabla \cdot \nabla c_i = D_i \nabla^2 c_i.$$

The diffusion equation is a partial differential equation for $c_i(x, y, z, t)$, i.e. for the evolution of the concentration in time and space. In general, it is difficult to solve. There are books that solve the diffusion equation for many different situations and geometries; they also give methods for solving these equations for some classes of problems. The archetype of these books is Crank's *The Mathematics of Diffusion* [4]. If you ever need to solve a diffusion equation, it's a good idea to start with one of these books before expending a large amount of effort. If you're lucky, the solution to your problem may be there waiting for you. And if you can't find an analytic solution, there is always the option of solving the equation numerically, which gives you a great deal of flexibility in terms of the shape of the container, among other things.

In fact, let's look at a couple of those cases for which we can look up a solution.

Example 11.1. *Suppose that we put a small drop of material (e.g. a dye) in the center of a very long, narrow tube. We want to know how the concentration profile evolves over time, i.e. we want to solve the diffusion equation for this case. Because the tube is narrow, we can treat the system as being one-dimensional; the position along the tube defines an x coordinate and the drop is placed at the origin. A drop of negligible width can be modeled using a* **Dirac delta function** *with the following properties:*

1. $\delta(x) = 0$ *everywhere except at* $x = 0$.
2. $\int_{-a}^{a} \delta(x)\,\mathrm{d}x = 1$ *for any* $a > 0$.

Note that in order for $\delta(x)$ *to have property (2), the delta function must have units that are the inverse of those of x. If x represents a spatial coordinate with units of length, then* $\delta(x)$ *has units of* length^{-1}.

The concentration profile of a drop of negligible width containing n moles of a substance placed in the center of the tube at time zero could then be approximately described by

$$c(x, t = 0) = \frac{n}{A}\delta(x),$$

where A is the cross-sectional area of the tube. Note that we are dropping the subscript i, since we are only considering one diffusing substance.

Although this concentration profile is an idealization, it was chosen because the solution of the diffusion equation with this initial condition is known:

$$c(x, t) = \frac{n}{A\sqrt{4\pi Dt}} \exp\left(-\frac{x^2}{4Dt}\right). \tag{11.10}$$

This is a Gaussian distribution with a time-dependent variance of $2Dt$. *Solutions are plotted at different times in figure 11.3. Note that diffusion isn't very fast. On a timescale of minutes, material moves by diffusion over length scales of millimeters.*

Example 11.2. *What if we wanted to solve an analogous problem in three dimensions? We would use a three-dimensional version of a delta function to describe a tiny drop placed in the center of a container. Provided the container was large enough so that we didn't need to worry about interactions between the material and the walls over the observational timescale, the solution starting from a spherical droplet should be spherically symmetric. We would write the Laplacian in spherical polar coordinates (equation (11.2)) and, because of the spherical symmetry of the solution, both* $\partial c/\partial\theta$ *and* $\partial^2 c/\partial\phi^2$ *would be zero. The diffusion equation therefore reduces to*

$$\frac{\partial c}{\partial t} = D\frac{1}{r^2}\frac{\partial}{\partial r}\left(r^2\frac{\partial c}{\partial r}\right).$$

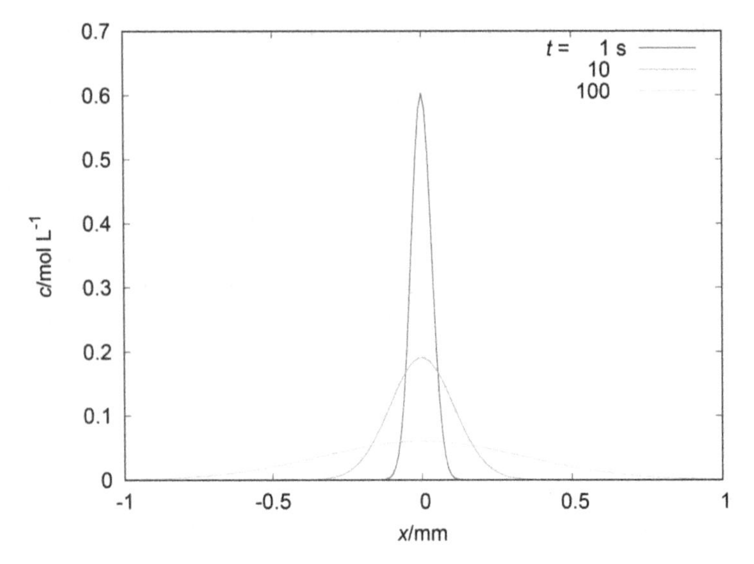

Figure 11.3. Solutions of the one-dimensional diffusion equation as a function of time starting from delta-function initial conditions with $D = 5.7 \times 10^{-10}$ m^{2}s^{-1} (the diffusion coefficient of sucrose in water at 20 °C), $A = 1.96 \times 10^{-5}$ m^2 (corresponding to a pipe with a radius of 2.5 mm), and $n = 1$ nmol.

This problem has also been solved. The solution is

$$c(r,\, t) = \frac{n}{(4\pi Dt)^{3/2}} \exp\left(\frac{-r^2}{4Dt}\right).$$

Since $r^2 = x^2 + y^2 + z^2$, this solution could also be written

$$c(x,\, y,\, z,\, t) = n\left[\frac{1}{\sqrt{4\pi Dt}} \exp\left(\frac{-x^2}{4Dt}\right)\right]\left[\frac{1}{\sqrt{4\pi Dt}} \exp\left(\frac{-y^2}{4Dt}\right)\right]\left[\frac{1}{\sqrt{4\pi Dt}} \exp\left(\frac{-z^2}{4Dt}\right)\right].$$

Note that, give or take a normalization constant associated with the different geometries, the solution of the diffusion equation in three dimensions looks like a product of one-dimensional solutions (equation (11.10)).

In both of the foregoing examples, the variance of the Gaussian concentration profile in any given Cartesian direction is $2Dt$. Note that the factor of A in equation (11.10) was only there to convert the lineal density (molecules per unit length) into a volumic density (concentration). Let's get rid of that and consider true one-dimensional diffusion. If we divide the lineal density by n, we get a density per molecule, i.e. a probability density that a particular molecule will be found near a particular x. We can calculate averages from probability densities. In particular,

$$\langle x^2 \rangle = \int_{-\infty}^{\infty} x^2 \frac{1}{\sqrt{4\pi Dt}} \mathrm{e}^{-x^2/4Dt} \mathrm{d}x = 2Dt.$$

This is called the mean-squared displacement. The square root of this quantity, the root-mean-squared displacement, is a measure of how far a typical molecule travels in time t. Note that this means that the distance traveled by a typical molecule increases according to the square root of the time elapsed.

11.3.3 Stokes–Einstein theory

We previously found that the diffusion coefficient could be related to a frictional coefficient through equation (11.8). We can continue our peculiar journey through this theory using equations from classical hydrodynamics for the frictional coefficient. For example, for a sphere moving through a fluid, hydrodynamics gives

$$f_i = 6\pi r_i \eta,$$

where η is the viscosity of the solvent. If we assume that this equation holds for molecules just as it does for macroscopic spheres, we get the **Stokes–Einstein** formula:

$$D_i = \frac{k_B T}{6\pi r_i \eta}.$$

The Stokes–Einstein formula lets us estimate the radius of a molecule in solution from its diffusion coefficient, or vice versa. There are many caveats to the use of this formula, not the least of which is that most molecules are not spherical. The frictional coefficient is larger for an ellipsoid than for a sphere of equal volume. The two are related by a factor that depends on the ratio of the major axis to the minor axis and differs for prolate and oblate ellipsoids. Tables of the frictional coefficient ratio vs the aspect ratio of the molecule are available and, when combined with measurements of molecular volume, can be used to determine a rough shape for a molecule in solution (see [5], chapter 6). For our purposes, though, the Stokes–Einstein equation will suffice.

Example 11.3. *Buckminsterfullerene (C_{60}) has a diameter of 10.18* Å. *Benzonitrile has a viscosity of 1.24 mPa s at* 25 °C. *Given that C_{60} is approximately spherical, we should be able to use Stokes–Einstein theory to estimate the diffusion coefficient of this molecule in benzonitrile (or in any other solvent for that matter). At* 25 °C,

$$\begin{aligned} D &= \frac{k_B T}{6\pi r \eta} \\ &= \frac{(1.380\,649 \times 10^{-23}\ \mathrm{J\ K^{-1}})(298.15\ \mathrm{K})}{6\pi(5.09 \times 10^{-10}\ \mathrm{m})(1.24 \times 10^{-3}\ \mathrm{Pa\ s})} \\ &= 3.46 \times 10^{-10}\ \mathrm{m^2\ s^{-1}}. \end{aligned}$$

Note that we needed the radius and not the diameter for this calculation. Also note the SI units of viscosity.

This diffusion coefficient has been measured. The experimental value is $(4.1 \pm 0.3) \times 10^{-10}\ \mathrm{m^2\ s^{-1}}$, *a bit higher than the Stokes–Einstein value. This is an anomalous result. On occasion, the calculated Stokes–Einstein value is larger than the experimental value because the experimental value includes some solvent that is dragged around with the solute. Here, it seems likely that the solvent organizes itself around the buckminsterfullerene molecule in a way that facilitates the latter's diffusion, reducing the effective drag.*

11.3.4 Electrodiffusion

Ions in solution experience medium-range forces due to the charges they carry. Moreover, the motion of ions can be affected by external electric fields. Accordingly, to describe the diffusion of ions, we need to take electrostatics into account.

Recall that the force on an ion of charge z_i (in elementary units) in an electric field $\mathbf{E}$ is

$$\mathbf{F}_i = z_i e \mathbf{E}.$$

If the only force acting on an ion were the electric force, then this force would be balanced by the drag force when the ion reached a terminal velocity. Thus,

$$f_i \mathbf{u}_i = z_i e \mathbf{E}$$

or

$$\mathbf{u}_i = z_i e \mathbf{E} / f_i \, .$$

We define the **mobility** of an ion as

$$\upsilon_i = u_i / E = |z_i| e / f_i \tag{11.11}$$

so that

$$\mathbf{u}_i = \operatorname{sgn}(z_i) \upsilon_i \mathbf{E}$$

where $\operatorname{sgn}(z_i)$ is the sign of the charge. The flux of ion i due solely to the electric field is

$$\mathbf{J}_i^{(E)} = c_i \mathbf{u}_i = \operatorname{sgn}(z_i) c_i \upsilon_i \mathbf{E}.$$

But of course, an ion can also diffuse. As a rule, the effects of various transport mechanisms can be added to give the overall flux, unless these mechanisms are coupled. Thus, the flux of ions of type i due to diffusion and the effect of an electric field is given by

$$\mathbf{J}_i = -D_i \nabla c_i + \operatorname{sgn}(z_i) c_i \upsilon_i \mathbf{E}. \tag{11.12}$$

If we now apply the transport equation (11.5) to this flux, we get the **diffusion-conduction equation**:

$$\frac{\partial c_i}{\partial t} = D_i \nabla^2 c_i + \operatorname{sgn}(z_i) \upsilon_i \nabla \cdot (c_i \mathbf{E}).$$

The diffusion coefficient and mobility are related quantities. If we compare equations (11.8) and (11.11), we get

$$v_i = \frac{|z_i|e}{k_B T} D_i. \tag{11.13}$$

We now want to consider an ion of type i in a spherically symmetric electric potential $V = V(r)$ with $V(r \to \infty) = 0$. This could, for example, be the potential due to one particular ion in a solution. We are going to look for steady (time-independent) solutions of the diffusion-conduction equation for this case. These solutions will necessarily also have a spherical symmetry, i.e. $c_i = c_i(r)$. In a steady state, there should be no net flux anywhere in the system, i.e. $\mathbf{J} = \mathbf{0}$.

The electric field and electric potential are related by $\mathbf{E} = -\nabla V$. Therefore, equation (11.12) can be written

$$\mathbf{J}_i = -D_i \nabla c_i - \mathrm{sgn}(z_i) c_i v_i \nabla V.$$

For spherically symmetric functions, the gradient, equation (11.1), reduces to a simple (partial) derivative by r. Thus, the steady-state problem reduces to

$$\mathbf{J}_i = \mathbf{0} = -D_i \frac{\mathrm{d}c_i}{\mathrm{d}r} - \mathrm{sgn}(z_i) c_i v_i \frac{\mathrm{d}V}{\mathrm{d}r}.$$

Since the steady-state solutions won't depend on time, so that r is the only remaining variable, I replaced the partial derivatives by ordinary derivatives. This problem can be solved by separation of variables. Here is the separation:

$$\frac{\mathrm{d}c_i}{c_i} = -\frac{\mathrm{sgn}(z_i) v_i}{D_i} \mathrm{d}V.$$

Note that, according to equation (11.13), the ratio v_i/D_i which appears here is $|z_i|e/k_B T$. Therefore

$$\frac{\mathrm{d}c_i}{c_i} = -\frac{z_i e}{k_B T} \mathrm{d}V.$$

$$\therefore \int_{c_i^\circ}^{c_i(r)} \frac{\mathrm{d}c_i}{c_i} = -\frac{z_i e}{k_B T} \int_0^{V(r)} \mathrm{d}V.$$

The limits of integration are based on the assumption that, far away from the center of the field where $V(r) \to 0$, the concentration tends toward its bulk value, c_i°.

$$\therefore \ln\left(\frac{c_i(r)}{c_i^\circ}\right) = -\frac{z_i e}{k_B T} V(r) = -\frac{U(r)}{k_B T},$$

where $U(r)$ is the electrostatic potential energy of ion i in the potential $V(r)$.

$$\therefore c_i(r) = c_i^\circ \exp\left(-\frac{U(r)}{k_B T}\right), \tag{11.14}$$

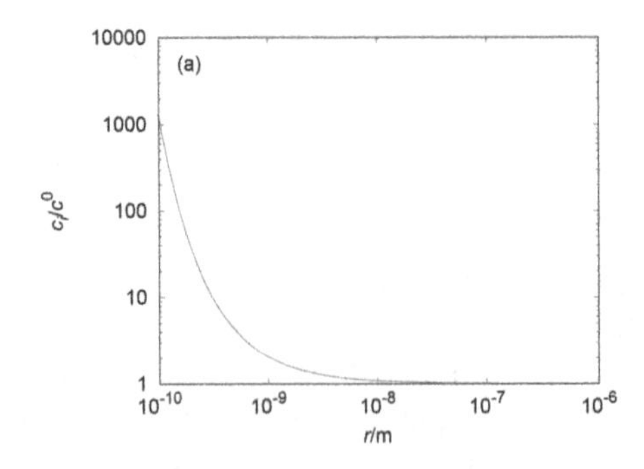

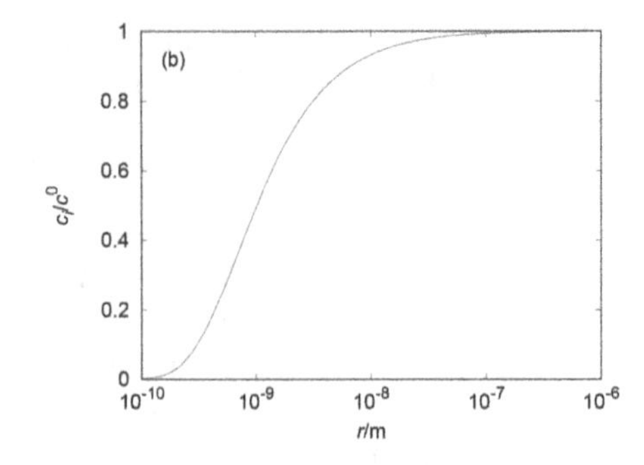

Figure 11.4. The distribution of ions around a charge. (a) $z_1 = -z_2 = 1$; (b) $z_1 = z_2 = 1$. In both panels, $T = 298.15$ K and the distribution was computed in water (relative permittivity 78.37).

which is a Boltzmann distribution! A Boltzmann distribution naturally arose from the balance between electrostatic forces and diffusion. Boltzmann distributions tend to develop in situations in which energetics compete with entropy. In this case, we have a position-dependent potential energy due to the electric field of an ion, and diffusion, which tends to spread matter out, an entropically favorable process.

So what does this Boltzmann distribution look like? figure 11.4 shows this distribution for (a) charges of opposite sign and (b) charges with the same sign. As you might have guessed, ions of opposite charge attract, and there is thus a higher concentration near the central ion than farther away. On the other hand, repulsion between ions of opposite charge results in a very low concentration near the central ion, and a higher concentration farther away. In a solution with many ions of both kinds, these effects work together to organize the average relative positioning of ions.

11.4 The theory of diffusion-influenced reactions

In this section, we will see two distinct versions of the theory of diffusion-influenced reactions. These theories are complementary, yielding slightly different insights.

11.4.1 Version 1: based on the physics of diffusion

In the gas phase, a collision is a single event with a very short lifetime. In solution, however, once two molecules have come into direct contact with each other, they may stay in contact for a long time because the solvent molecules that surround them keep them from drifting away freely. These solvent molecules form a **cage**, which has to be partially disassembled in order for the two solute molecules to move away from each other. This means that encounters between molecules in solution last much longer than collisions in the gas phase. A pair of molecules that are in direct contact and surrounded by solvent is called an **encounter pair**.

If we want to describe reactions in solution, we therefore need to take into account:

1. the rate at which molecules diffuse into close proximity, i.e. the rate at which encounter pairs are formed;

2. the lifetime of encounter pairs; and
3. the intrinsic reaction rate for two molecules in direct contact.

Since both diffusion and reaction rates are important, we describe typical reactions in solution as **diffusion influenced**.

In the rest of this section, we will focus on a generic bimolecular reaction

$$A + B \xrightarrow{k} \text{product(s)}$$

which proceeds at the rate

$$v = k[\mathrm{A}][\mathrm{B}].$$

You will likely notice that the theory of diffusion-influenced reactions described below has many similarities with collision theory from gas-phase kinetics. This is no coincidence given that in both theories we need to focus on the rate at which molecules encounter each other.

For simplicity, we will assume spherical molecules. This simplifies the description of an encounter pair, which will be assumed to be in contact when their centers are a distance $R_{\mathrm{AB}} = R_{\mathrm{A}} + R_{\mathrm{B}}$ from each other. We focus on one particular, stationary A molecule and assume the solution is sufficiently dilute that the distribution of B molecules around one A molecule does not affect the distribution around the others. We can compensate for treating our A molecule as stationary by replacing the diffusion coefficient of the mobile molecules B by the relative diffusion coefficient

$$D_{\mathrm{AB}} = D_{\mathrm{A}} + D_{\mathrm{B}}.$$

If we 'turn on' the reaction at $t = 0$, some B molecules may already be at a distance R_{AB} from A. This leads to time-dependent rate constants (see [1], section 18.3). This effect is short-lived (nanoseconds), so we focus instead on the steady regime that follows this transient. Most experiments, especially those that involve mixing, can't be thought of as having a reaction that suddenly turns on at a specific moment in time anyway. It may, however, be possible to observe this effect in some ultra-rapid flash photolysis experiments.

If we assume that the concentration of encounter pairs is in a steady state, the rate at which B molecules reach a distance R_{AB} from the center of an A molecule balances the rate of reaction. Of course, a reaction removes our A molecule, so that conceptually, what we are doing is either replacing the lost A molecule immediately by another or, if you prefer, shifting our attention to another A molecule once the molecule we were initially watching has reacted.

Now consider the quantity $v/[\mathrm{A}] = k[\mathrm{B}]$, which has units of (molecules) s^{-1} and is the rate of reaction per molecule of A. The flux at $r = R_{\mathrm{AB}}$ is the rate of arrival of molecules of B at a sphere of radius R_{AB} centered on a given A molecule per unit area. Therefore, the steady-state balance between arrival at $r = R_{\mathrm{AB}}$ and reaction can be written

$$v/[\mathrm{A}] = k[\mathrm{B}] = -4\pi R_{\mathrm{AB}}^2 L J_{\mathrm{B}}(r = R_{\mathrm{AB}}).$$

In this equation, $4\pi R_{AB}^2$ is the area of a sphere of radius R_{AB}. This area times the flux of B molecules through this surface, $J_B(r = R_{AB})$, is therefore the number of moles of B arriving at $r = R_{AB}$ per unit time. The factor of L converts this molar flux to a flux of molecules. The units of the right-hand side of this equation are therefore the same as those of the left, i.e. (molecules) s^{-1}.

When a steady state is reached, the concentration of B in any given region of space should be constant. Because molecules of B react at $r = R_{AB}$, the net flux is directed inward towards A, where reaction depletes the concentration of B molecules. If we imagine two concentric spheres of arbitrary radii r_1 and r_2, this means that the number of molecules entering the space between these spheres through the outer sphere must be matched by the number of molecules leaving this space through the inner sphere (figure 11.5). Since there is nothing special about r_1 and r_2, the rate at which molecules pass through a sphere of any r must be the same. Therefore,

$$k[\mathrm{B}] = -4\pi r^2 L J_B(r).$$

The steady-state distribution of B around a given A should be (on average) spherical. Thus,

$$J_B = -\left[D_{AB}\frac{d[\mathrm{B}]_r}{dr} + \frac{z_B e}{k_B T}D_{AB}[\mathrm{B}]_r\frac{dV}{dr}\right]$$

where $[\mathrm{B}]_r$ is the concentration of B at distance r from an A molecule. Using $U(r) = z_B e V(r)$, we can also write

$$J_B = -D_{AB}\left[\frac{d[\mathrm{B}]_r}{dr} + \frac{1}{k_B T}[\mathrm{B}]_r\frac{dU}{dr}\right].$$

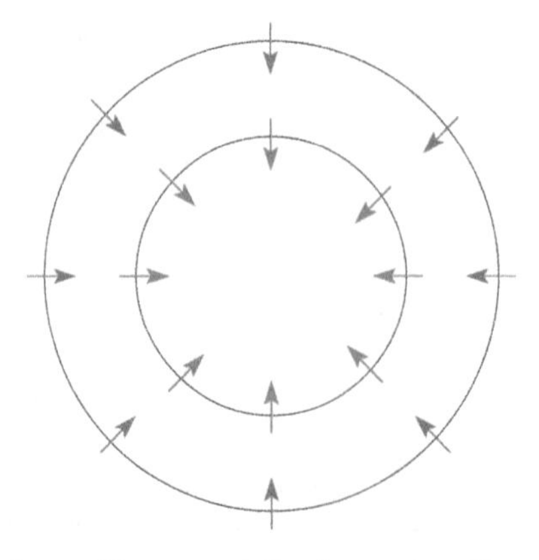

Figure 11.5. Two concentric spheres with representative flux vectors. Since concentration is a number per volume, in order for the concentration to remain constant in the shell between the two spheres once the system has reached a steady state, the number of molecules entering the outer sphere must exactly balance the number of molecules leaving the shell through the inner sphere.

Using this expression for the flux, the steady-state condition becomes

$$k[\mathrm{B}] = 4\pi r^2 L D_{\mathrm{AB}} \left[\frac{\mathrm{d}[\mathrm{B}]_{\mathrm{r}}}{\mathrm{d}r} + \frac{1}{k_{\mathrm{B}}T}[\mathrm{B}]_{\mathrm{r}} \frac{\mathrm{d}U}{\mathrm{d}r} \right]. \tag{11.15}$$

The $k[\mathrm{B}]$ term on the left-hand side derives from the rate law for the reaction. When we write a rate law, a concentration such as [B] represents the average concentration in the solution. This is also the concentration we expect to find far from any given A molecule. In other words,

$$\lim_{r\to\infty}[\mathrm{B}]_{\mathrm{r}} = [\mathrm{B}].$$

We will now employ a trick. Using the chain rule, we can show that

$$\frac{\mathrm{d}}{\mathrm{d}r}\left[[\mathrm{B}]_{\mathrm{r}} \exp\left(\frac{U(r)}{k_{\mathrm{B}}T}\right)\right] = \exp\left(\frac{U(r)}{k_{\mathrm{B}}T}\right)\left[\frac{\mathrm{d}[\mathrm{B}]_{\mathrm{r}}}{\mathrm{d}r} + \frac{1}{k_{\mathrm{B}}T}[\mathrm{B}]_{\mathrm{r}}\frac{\mathrm{d}U}{\mathrm{d}r}\right]. \tag{11.16}$$

Note that the quantity in brackets on the right-hand side of this equation is exactly the same as the quantity in brackets in equation (11.15). We can therefore use equation (11.16) to eliminate the bracketed term from equation (11.15). We get

$$k[\mathrm{B}] = 4\pi r^2 L D_{\mathrm{AB}} \exp\left(-\frac{U(r)}{k_{\mathrm{B}}T}\right)\frac{\mathrm{d}}{\mathrm{d}r}\left[[\mathrm{B}]_{\mathrm{r}} \exp\left(\frac{U(r)}{k_{\mathrm{B}}T}\right)\right].$$

This can be rearranged to

$$\frac{\mathrm{d}}{\mathrm{d}r}\left[[\mathrm{B}]_{\mathrm{r}} \exp\left(\frac{U(r)}{k_{\mathrm{B}}T}\right)\right] = \frac{k[\mathrm{B}]}{4\pi r^2 L D_{\mathrm{AB}}} \exp\left(\frac{U(r)}{k_{\mathrm{B}}T}\right).$$

We can solve this equation by separation of variables subject to the boundary conditions $[\mathrm{B}]_{\mathrm{r}} \to [\mathrm{B}]$ and $U(r) \to 0$ as $r \to \infty$:

$$\int_{r=R_{\mathrm{AB}}}^{\infty} \mathrm{d}\left[[\mathrm{B}]_{\mathrm{r}} \exp\left(\frac{U(r)}{k_{\mathrm{B}}T}\right)\right] = \int_{R_{\mathrm{AB}}}^{\infty} \frac{k[\mathrm{B}]}{4\pi r^2 L D_{\mathrm{AB}}} \exp\left(\frac{U(r)}{k_{\mathrm{B}}T}\right) \mathrm{d}r,$$

$$\therefore \left[[\mathrm{B}]_{\mathrm{r}} \exp\left(\frac{U(r)}{k_{\mathrm{B}}T}\right)\right]_{R_{\mathrm{AB}}}^{\infty} = \frac{k[\mathrm{B}]}{4\pi L D_{\mathrm{AB}}} \int_{R_{\mathrm{AB}}}^{\infty} \frac{1}{r^2} \exp\left(\frac{U(r)}{k_{\mathrm{B}}T}\right) \mathrm{d}r.$$

$$\therefore [\mathrm{B}] - [\mathrm{B}]_{R_{\mathrm{AB}}} \exp\left(\frac{U(R_{\mathrm{AB}})}{k_{\mathrm{B}}T}\right) = \frac{k[\mathrm{B}]}{4\pi L D_{\mathrm{AB}}\beta} \tag{11.17}$$

where

$$\beta^{-1} = \int_{R_{\mathrm{AB}}}^{\infty} \frac{1}{r^2} \exp\left(\frac{U(r)}{k_{\mathrm{B}}T}\right) \mathrm{d}r.$$

We now assume that there is an intrinsic rate constant, k_{R}, for reaction between A and B when the two are in contact such that

$$v = k_{\mathrm{R}}[\mathrm{A}][\mathrm{B}]_{R_{\mathrm{AB}}} = k[\mathrm{A}][\mathrm{B}]. \tag{11.18}$$

By comparing the two expressions for v, we get

$$[\mathrm{B}]_{R_{\mathrm{AB}}} = k[\mathrm{B}]/k_{\mathrm{R}}.$$

Our next step is to substitute for $[\mathrm{B}]_{R_{\mathrm{AB}}}$ in equation (11.17) and solve for k. The result is

$$k = \frac{4\pi L D_{\mathrm{AB}}\beta k_{\mathrm{R}}}{k_{\mathrm{R}} + 4\pi L D_{\mathrm{AB}}\beta \exp\left(\dfrac{U(R_{\mathrm{AB}})}{k_{\mathrm{B}}T}\right)}. \tag{11.19}$$

This is the key equation of this version of the theory of diffusion-influenced reactions. It allows us to calculate the rate constant for a bimolecular reaction in solution, provided we know the intrinsic rate constant k_{R}, the potential energy of an A–B pair $U(r)$, and the diffusion coefficients and radii of A and B. (If necessary, we can get by with just the diffusion coefficients or just the radii, since we can use the Stokes–Einstein equation to estimate a D from an R or vice versa.)

Now suppose that k_{R} is very large, i.e. that A and B react nearly every time they meet in solution. We say that such a reaction is **diffusion controlled**. Turning to equation (11.19), if k_{R} is very large, the rate constant reduces to

$$k \approx 4\pi L D_{\mathrm{AB}}\beta \equiv k_{\mathrm{D}}. \tag{11.20}$$

k_{D} is the **diffusion-controlled rate constant**. The diffusion-influenced rate constant can consequently be written

$$k = \frac{k_{\mathrm{D}}k_{\mathrm{R}}}{k_{\mathrm{R}} + k_{\mathrm{D}} \exp\left(\dfrac{U(R_{\mathrm{AB}})}{k_{\mathrm{B}}T}\right)}. \tag{11.21}$$

We see now why we speak of 'diffusion-influenced reactions.' The rate constant depends both on a pure reaction-rate constant, k_{R} and on the diffusion-controlled rate constant k_{D}. We just saw that if k_{R} is very large, $k \to k_{\mathrm{D}}$, i.e. diffusion dominates the rate of reaction. If, on the other hand, k_{R} is very small, then

$$k \approx k_{\mathrm{R}} \exp\left(\frac{-U(R_{\mathrm{AB}})}{k_{\mathrm{B}}T}\right),$$

where we recognize the Boltzmann factor that arises from the steady-state balance between the electrostatic interaction and diffusion. (Compare equation (11.14).) Diffusion therefore plays a role in the rates of reaction in solution regardless of the regime we consider.

Let us now consider a different limit, that of weak intermolecular forces. Here, $U(r) \approx 0$, except when A and B are very close, and

$$\beta^{-1} = \int_{R_{\mathrm{AB}}}^{\infty} \frac{1}{r^2} \exp\left(\frac{U(r)}{k_{\mathrm{B}}T}\right) \mathrm{d}r \approx \int_{R_{\mathrm{AB}}}^{\infty} \frac{1}{r^2}\, \mathrm{d}r = \frac{1}{R_{\mathrm{AB}}},$$

or $\beta = R_{\mathrm{AB}}$. The diffusion-controlled rate constant becomes

$$k_D = 4\pi L D_{AB} R_{AB} \tag{11.22}$$

and the diffusion-influenced rate constant is simply

$$k = \frac{k_D k_R}{k_R + k_D}.$$

This is the solution-phase equivalent of hard-sphere reactive scattering theory.

We can also consider the Coulomb interaction between two ions:

$$U(r) = \frac{z_A z_B e^2}{4\pi\varepsilon r}$$

where ε is the permittivity of the solvent[7]. For this potential, we get

$$\beta = \frac{z_A z_B e^2}{4\pi\varepsilon k_B T\left[\exp\left(\frac{z_A z_B e^2}{4\pi\varepsilon k_B T R_{AB}}\right) - 1\right]}.$$

If we divide both sides of this equation by R_{AB}, we get

$$\frac{\beta}{R_{AB}} = \frac{\zeta}{e^{\zeta} - 1},$$

where

$$\zeta = \frac{z_A z_B e^2}{4\pi\varepsilon k_B T R_{AB}}.$$

If you substitute in some typical numbers, you will find that $|\zeta|$ tends to be larger than one and can be moderately large if the charges are greater than one. Now consider the case $z_A z_B > 0$ (charges of the same sign). Because ζ is largish,

$$\frac{\beta}{R_{AB}} \approx \frac{\zeta}{e^{\zeta}},$$

which will be a small quantity. Thus, β will be smaller than R_{AB}. Comparing the general equation for k_D, equation (11.20), to the weak-forces limit (11.22), we see that a repulsive interaction reduces the value of k_D compared to the situation in which there is no interaction. Conversely, if ζ is relatively large and negative,

$$\frac{\beta}{R_{AB}} \approx -\zeta,$$

so β will be larger than R_{AB}. As we would expect, attractive forces accelerate diffusional encounters.

[7] Depending on the textbook you studied, you may have seen this equation in slightly different forms. In some textbooks, you will see $\varepsilon_r\varepsilon_0$ instead of ε, where ε_r is the relative permittivity. In one or two books I have seen, people write $\varepsilon\varepsilon_0$, so the relative permittivity is symbolized by a bare ε, which can be confusing. Some books have tried to break out of all of these ε's by using κ for the relative permittivity. I use ε only for the permittivity, and ε_r for the relative permittivity, both of which follow IUPAC recommendations.

11.4.2 Version 2: based on classical kinetics methods

Let us separate an elementary bimolecular reaction into two steps: (i) the formation of the encounter pair, denoted by {AB}, and (ii) the reaction itself:

$$\mathrm{A} + \mathrm{B} \underset{k_{-D}}{\overset{k_D}{\rightleftharpoons}} \{\mathrm{AB}\} \overset{k_r}{\rightarrow} \text{product}(s).$$

It is important to understand that the encounter pair is neither a chemical intermediate nor a transition state. It is simply a pair of molecules that have come into contact through diffusion. The encounter pair has a very short lifetime (to be estimated later), which makes it a perfect target for application of the steady-state approximation:

$$\frac{\mathrm{d}[\{\mathrm{AB}\}]}{\mathrm{d}t} = k_{\mathrm{D}}[\mathrm{A}][\mathrm{B}] - (k_{-\mathrm{D}} + k_{\mathrm{r}})[\{\mathrm{AB}\}] \approx 0.$$

$$\therefore [\{\mathrm{AB}\}] \approx \frac{k_{\mathrm{D}}}{k_{-\mathrm{D}} + k_{\mathrm{r}}}[\mathrm{A}][\mathrm{B}].$$

$$\therefore v = k_{\mathrm{r}}[\{\mathrm{AB}\}] \approx \frac{k_{\mathrm{D}}k_{\mathrm{r}}}{k_{-\mathrm{D}} + k_{\mathrm{r}}}[\mathrm{A}][\mathrm{B}].$$

By comparison with the rate law for the elementary reaction, we conclude that the mass-action rate constant is given by

$$k = \frac{k_{\mathrm{D}}k_{\mathrm{r}}}{k_{-\mathrm{D}} + k_{\mathrm{r}}}. \tag{11.23}$$

This is the key equation of this version of the theory of diffusion-influenced reactions. It has many of the same features as the previous version of the theory. If we consider the limit of large k_{r}, k tends to the limit k_{D}, i.e. the reaction is diffusion controlled. If, on the other hand, k_{r} is small, we get

$$k \to k_{\mathrm{r}}\frac{k_{\mathrm{D}}}{k_{-\mathrm{D}}} \equiv k_{\mathrm{r}}K_{\mathrm{D}}, \tag{11.24}$$

where K_{D} is the equilibrium constant for formation of the encounter pair[8]. The case in which k_{r} is small is described as **activation controlled**. In this case, because $k_{-\mathrm{D}} \gg k_{\mathrm{r}}$, the encounter complex is formed and dissolved many times before a reaction can occur.

While the two theories have some similarities, there are both obvious and not-so-obvious differences as well, to wit:

- In version 1 of the theory, we obtained equation (11.21), which gives the rate constant in terms of a *second-order* rate constant k_{R} for the reaction of A with molecules of B at a center-to-center distance of R_{AB}. (See equation (11.18), which defines k_{R}.)
- We now have equation (11.23), which involves a *first-order* rate constant k_{r} for the formation of products from an encounter pair.

[8] For those of you with backgrounds in biochemistry, this is *not* a dissociation constant.

These two theories provide different ways of describing the same thing, so there must be a relationship between them. Let us set the two equations for k equal to each other and work from there. Note that k_D has essentially the same meaning in both theories.

$$\frac{k_D k_R}{k_R + k_D \exp\left(\frac{U(R_{AB})}{k_B T}\right)} = \frac{k_D k_r}{k_{-D} + k_r}.$$

$$\therefore \frac{1}{1 + \frac{k_D}{k_R}\exp\left(\frac{U(R_{AB})}{k_B T}\right)} = \frac{1}{1 + k_{-D}/k_r}.$$

$$\therefore k_R = k_r K_D \exp\left(\frac{U(R_{AB})}{k_B T}\right)$$

or

$$k_r = k_R K_D^{-1} \exp\left(-\frac{U(R_{AB})}{k_B T}\right). \tag{11.25}$$

Here, K_D^{-1} has units of concentration and has the interpretation of a characteristic concentration scale for the formation of the encounter pair, in the following sense: suppose that B is the limiting reactant. If $[A] \approx K_D^{-1}$, then approximately half of the B will be found in encounter pairs. (We will see later that K_D^{-1} is likely to be very large compared to reactant concentrations under almost all realistic conditions. That being the case, $[\{AB\}]$ will be a small fraction of the total available B.) If we now compare equations (11.14) and (11.25), we see that k_r is the pseudo-first-order rate constant that would be obtained if we had a Boltzmann distribution of the excess reactant at concentration K_D^{-1} surrounding each molecule of the limiting reactant.

Returning now to the theory being developed in this section, for the special case of molecules that experience similar intermolecular forces with the solvent as with each other, we can estimate K_D by a statistical argument. This also leads to an estimate of k_{-D}, since we know how to calculate k_D.

Focus again on a particular molecule of A. Suppose that the reactive site that must be accessed by B makes contact with $\mathcal{N}$ neighboring molecules in solution. $\mathcal{N}$ is thus the coordination number of this site. Let [S] be the mole density of the solvent. If $[B] \ll [S]$, then the probability that any given molecule of S has been replaced by a B in the first solvation sphere of A is $[B]/[S]$. Under the same assumption, the probability that any one of the $\mathcal{N}$ molecules solvating the reactive site is a molecule of B is $\mathcal{N}[B]/[S]$. Another way to think about it is that $\mathcal{N}[B]/[S]$ is the fraction of A molecules whose reactive sites are in contact with a B molecule, i.e. the fraction of A molecules that have formed an encounter pair with a B. Therefore,

$$[\{AB\}] \approx \mathcal{N}\frac{[B]}{[S]}[A].$$

Since

$$K_D = \frac{[\{AB\}]}{[A][B]},$$

we get

$$K_D = \frac{\mathcal{N}}{[S]}.$$

Example 11.4. *Bromphenol blue (BPB) is a pH indicator. It reacts with hydroxide ions, changing from a blue quinoid form to a colorless carbinol structure:*

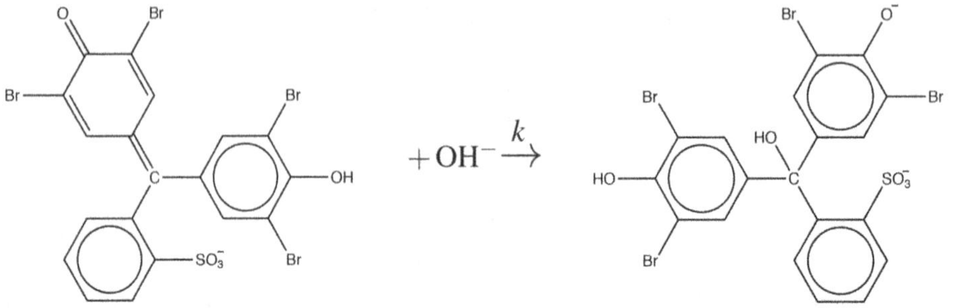

$$+\,OH^- \xrightarrow{k}$$

quinoid (blue) *carbinol (colorless)*

The rate constant for this reaction is $k = 9.30 \times 10^{-4}$ L mol^{-1}s^{-1} *in water at 25 °C. We want to use this experimental value of the rate constant to estimate* k_r *and thus to determine whether this reaction is diffusion controlled or activation controlled.*

Here are some useful data: $D_{OH^-} = 5.30 \times 10^{-9}$ m^{2}s^{-1}, $D_{BPB} = 4.4 \times 10^{-10}$ m^{2}s^{-1}, and $\varepsilon_r(H_2O) = 78.37$ *(all at 25 °C). We also need an estimate of* R_{AB}. *This is the distance at which the two reactants, bromphenol blue and hydroxide, are in contact with each other. Note that the hydroxide reacts at the central carbon. A reasonable estimate of this distance might be the sum of the van der Waals radii of carbon and oxygen. Mantina* et al *give radii of 1.70 and 1.52 Å, respectively for carbon and oxygen [6]. Thus, we have* $R_{AB} = 3.22$Å.

We start by calculating k_D. *The two reactants are both anions with a single negative charge.*

$$\varepsilon = \varepsilon_r\varepsilon_0 = 6.939 \times 10^{-10}\ \mathrm{C^2J^{-1}m^{-1}}.$$

$$\beta = \frac{z_A z_B e^2}{4\pi\varepsilon k_B T\left[\exp\left(\dfrac{z_A z_B e^2}{4\pi\varepsilon k_B T R_{AB}}\right) - 1\right]}$$

$$= 8.704 \times 10^{-11}\ \mathrm{m}.$$

$$\therefore k_D = 4\pi L D_{AB}\beta$$

$$= 4\pi(6.022\,141 \times 10^{23}\ \mathrm{mol^{-1}})[(5.30 + 0.44) \times 10^{-9}\ \mathrm{m^2s^{-1}}] \times (8.704 \times 10^{-11}\ \mathrm{m})$$

$$= 3.78 \times 10^{6}\ \mathrm{m^3mol^{-1}s^{-1}} \equiv 3.78 \times 10^{9}\ \mathrm{L\ mol^{-1}s^{-1}}.$$

Before we go any further, we can already tell that this reaction will be activation controlled, given that the observed rate constant is much, much smaller than k_D.

We also need an estimate of K_D. The quinoid form of bromphenol blue has a trigonal planar geometry around the central carbon, where the reaction with hydroxide takes place. It is likely that two water molecules solvate this site, one from each side, i.e. $\mathcal{N} = 2$. Given the number of electron-withdrawing groups in the vicinity of this carbon, it is likely that it carries a partial positive charge. Thus, we would expect water molecules to present themselves to this site oxygen-first. This is the same approach as that necessary for the hydroxide ion to react here. There is therefore a reasonable similarity between the ways in which hydroxide and water coordinate to this site, which is required for the method described above to work. Since the mole density of water at 25 °C *is* 55.33 mol L^{-1}, *we have*

$$K_D \approx \frac{\mathcal{N}}{[H_2O]} = \frac{2}{55.33 \text{ mol L}^{-1}} = 4 \times 10^{-2} \text{ L mol}^{-1}.$$

Given that we have already decided that the reaction is activation controlled, we can calculate k_r using equation (11.24):

$$k_r = \frac{k}{K_D} = \frac{9.30 \times 10^{-4} \text{ L mol}^{-1}\text{s}^{-1}}{4 \times 10^{-2} \text{ L mol}^{-1}} = 3 \times 10^{-2} \text{ s}^{-1}.$$

Although it wasn't the original objective of this example, it is instructive to calculate k_{-D}:

$$k_{-D} = \frac{k_D}{K_D} = \frac{3.78 \times 10^{9} \text{ L mol}^{-1}\text{s}^{-1}}{4 \times 10^{-2} \text{ L mol}^{-1}} = 1 \times 10^{11} \text{ s}^{-1}.$$

As a way of putting this number into perspective, note that this value of k_{-D} implies a half-life of about 7 ps for the encounter pair.

It is worth thinking about why k_r is so small in this reaction. One way to think about small k_r values is that they imply that the transition state is hard to reach. In this particular reaction, the quinoid reactant and carbinol product have different geometries about the central carbon. In order to reach the transition state, it is likely that the trigonal planar quinoid has to bend a long way out of plane as the hydroxide comes in. Large rearrangements like this one typically have large free energy barriers, hence the small value of the intrinsic rate constant k_r.

Further reading

There is no better reference on diffusion-influenced reactions, to my knowledge, than Stephen Rice's book:

- Rice S A 1985 *Diffusion-Limited Reactions*, vol 25 (*Comprehensive Chemical Kinetics Series*) (Amsterdam: Elsevier)

Exercise

11.1 Copper(II) ions have a mobility in water at 25 °C of 5.56×10^{-8} m^2 s^{-1}V^{-1}.
(a) Calculate the diffusion coefficient of a copper(II) ion in water at this temperature.

(b) The viscosity of water at 25 °C is 8.91×10^{-4} Pa s. Calculate the radius of the ion.

(c) The ionic radius of copper(II) ions obtained from crystallographic data is 72 pm. Does your value agree with this one? If not, why do you think it might differ?

11.2 Typical diffusion coefficients for small molecules in water are of the order of 10^{-9} m^2s^{-1}.

(a) What value of the radius does this diffusion coefficient imply? Assume that the solvent is water at 25 °C.

(b) Calculate the diffusion-limited rate constant for typical neutral molecules in water at 25 °C. Report your final answer in L $mol^{-1}s^{-1}$.

(c) What would the diffusion-limited rate constant be if the two reactants were ions with charges of +1 and −2? The permittivity of water at 25 °C is 6.939×10^{-10} $C^2N^{-1}m^{-2}$.

(d) What if the charges were +1 and +2?

(e) Estimate the half-life at 25 °C of a typical encounter pair for neutral molecules in which one reactant has a coordination number of six. The mole density of water at this temperature is 55.33 mol L^{-1} and its viscosity is 8.91×10^{-4} Pa s.

11.3 In this question, you will compare the collision/diffusion-limited rate constants in the gas phase and in solution. Suppose that we have a bimolecular reaction A + B → products. Assume the two reactants are uncharged hard spheres with the following properties:

	r/Å	M/g mol^{-1}
A	3.8	150
B	5.0	332

Assume the reaction is collision-limited in the gas phase and diffusion-controlled in solution. Predict the rate constants in the gas phase and in ethylene glycol solution for this reaction at 100 °C. At this temperature, the viscosity of ethylene glycol is 1.975 mPa s. In which medium (gas or solution) is the reaction fastest? Is the difference large? Is this what you would have expected?

11.4

(a) Sketch the reaction profile for the recombination of radicals. What is the activation energy?

(b) In the reverse reaction (dissociation into radicals), what is the activation energy? You can either give your answer in words or show it on your reaction profile.

(c) The products in a radical recombination reaction hold excess energy which they need to shed. Briefly explain why, with reference to your potential energy diagram.
(d) How many normal modes of vibration does ethane have?
(e) The radical recombination of methyl radicals (CH_3) into ethane has been thoroughly studied. Once the two radicals have recombined, we have an excited species whose energy exceeds the activation energy, to which RRK theory should apply. Intramolecular vibrational relaxation rapidly moves energy around the molecule. Eventually, this energy should find its way into the bond that has been formed, resulting in the redissociation of the molecule. Which of the rate constants from RRK theory governs the rate of redissociation?
(f) For an RRK treatment of the redissociation of ethane following the recombination of methyl radicals, what would be a sensible value of s?
(g) The carbon–carbon bond dissociation energy in ethane is 377 kJ mol^{-1}, and the carbon–carbon bond stretching frequency is 2.98×10^{13} Hz. For a molecule exceeding the dissociation energy by 20 kJ mol^{-1}, estimate the rate constant for redissociation and then the half-life of the newly formed ethane molecule.
(h) De-excitation often occurs via collision. Suppose that recombination is studied in a helium bath gas. Calculate the collisional rate constants of helium and ethane at 300 K. The van der Waals radii of helium and ethane are 1.4 and 4.0 Å, respectively.
(i) Assuming that collision is 100% efficient in de-excitation, at what concentration of the bath gas would the rates of de-excitation and redissociation be equal? Give your answer in mol L^{-1}. What does your answer tell you about this system?
(j) This reaction can be studied in solution using pulse radiolysis, a technique in which an intense flash of radiation causes the molecules to dissociate. The reassociation of the radicals can then be followed spectroscopically. What is the major difference between this reaction in the gas phase and in solution?
(k) Assuming that the radius of a methyl radical is 2.0 Å (half that of ethane), estimate the diffusion-limited rate constant for the recombination of methyl radicals in water at 300 K. The viscosity of water at this temperature is 8.47×10^{-4} Pa s. The experimental value of the rate constant is $(1.24 \pm 0.20) \times 10^{9}$ L mol^{-1}s^{-1}. Would you say that this reaction is diffusion-limited?

References

[1] Roussel M R 2012 *A Life Scientist's Guide to Physical Chemistry* (Cambridge: Cambridge University Press)
[2] Castets V, Dulos E, Boissonade J and De Kepper P 1990 Experimental evidence of a sustained standing Turing-type nonequilibrium chemical pattern *Phys. Rev. Lett.* **64** 2953–6

[3] Roussel M R 2019 *Nonlinear Dynamics: A Hands-On Introductory Survey (IOP Concise Physics)* (San Rafael, CA: Morgan & Claypool)
[4] Crank J 1956 *The Mathematics of Diffusion* (Oxford: Clarendon)
[5] Tinoco I Jr, Sauer K and Wang J C 1995 *Physical Chemistry: Principles and Applications in Biological Sciences* 3rd edn (Englewood Cliffs, NJ: Prentice-Hall)
[6] Mantina M, Chamberlin A C, Valero R, Cramer C J and Truhlar D G 2009 Consistent van der Waals radii for the whole main group *J. Phys. Chem.* **113** 5806–12

IOP Publishing

Foundations of Chemical Kinetics
A hands-on approach
Marc R Roussel

Chapter 12

Transition-state theory in solution

12.1 Should we be using transition-state theory for reactions in solution?

In the previous chapter, we noted that bimolecular reactions in solution involve solvent caging. The encounter pair is held together (briefly) by a solvent cage. There is a barrier to forming this species and to breaking it up due to the necessary rearrangement of the solvent around A and B. One way to think about the effect of solvent caging is the potential energy profile shown in figure 12.1. This picture of a bimolecular reaction in solution implies the following mechanistic description:

$$\mathrm{A} + \mathrm{B} \overset{K_D}{\rightleftharpoons} \{\mathrm{AB}\} \overset{K_{\mathrm{tot}}^{\ddagger}}{\rightleftharpoons} \mathrm{TS} \xrightarrow{k^{\ddagger}} \text{products}.$$

Note that there is a low-lying transition state (not labeled in the figure) between A + B and {AB}. According to this scheme, we should apply transition-state theory (TST) to the encounter pair and not to the separated reactants, since it is the encounter pair that is separated from the products by the reaction's transition state, and not the separated molecules of A and B. However, consider the following:

$$[\{\mathrm{AB}\}] = K_{\mathrm{D}}[\mathrm{A}][\mathrm{B}]$$

and

$$[\mathrm{TS}] = K_{\mathrm{tot}}^{\ddagger}[\{\mathrm{AB}\}].$$
$$\therefore [\mathrm{TS}] = K_{\mathrm{tot}}^{\ddagger}K_{\mathrm{D}}[\mathrm{A}][\mathrm{B}] = K_{\mathrm{eff}}^{\ddagger}[\mathrm{A}][\mathrm{B}],$$

which defines the effective A + B to TS equilibrium constant $K_{\mathrm{eff}}^{\ddagger}$. Now since

$$K^{\ddagger} = \exp\left(\frac{-\Delta G^{\circ}(\{\mathrm{AB}\} \rightleftharpoons \mathrm{TS})}{RT}\right)$$

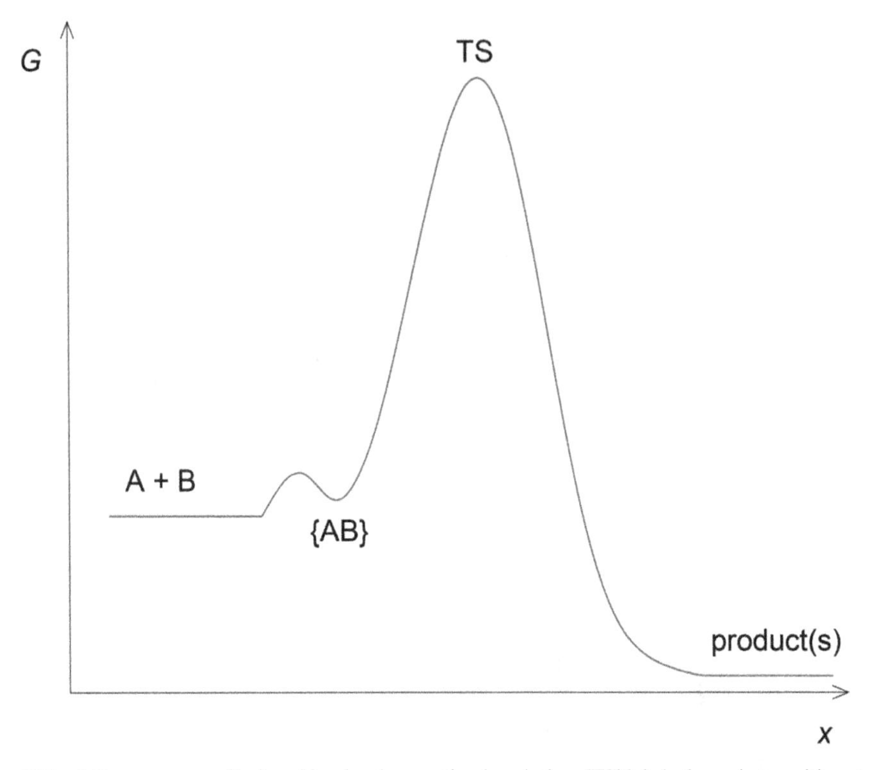

Figure 12.1. A free-energy profile for a bimolecular reaction in solution. 'TS' labels the main transition state of the reaction.

and

$$K_{\mathrm{D}} = \exp\left(\frac{-\Delta G^{\circ}(\mathrm{A} + \mathrm{B} \rightleftharpoons \{\mathrm{AB}\})}{RT}\right),$$

it follows that

$$K_{\mathrm{eff}}^{\ddagger} = \exp\left(\frac{-\Delta G^{\circ}(\mathrm{A} + \mathrm{B} \rightleftharpoons \mathrm{TS})}{RT}\right).$$

The equilibrium constant that connects the reactants to the transition state, $K_{\mathrm{eff}}^{\ddagger}$, is therefore exactly what we would have calculated if we had ignored the encounter pair. This is because TST is an equilibrium theory, and 'intermediate species' (although {AB} is not a chemical intermediate) can always be squeezed out in equilibrium theories. We can therefore apply TST to reactions in solution as if the encounter pair did not exist.

12.2 Rate constants in solution according to transition-state theory

It is possible to perform TST calculations for reactions in solution, but it's not as straightforward as in the gas phase. The problem is that because the solvent is a condensed phase in which the solvent molecules are constantly in contact with the solutes, we really need to think about a system that includes the solvent as well as the reactants. The problem is illustrated in figure 12.2. The energy of the system (solvent + solutes) is probably extremely similar for the two slightly different configurations of the solvent cage shown in this figure. There may be slightly different solvent–solute and solvent–solvent contacts in these two configurations, but these would have only a small effect on the energy. Consequently, at temperatures at which the solvent is a liquid, the system would sample a very large number of configurations of similar energy. Instead of a nice simple potential energy surface (PES), we now have to think about a very high-dimensional PES that includes solvent degrees of freedom and corresponding deformations in the solutes. In addition, this PES has a very large number of shallow minima—picture something like an egg carton, figure 12.3 but much less regular in structure—with the system hopping between minima at high frequency. In other words, the potential energy profile shown in figure 12.1 doesn't convey the full complexity of a reaction in solution. Accordingly, the simple approach to calculating rate constants in TST seen in section 7.2 becomes

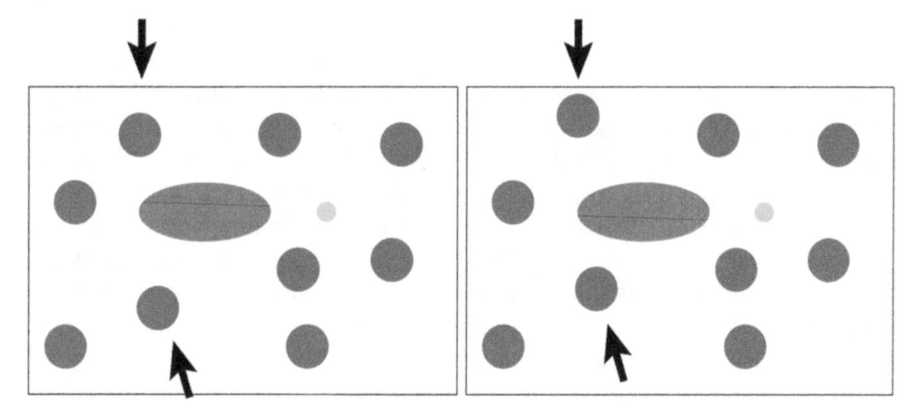

Figure 12.2. Two configurations of the solvent creating a solvent cage around two solutes. The solvent molecules are represented by blue circles, and the two solutes are represented by the red ellipse and the small green circle, respectively. The two configurations are identical except that two solvent molecules, pointed to by arrows, have moved.

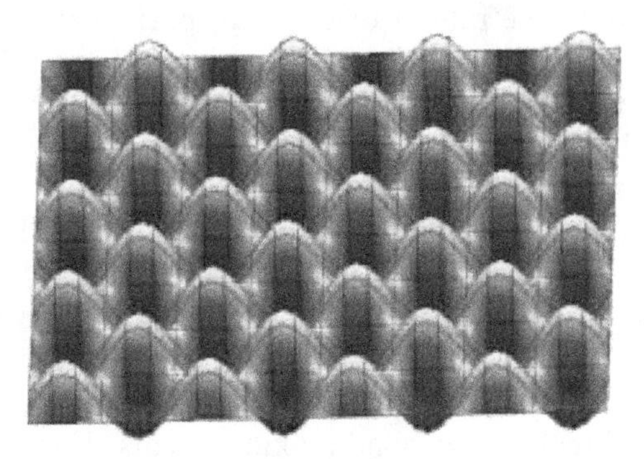

Figure 12.3. An 'egg carton' potential.

problematic because there is no single configuration of solvent and solutes that we can use to calculate the necessary partition functions, and, in any event, it would be a huge undertaking to calculate a partition function for a system with many solvent molecules.

The thermodynamic formalism of TST now comes to our rescue. It gave us equation (7.11a) for first-order reactions, or

$$k_0^{(2)} = \frac{k_B T}{hc^\circ} e^{-\Delta^\ddagger G^\circ/RT} \tag{12.1}$$

for second-order reactions. The subscripted nought indicates that this is the ideal-solution rate constant. (Nonideal effects will be discussed starting in section 12.3.) Molecular dynamics simulations can sample representative conformations of the solvent and solutes, from which we can obtain free energies. The basic idea is to start with a reaction coordinate. Any reasonable coordinate will do, including simple geometric coordinates (e.g. the difference between the length of a bond to be made and a bond to be broken) or the distance along the intrinsic reaction coordinate calculated from a gas-phase calculation. The system is constrained to a sequence of values of the reaction coordinate that takes the system from the reactant to the product state, including, of course, the classical (saddle-point) transition state. The free energy is calculated by sampling conformations of the remaining degrees of freedom, including the solvent degrees of freedom. This gives us a free-energy profile of the reaction. We can then apply the same kinds of ideas that we saw previously for gas-phase TST but using the free-energy profile: we can obtain a variational TST by setting a transition-state dividing surface across the reaction path in such a way as to minimize the computed rate constant. And we can compute tunneling corrections from the sampled paths. Although there are many more details to these calculations, roughly speaking, this lets us calculate $\Delta^\ddagger G^\circ$. This is, as you can imagine, an extremely compute-intensive process that likely requires computer power beyond what is available to you in a course, so we won't try it ourselves.

Note that for larger molecules, sampling important conformations of the reactants may require some additional knowledge, which would be used to start the simulations in different conformations. In an ideal world, we would try many different initial conditions, and use the free-energy calculations to decide which ones contribute significantly to the rate of reaction. In particular, enzymes are not only large, they tend to be somewhat flexible. If experimental work has suggested that a flexible loop might be important for binding or catalysis, it might be necessary to sample different conformations of that loop to find the conformations that facilitate binding, accommodation of the substrate, the actual catalytic event, and possibly also product dissociation. Expertise in biochemistry is clearly important in such calculations, along with knowledge of TST.

As a rule, calculations such as those described above use a hybrid method that combines quantum mechanics for the parts of the system directly involved in the reaction and classical mechanics for those parts that are outside the immediate vicinity of any bond-breaking or bond-making events. For an enzyme-catalyzed reaction, for example, the solvent and most of the enzyme might be treated with a fast classical molecular dynamics method, while the enzyme's active site and the reactant(s) would be treated quantum mechanically.

12.3 What transition-state theory tells us about reactions in solution

Applying TST directly to reactions in solution is tricky and requires significant computer power. However, TST provide a framework that yields insights into the factors that affect the rates of reactions in solution and even allows us to rationalize, sometimes even to predict, the effects of changes in the reaction medium on the rate of reaction.

We start with a second look at the thermodynamic formalism of TST first seen in section 7.1, this time focusing on the appearance of activity coefficients in the TST equations, which we neglected in our study of gas-phase kinetics.

TST decomposes a second-order reaction $A + B \xrightarrow{k^{(2)}}$ products into

$$A + B \overset{K^{\ddagger}_{\text{tot}}}{\rightleftharpoons} \text{TS} \xrightarrow{k^{\ddagger}} \text{products},$$

such that $v = k^{\ddagger}[\text{TS}]$. (Recall that we use $K^{\ddagger}_{\text{tot}}$ because, at this stage, this is the 'total' equilibrium constant including all modes of the transition state. Eventually, the reactive mode will be extracted from this equilibrium constant.) When the transition state is in equilibrium with the reactants, we have

$$K^{\ddagger}_{\text{tot}} = \frac{a_{\text{TS}}}{a_{\text{A}}\, a_{\text{B}}} = \frac{\gamma_{\text{TS}}}{\gamma_{\text{A}}\, \gamma_{\text{B}}} \frac{[\text{TS}]c^{\circ}}{[\text{A}][\text{B}]}.$$

We use this equation to isolate [TS], which is then substituted into the rate equation:

$$[\text{TS}] = \frac{K^{\ddagger}_{\text{tot}}}{c^{\circ}} \frac{\gamma_{\text{A}}\, \gamma_{\text{B}}}{\gamma_{\text{TS}}} [\text{A}][\text{B}].$$

$$\therefore\ v = k^{\ddagger} \frac{K^{\ddagger}_{\text{tot}}}{c^{\circ}} \frac{\gamma_{\text{A}}\, \gamma_{\text{B}}}{\gamma_{\text{TS}}} [\text{A}][\text{B}].$$

$$\therefore\ k^{(2)} = k^{\ddagger} \frac{K^{\ddagger}_{\text{tot}}}{c^{\circ}} \frac{\gamma_{\text{A}}\, \gamma_{\text{B}}}{\gamma_{\text{TS}}}.$$

For an ideal solution, all of the activity coefficients are equal to unity, and we get the limiting value

$$k_0^{(2)} = k^{\ddagger} \frac{K^{\ddagger}_{\text{tot}}}{c^{\circ}} = \frac{k_{\text{B}}T}{hc^{\circ}} K^{\ddagger}.$$

For emphasis: $k_0^{(2)}$ is the rate constant measured (or calculated) under ideal conditions, which is to say a rate constant measured in a dilute solution or, better still, extrapolated to dilute conditions following measurements in which we vary the concentrations of reactants. Thus $k_0^{(2)}$ and $k^{(2)}$ are related by

$$k^{(2)} = k_0^{(2)} \frac{\gamma_A \gamma_B}{\gamma_{TS}}$$

under nonideal conditions. This equation tells us how to incorporate nonideal effects into TST.

We can repeat this derivation for a first-order reaction R → product(s). In this case, we get

$$k^{(1)} = k_0^{(1)} \frac{\gamma_R}{\gamma_{TS}}. \tag{12.2}$$

Before proceeding further, let's think about the factors that might affect the values of the activity coefficients of a solute. Solutes behave ideally when they are in dilute solution, i.e. when the forces between solutes are negligible. So what affects the importance of forces between the solutes?

- The concentration is clearly important because it determines the typical distances between solute molecules and how often two solute molecules are likely to meet.
- The nature of the forces between the solutes is important. Some forces act only at short range, while others act at significant distances. Some forces are very strong, others much weaker. For example, hydrogen bonding only acts at a short distance, so hydrogen bonding between species involved in the reaction is only likely to be important at very high solute concentrations; on the other hand, the forces due to hydrogen bonding are relatively strong, so when these interactions start to become more important, they are likely to cause a very rapid change in the activity coefficients. The screened Coulomb force between ions, however, is both strong and acts over larger distances, so nonideal behavior is expected at relatively low concentrations in electrolyte solutions.
- The sizes of the solutes also matter, although the dependence on size can be complicated. For larger solutes (e.g. polymers), the main consideration is that solute–solute interactions become more frequent as the fraction of the volume occupied by the solutes increases. However, other effects can come into play. In Debye–Hückel theory, for example, the fact that ions occupy space leads to smaller deviations from ideal behavior with increasing ionic radii, in essence because this increases the minimum distance between ionic centers.

12.3.1 Unimolecular reactions

Suppose we are studying a unimolecular reaction. Given that the transition state has to contain all of the same atoms as the reactant, we would expect the forces between the reactant and other solutes to be about the same as the forces between the transition state and other solutes. This is not guaranteed, since the transition state could, for example, have a dipole moment that is very different from that of the reactant, but this only happens in some reactions. Moreover, except for large-scale rearrangements of macromolecules, which arguably are not elementary processes in the usual sense of the word, we don't expect much of a change in the volume occupied as a molecule transits towards its transition state. This line of reasoning implies that, in most cases, $\gamma_{TS} \approx \gamma_R$. Equation (12.2) then predicts that the rate constant should depend very weakly on the factors that affect the activity coefficients (ionic strength, cosolute concentrations, etc).

Now imagine that we slowly decrease the density of the solvent, holding the solute concentration, temperature, etc constant. (This is, obviously, a gedanken experiment.) Eventually, we would no longer describe the reaction as occurring in solution but in the gas phase. Throughout this operation, the argument given above would continue to hold, and we would therefore expect very little change in the ratio of γ_R/γ_{TS}, which should remain close to unity throughout. The solvent might have effects on $k_0^{(1)}$, among other things, because of the dielectric constant of the solvent, whose effects we will discuss later. Subtle effects associated with the reorganization of the solvent around the molecule as it moves to the transition state can also affect the value of $k_0^{(1)}$. These effects should be modest given that, as a rule, most of the work necessary to reach the transition state involves intramolecular interactions that should be relatively little affected by the solvent. The surprising conclusion is that *rate constants for unimolecular reactions in solution and in the gas phase should be about the same.*

There is one caveat to this: recalling our discussion of gas-phase unimolecular reactions in chapter 8, we know that the effective first-order rate constant depends on the pressure. In solution, molecules are constantly exchanging energy with the solvent, which is the equivalent of the bath gas. Thus, the rate constant in solution should be equivalent to the high-pressure limit of the gas-phase rate constant.

Example 12.1. *Branton and coworkers studied the following reaction in both the gas phase and in dimethyl phthalate solution* [1]:

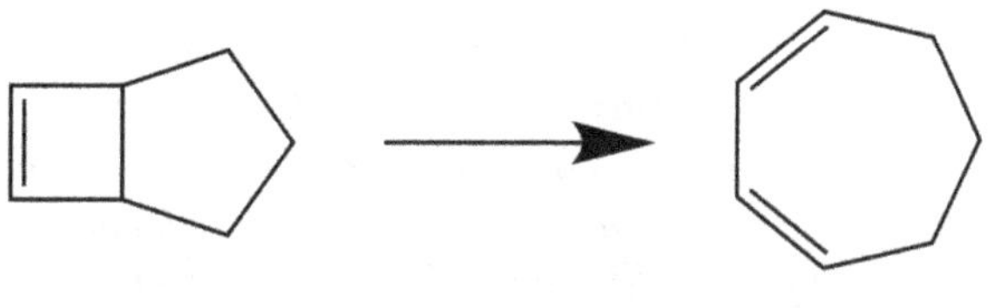

(12.3)

The results of their experiments, presented as an Arrhenius plot, are shown in figure 12.4. *It is obvious before doing any further work that the reaction has essentially the same activation energy (the same slope of the Arrhenius plot) in both cases. The rate of reaction is slightly larger in solution, which results in a somewhat larger pre-exponential factor (related to the intercept) in solution than in the gas phase.*

Table 12.1 *shows the Arrhenius parameters and the entropy of activation computed from the Arrhenius plots for this reaction. The entropy of activation was computed from the equation*

$$A = \frac{k_B T}{h} e^{1+\Delta^{\ddagger} S^{\circ}/R},$$

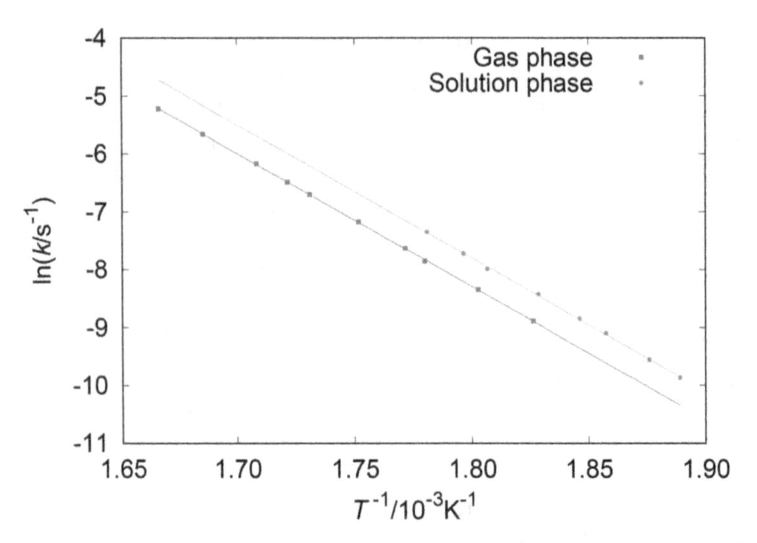

Figure 12.4. Kinetic data produced by Branton and coworkers [1] for reaction (12.3).

Table 12.1. Arrhenius parameters and the entropy of activation corresponding to the Arrhenius plots in figure 12.4. The entropy of activation was calculated assuming a temperature of 550 K, roughly the middle of the range over which the gas-phase and solution-phase data overlap.

	E_a/kJ mol^{-1}	A/10^{14}s^{-1}	$\Delta^{\ddagger}S°$/J K^{-1}mol^{-1}
Gas	191.1 ± 0.6	2.28 ± 0.30	16.55 ± 0.13
Solution	191.3 ± 2.0	4.0 ± 1.8	21.2 ± 0.5

which is valid for unimolecular gas-phase reactions as well as for reactions in solution. As noted above, there is no significant difference in the activation energies in the gas phase and in solution. There is a larger difference in the pre-exponential factors due to a small difference in the entropies of activation. (Differences between gas and solution due to different enthalpies of solvation of the reactant and transition state would affect the activation energy.) Since the difference is only apparent in the entropy of activation, this suggests that solvent reorganization may be largely responsible for the differences between the gas-phase and solution rate constants.

The small differences observed in this example do not negate the fact that the rate constants in solution and in the gas phase are comparable for this reaction.

Let's consider for a moment an enzyme-catalyzed reaction, say one that forms a new chemical bond. The catalysis step might be represented as EAB → EP, where A and B are the substrates (reactants) and P is the product. The arguments presented in this section suggest that gas-phase calculations might do a reasonable job of estimating the rate constant for this step *provided* the enzyme is reasonably rigid. If the enzyme undergoes large-amplitude motions during the catalysis step, then reaching the transition state might involve a large change in the activity coefficient.

If the enzyme is flexible, we're back to the problem of the egg carton potential of section 12.2. Either way, we would have to deploy the machinery described in section 12.2 to calculate a rate constant. Most enzymes are at least somewhat flexible, but there are some that are essentially rigid, sometimes known by biochemists as 'rocks' (although the latter term sometimes has other meanings).

12.3.2 The effect of solvation

Example 12.1 shows that solvation, while it may not always lead to dramatic effects, does have some effects on reactions in solution vs the gas phase. These effects are probably larger for bimolecular reactions than for unimolecular reactions, so we turn our attention to the former. Equation (12.1) gives us the TST rate constant for a second-order reaction in the thermodynamic formulation. This equation should apply both in solution and in the gas phase. We use subscripts to differentiate between reactions in these two phases: $k_{0,\mathrm{s}} \leftrightarrow \Delta^{\ddagger}G_{\mathrm{s}}^{\circ}$ for reactions in solution and $k_{0,\mathrm{g}} \leftrightarrow \Delta^{\ddagger}G_{\mathrm{g}}^{\circ}$ for reactions in the gas phase. We would like to determine how these rate constants are related for a given reaction.

The standard way to compare reactions in solution and in the gas phase is to develop a thermodynamic cycle. In TST, since the rate constant depends on the properties of the transition state, we consider the following cycle:

$$\begin{array}{ccc} \mathrm{A}_{(\mathrm{sol})} + \mathrm{B}_{(\mathrm{sol})} & \xrightarrow{\Delta^{\ddagger}G_s^{\circ}} & \mathrm{TS}_{(\mathrm{sol})} \\ {\scriptstyle -\Delta_{\mathrm{solv}}G^{\circ}(\mathrm{A}) - \Delta_{\mathrm{solv}}G^{\circ}(\mathrm{B})}\downarrow & & \uparrow{\scriptstyle \Delta_{\mathrm{solv}}G^{\circ}(\mathrm{TS})} \\ \mathrm{A}_{(\mathrm{g})} + \mathrm{B}_{(\mathrm{g})} & \xrightarrow{\Delta^{\ddagger}G_g^{\circ}} & \mathrm{TS}_{(\mathrm{g})} \end{array}$$

In this cycle, $\Delta_{\mathrm{solv}}G^{\circ}$ is the standard free energy of solvation of a particular species. We can read the relationship between $\Delta^{\ddagger}G_s^{\circ}$ and $\Delta^{\ddagger}G_g^{\circ}$ directly from this cycle:

$$\begin{aligned} \Delta^{\ddagger}G_s^{\circ} &= \Delta^{\ddagger}G_g^{\circ} + \Delta_{\mathrm{solv}}G^{\circ}(\mathrm{TS}) - [\Delta_{\mathrm{solv}}G^{\circ}(\mathrm{A}) + \Delta_{\mathrm{solv}}G^{\circ}(\mathrm{B})] \\ &= \Delta^{\ddagger}G_g^{\circ} + \Delta^{\ddagger}\Delta_{\mathrm{solv}}G^{\circ} \end{aligned}$$

where

$$\Delta^{\ddagger}\Delta_{\mathrm{solv}}G^{\circ} = \Delta_{\mathrm{solv}}G^{\circ}(\mathrm{TS}) - [\Delta_{\mathrm{solv}}G^{\circ}(\mathrm{A}) + \Delta_{\mathrm{solv}}G^{\circ}(\mathrm{B})]. \tag{12.4}$$

If we substitute $\Delta^{\ddagger}G_s^{\circ}$ into equation (12.1), we get

$$k_{0,s} = \frac{k_{\mathrm{B}}T}{c^{\circ}h}\exp\left(-\frac{\Delta^{\ddagger}G_g^{\circ}}{RT}\right)\exp\left(-\frac{\Delta^{\ddagger}\Delta_{\mathrm{solv}}G^{\circ}}{RT}\right)$$

$$\therefore k_{0,s} = k_{0,g}\exp\left(-\frac{\Delta^{\ddagger}\Delta_{\mathrm{solv}}G^{\circ}}{RT}\right). \tag{12.5}$$

We see that the rate constant in solution is related to the gas-phase rate constant very directly by a term that depends on the difference in solvation free energies between the transition state and the reactants. Solvation free energies can be estimated computationally, so equation (12.5) could be used to estimate a solution-phase rate constant given a rate constant in the gas phase, or vice versa. We can also use experimental data to estimate $\Delta^{\ddagger}\Delta_{\text{solv}}G^{\circ}$ from this equation.

Additionally, note that while we derived equation (12.5) by assuming a second-order reaction, the same equation would hold for any order of reaction.

Example 12.2. *For the Diels-Alder reaction of acrolein with cyclopentadiene*

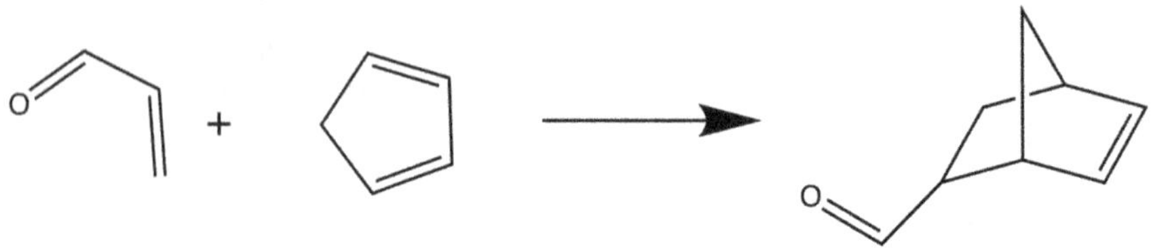

Wassermann measured the following rate constants in the gas phase and in solution [2]:

In the gas phase: $A = (1.6 \pm 1.8) \times 10^6$ L mol^{-1}s^{-1}, $E_a = 64 \pm 4$ kJ mol^{-1}

In benzene: $A = (1.3 \pm 0.9) \times 10^6$ L mol^{-1}s^{-1}, $E_a = 57.3 \pm 2.1$ kJ mol^{-1}

We want to use these data to learn something about the relative solvation of the transition state. To do this, we need to compare the rate constants of the two states at the same temperature. Let's use the boiling point of benzene (353.2 K), which is the highest temperature at which the experiment could be done in the solvent without increasing the pressure. At this temperature,

$$k_{0,s} = 4.4 \times 10^{-3}\ \text{L mol}^{-1}\text{s}^{-1},$$

$$k_{0,g} = 5.5 \times 10^{-4}\ \text{L mol}^{-1}\text{s}^{-1}.$$

We can rearrange equation (12.5) to

$$\begin{aligned}\Delta^{\ddagger}\Delta G^{\circ} &= RT\ln(k_{0,g}/k_{0,s}) \\ &= (8.314\,463\ \text{J K}^{-1}\text{mol}^{-1})(353.2\ \text{K})\ln\left(\frac{5.5\times10^{-4}\ \text{L mol}^{-1}\text{s}^{-1}}{4.4\times10^{-3}\ \text{L mol}^{-1}\text{s}^{-1}}\right) \\ &= -6.1\ \text{kJ mol}^{-1}.\end{aligned}$$

Given the definition of $\Delta^{\ddagger}\Delta G^{\circ}$ (equation (12.4)), this negative value tells us that the transition state is better solvated than the reactants. We could then proceed to ask why that might be by determining the structure of the transition state (e.g. computationally) and then asking how this structure would be solvated compared to the reactants.

12.3.3 The effect of pressure

Equation (11.3) can be used to predict the effect of pressure on a chemical reaction. Imagine applying this equation to each of the reactants and products of a reaction, and then subtracting to obtain $\Delta \mathrm{d}G = \mathrm{d}\Delta G$. This would give us

$$\mathrm{d}\Delta G = \Delta V\,\mathrm{d}p - \Delta S\,\mathrm{d}T. \tag{12.6}$$

Since the rate constant depends on $\Delta^{\ddagger}G^{\circ}$, this equation implies that rate constants depend on pressure. Equation (12.6) implies

$$\left.\frac{\partial \Delta G}{\partial p}\right|_T = \Delta V.$$

Taking a logarithm of equation (12.1), we get

$$\ln k_0^{(2)} = \ln\left(\frac{k_B T}{c^{\circ} h}\right) - \frac{\Delta^{\ddagger}G^{\circ}}{RT}.$$

We now take a derivative of this equation with respect to p:

$$\left.\frac{\partial}{\partial p}\ln k_0^{(2)}\right|_T = -\frac{1}{RT}\left.\frac{\partial \Delta^{\ddagger}G^{\circ}}{\partial p}\right|_T = -\frac{\Delta^{\ddagger}V^{\circ}}{RT}, \quad (12.7)$$

where $\Delta^{\ddagger}V^{\circ}$ is the change in the molar volume (the change in volume of solution per mole of reaction) on accessing the transition state. If $\Delta^{\ddagger}V^{\circ} > 0$, the rate constant decreases as p increases. The reverse is true if $\Delta^{\ddagger}V^{\circ} < 0$. Note that although we started out by considering a second-order reaction, this equation would hold for reactions of any order.

Example 12.3. *Ethylene oxide (oxirane,* C_2H_4O*) undergoes acid-catalyzed hydrolysis to ethylene oxide. The overall reaction is as follows:*

$$\text{(oxirane)} + H_2O \longrightarrow HO{-}CH_2CH_2{-}OH$$

The rate law for this reaction is

$$v = k[H^+][C_2H_4O],$$

implying that the rate-determining step involves protonation of the ethylene oxide. Baliga and coworkers obtained the following data for this reaction in water at 25.00 °C [3].

p/bar	1	500	1000	2000	3000
$k/10^{-3}$ L mol^{-1}s^{-1}	9.30	10.9	12.7	16.9	20.4

Equation (12.7) tells us that the slope of a graph of $\ln k$ *vs* p *is* $-\Delta^{\ddagger}V^{\circ}/RT$. *As usual, it is good practice to work in SI units where it matters. In this case, the pressures should be converted to Pa before graphing. If we fit a straight line to the full data set, we get the graph shown in figure 12.5(a). The fit is not particularly good, with obvious curvature of the data around the line of best fit. However, you might notice that the first four points do*

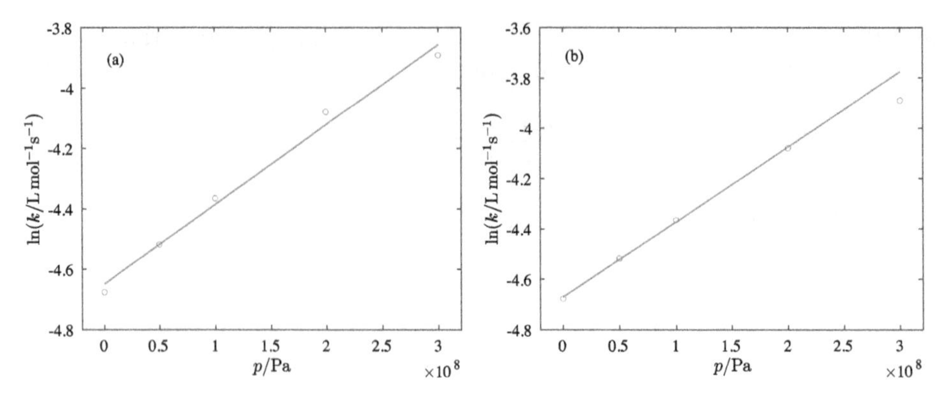

Figure 12.5. The dependence of the rate constant on pressure for the acid-catalyzed hydrolysis of ethylene oxide. (a) A fit of the entire data set. (b) A fit of the first four points only.

in fact fit a line quite well. If we fit only these first four points, we get the graph shown in figure 12.5*(b). It is clear that something changes in a significant way between 2000 and 3000 bar. There are many possibilities, including a change in the solvent structure at this very high pressure.*

The slope of the line in figure 12.5*(b) is* 2.98×10^{-9} Pa^{-1}*. From this slope, we can calculate* $\Delta^{\ddagger} V^{\circ} = -7.38 \times 10^{-6}$ $\text{m}^3\text{mol}^{-1}$*. Volumes of activation are normally given in* $\text{cm}^3\text{mol}^{-1}$*. In these units,* $\Delta^{\ddagger} V^{\circ} = -7.38$ $\text{cm}^3\text{mol}^{-1}$*. Personally, I find it easier to think about these quantities if I convert them to units that make sense for a single reaction*[1]: $\Delta^{\ddagger} V^{\circ} = -12.3$ Å^3 event^{-1}*. From the density of water, we can calculate that each water molecule occupies about* 30 Å^3*, so* $\Delta^{\ddagger} V^{\circ}$ *represents a decrease in volume equivalent to about 40% of the volume of a molecule of water. This presumably means that the solvent reorganizes around the solutes in a way that allows for more efficient packing of the transition state and solvent vs the reactants and solvent.*

12.4 Kramers' theory

Our discussion of TST in solution is missing something: drag. Reaching the transition state requires that atoms move, and when they do, they encounter drag from the solvent. Drag is implicitly taken into account in the calculations described in section 12.2 by sampling different solvent conformations and determining how they affect the free energy along the reaction coordinate. However, a classic theory developed by Kramers affords us some additional insights into the effect of drag on the rate of a chemical reaction. To get there, we first need to talk about a microscopic picture of molecular diffusion.

[1] *Note that the 'mol' in the units of any reaction statistic, including* $\Delta^{\ddagger} V^{\circ}$*, represents moles of reaction, i.e. a multiple of the number of reaction events.*

12.4.1 Langevin equations

In our previous treatment of diffusion, we focused on a macroscopic observable: concentration. What if we wanted to model the diffusion of a single molecule? We could, in essence, apply $F = ma$ to one particle, provided we can come up with a model for the effect of the solvent. We can think of the solvent as exerting two forces on a moving particle:

1. The drag force, which we have already discussed.
2. Collisions between solvent molecules and the solute also generate an undirected, random force.

The separation of the force into drag and a random force is somewhat artificial: both arise from collisions with the solvent. This separation recognizes that, if the particle has velocity v, this creates an asymmetry in the interaction with the solvent that can be separated from the (on average) isotropic term due to random motion of the solvent.

Let us set up the equations of motion in one dimension. We have the drag force $-fv$, a random force $F_r(t)$, and we can also consider an external force responsible for a potential energy $V(x)$. From $F = ma$, we get

$$m\frac{\mathrm{d}v}{\mathrm{d}t} = -\frac{\mathrm{d}V}{\mathrm{d}x} - fv + F_r(t), \tag{12.8}$$

where we used $a = \mathrm{d}v/\mathrm{d}t$. The position satisfies the differential equation

$$\frac{dx}{dt} = v. \tag{12.9}$$

These equations are a version of the **Langevin equation**. Equation (12.8) in particular is a **stochastic differential equation**, i.e. a differential equation with a random term. To do anything with this equation, we need more information about the randomly fluctuating force $F_r(t)$. The standard assumptions for $F_r(t)$ are the following:

- The time average of $F_r(t)$ is zero.
- The force varies rapidly in time so that its values at two **different** times t and t' are **uncorrelated**. Mathematically, we write

$$\langle F_r(t)F_r(t')\rangle = \Gamma\delta(t - t') \tag{12.10}$$

  where the angle brackets denote a time average (used here to compute a correlation), $\delta(\cdot)$ is a delta function, and Γ is a constant to be determined later. Since the delta function is zero when $t \neq t'$, this says that the random forces acting on the molecule at two different times, no matter how small $t - t'$ might be, are uncorrelated. This is the definition of white noise. Note also that Γ must have something to do with the typical magnitude of the force, sometimes called the noise intensity, based on the behavior of this equation for $t = t'$. We will make this more precise later.

The complete Langevin problem consists of equation (12.8) (and (12.9) if we want to track the position as well as the velocity), an initial condition, say $(x(0), v(0)) = (x_0, v_0)$, and the specification of the statistical properties of $F_r(t)$.

In the rest of this section, we will leave out the external potential. Equation (12.8) then becomes

$$m\frac{\mathrm{d}v}{\mathrm{d}t} = -fv + F_r(t). \tag{12.11}$$

We run into an issue right away. What would it mean to 'solve' an equation like (12.11)? This question provided the impetus for the development of a new branch of mathematics, stochastic calculus, in the middle of the 20th century. And if you have to develop a new branch of mathematics to answer a question, the answer is probably not that simple. On a superficial level, we can write a formal solution

$$v(t) = v_0 \exp(-ft/m) + \frac{1}{m}\exp(-ft/m)\int_0^t F_r(t')\exp(ft'/m)\,\mathrm{d}t', \tag{12.12}$$

but now we have the problem of interpreting the stochastic integral (the integral over the stochastic force F_r), which is very much the nub of the problem. We won't pursue this avenue. Instead, since we are interested in chemical problems, we will think about *many* molecules (an ensemble) satisfying the Langevin equation with the same initial condition. The random force $F_r(t)$ will be different for each realization while conserving the statistical properties given above. We will then ask what the *average* behavior of the ensemble is.

Let us therefore average both sides of equation (12.12):

$$\begin{aligned}\langle v(t)\rangle &= v_0 \exp(-ft/m) + \frac{1}{m}\exp(-ft/m)\int_0^t \langle F_r(t')\rangle \exp(ft'/m)\,\mathrm{d}t' \\ &= v_0 \exp(-ft/m)\end{aligned}$$

since $\langle F_r(t)\rangle = 0$. This isn't very exciting. All it says is that the average velocity decays exponentially to zero. We're averaging over many realizations, so regardless of the value of v_0, we will eventually have as many molecules traveling to the left as to the right, hence the asymptotic approach to zero. To put it another way, the first term on the right-hand side of equation (12.12) says that molecules 'forget' their initial velocity with a time constant m/f. The ensemble must also have this property.

If the average velocity converges to zero, then the variance is just $\langle v^2\rangle$. The latter quantity thus determines the spread of v values over time. To calculate this, we first need to square equation (12.12). A little bit of thought should lead you to conclude that the following is the correct equation for v^2:

$$\begin{aligned}v^2 = v_0^2 \exp(-2ft/m) &+ \frac{2v_0}{m}\exp(-2ft/m)\int_0^t F_r(t')\exp(ft'/m)\,dt' \\ &+ \frac{1}{m^2}\exp(-2ft/m)\int_0^t F_r(t')\exp(ft'/m)\,\mathrm{d}t'\int_0^t F_r(t'')\exp(ft''/m)\,\mathrm{d}t''.\end{aligned}$$

Note that we use different dummy variables in the two copies of the integral used to represent the integral's square. We can then combine these integrals:

$$v^2 = v_0^2 \exp(-2ft/m) + \frac{2v_0}{m} \exp(-2ft/m) \int_0^t F_r(t') \exp(ft'/m)\,dt'$$
$$+ \frac{1}{m^2} \exp(-2ft/m) \int_0^t dt' \int_0^t dt'' \exp[f(t' + t'')/m] F_r(t') F_r(t'').$$

Again, we average over an ensemble of particles with a common initial velocity.

$$\langle v^2 \rangle = v_0^2 \exp(-2ft/m) + \frac{\Gamma}{m^2} \exp(-2ft/m) \int_0^t dt' \exp(2ft'/m).$$

To get this result, we used equation (12.10).

Now note that

$$\lim_{t\to\infty} \langle v^2 \rangle = \frac{\Gamma}{2mf}.$$

Asymptotically, the variance of the velocity therefore depends on Γ, m, and f. Perhaps not too surprisingly, the variance increases with Γ, which we said earlier is something like a noise intensity. It also decreases with m and f, which again makes sense: f is related to a force that opposes motion, so a larger f should decrease typical velocities. And of course, a larger mass increases the inertia of the molecule.

The kinetic theory of matter tells us that the mean squared velocity is $\langle v^2 \rangle = k_B T/m$. Setting the Langevin equation value equal to the kinetic theory result, we get

$$\frac{\Gamma}{2f} = k_B T. \tag{12.13}$$

This is a version of the **fluctuation–dissipation theorem** because it relates the size of the fluctuations (controlled by Γ) to the rate of dissipation (f).

We also know that $D = k_B T/f$. Eliminating $k_B T$ between this equation and (12.13), we get

$$\Gamma = 2f^2 D.$$

Γ is therefore proportional to the diffusion coefficient. Thus, the larger the noise intensity, the faster diffusion occurs at a given temperature.

12.4.2 Simulation of the Langevin equation

Many ordinary differential equations (ODEs) cannot be solved analytically. Instead, we turn to numerical methods. The simplest numerical method for ODEs is Euler's method, in which we simply replace the derivatives by ratios of small changes, e.g. $dx/dt \approx \Delta x/\Delta t$. A differential equation $dx/dt = f(x, t)$ is thus approximated by $\Delta x = f(x, t)\Delta t$. At each step, we calculate $f(x, t)$, multiply it by Δt to get Δx, then increment x by Δx and t by Δt. The catch is that Δt needs to be fairly small for this to work.

The problem with stochastic differential equations, already hinted at, is that they don't obey the normal rules of calculus. In fact, depending on the problem, there are different versions of stochastic calculus that must be applied. The Langevin equation obeys Itô calculus. Itô calculus in turn implies slightly different rules for obtaining the solution of a stochastic differential equation than we would get by simply approximating dv/dt by $\Delta v/\Delta t$. Under the assumption that $F_r(t)$ is a white-noise term with the properties outlined above, the **Euler–Maruyama** scheme for the Langevin equation is

$$\Delta v = -\frac{fv}{m}\Delta t + \frac{\sqrt{\Gamma}}{m}\mathcal{N}(1)\sqrt{\Delta t}, \tag{12.14}$$

$$\Delta x = v\Delta t,$$

where $\mathcal{N}(1)$ is a normally distributed random number with unit variance. As in the Euler method for ODEs, Δv and Δx are added to the corresponding variables at each step to obtain their values at $t + \Delta t$.

As an example, suppose that we want to simulate the diffusion of sucrose in water for 10 ns at 20 °C. For glucose at this temperature, $D = 5.7 \times 10^{-10}$ m^{2}s^{-1}. It is easy to calculate that the mass of one glucose molecule is 5.68×10^{-25} kg.[2] We can use $f = k_B T/D$ and $\Gamma = 2fk_B T$ to calculate these two parameters given D.

We have the problem of choosing Δt. This is tricky. With Euler-class methods, if we choose too large a value of Δt, the method becomes unstable. A good starting point is to tell ourselves that we want $|\Delta v/v|$, the relative change in v, to be small. From equation (12.14), we see that Δv has two parts, $-(fv/m)\Delta t$ and the stochastic term. If both of these parts are sufficiently small relative to v, we should get reasonable solutions from the Euler–Maruyama scheme. For the first part, we have

$$\left|\frac{-\frac{fv}{m}\Delta t}{v}\right| = \frac{f}{m}\Delta t \ll 1,$$

$$\therefore \Delta t \ll m/f.$$

For the stochastic term, we first note that $\mathcal{N}(1)$ is of the order of unity most of the time. We therefore want $|\sqrt{\Gamma}\sqrt{\Delta t}/mv|$ to be small. However, the velocity v changes throughout the simulation. What we need is an estimate of a typical value of v. This is exactly what the root-mean-squared velocity $\sqrt{k_B T/m}$ gives us. Therefore, the second term is usually small when

[2] I am setting up the calculation in SI units because this is a consistent system, which makes it easier for me to get it right than if I get too cute with the units.

$$\left| \frac{\sqrt{\Gamma}\sqrt{\Delta t}}{mv} \right| \sim \frac{\sqrt{\Gamma}\sqrt{\Delta t}}{m}\sqrt{\frac{m}{k_B T}} = \sqrt{\frac{\Gamma}{m k_B T}}\sqrt{\Delta t} \ll 1.$$

Thus,

$$\Delta t \ll m k_B T / \Gamma,$$

or, since $\Gamma = 2f k_B T$,

$$\Delta t \ll \frac{m}{2f}.$$

Fortuitously, both conditions tell us that Δt should be much smaller than m/f. We can be more or less conservative about this. I have found that for this problem, $\Delta t = m/10f$ seems to be good enough, in the sense that the solutions don't do anything crazy.

When writing simulation codes, it is always a good idea to build in any tests you can think of. In this case, we should have $\langle v^2 \rangle = k_B T/m$. We should also have $\langle v \rangle \approx 0$. It will be easy to check whether these statistics are approximately correct at the end of the simulation.

Here is my Langevin simulation code:[3]

```
1  % Numerical solution of the Langevin equation in one dimension
2  % Marc R. Roussel (roussel@uleth.ca)
3
4  % Clear all variables.
5  clear
6
7  % Calculate frictional coefficient and Gamma
8  D = 5.7e-10; % sucrose, in SI units
9  m = 5.68e-25;    % sucrose, in kg
10 kB = 1.38e-23;
11 T = 293.15;
12 f = kB*T/D
13 Gamma = 2*f*kB*T
14
15 % Simulation for tmax time units
16 tmax = 10e-9; % in seconds
17 dt = m/(10*f); % Time step
18
19 % Initial conditions stored as first point of vectors for later plotting.
20 npts = 1;
```

[3] Matlab code Section12_4_2.m available from https://doi.org/10.1088/978-0-7503-5321-2.

```
21  t(npts) = 0;
22  v(npts) = 0;
23  x(npts) = 0;
24
25  % Simulate until tmax
26  while (t(npts) <= tmax)
27      % Calculate Delta_v
28      % Note: randn() generates Gaussian-distributed random numbers.
29      Delta_v = -f*v(npts)/m*dt + sqrt(Gamma)*randn(1)/m*sqrt(dt);
30      % Update velocity, position and time; store all for future plotting.
31      npts = npts + 1;
32      t(npts) = t(npts-1) + dt;
33      v(npts) = v(npts-1) + Delta_v;
34      % Note: Delta_x = v*dt
35      x(npts) = x(npts-1) + v(npts-1)*dt;
36  end
37
38  % Plot velocity and position vs time
39  figure('DefaultAxesFontSize',18)
40  plot(t,x)
41  xlim([0 tmax])
42  xlabel('t/s')
43  ylabel('x/m')
44
45  figure('DefaultAxesFontSize',18)
46  plot(t,v)
47  xlim([0 tmax])
48  xlabel('$t$/s','interpreter','latex')
49  ylabel('$v/$m s$^-1$','interpreter','latex')
50
51  % Check if <v> is approximately zero.
52  v_av = mean(v);
53  check_v_av = ['<v> = ',num2str(v_av),'.'];
54  disp(check_v_av)
55
56  % Check if <v^2> is approximately kB*T/m.
57  % Note: v.^2 squares each element of v
58  v2av = mean(v.^2);
59  compare_vals = ['<v^2> is ',num2str(v2av),' and kB T/m is ',...
60      num2str(kB*T/m),'.'];
61  disp(compare_vals)
```

The initialization should be reasonably self-explanatory. We start our discussion of the code with line 29, which implements the Euler–Maruyama equation (12.14). The `randn()` function generates matrices of normally distributed random numbers with unit variance. Here, we generate just one value. We use a counter called `npts` for the number of points of the trajectory stored. In lines 31 to 33, we increment the counter, time, and v. Line 35 increments x. In this case, Δx is sufficiently simple that we calculated it and incremented x in the same line. Note the reference to the velocity at the start of the step (`v(npts-1)`) rather than the updated velocity in this line, which is in the spirit of forward Euler methods[4]. In this particular situation, because the velocity evolves independently of x, we could equally well have used `v(npts)`.

After plotting the results, the code calculates $\langle v \rangle$ (line 52) and $\langle v^2 \rangle$ (line 58). It builds readable sentences outlining these results (lines 53 and 59–60) and prints them out using the `disp()` function. Lines 53 and 59–60 create character strings consisting of the literal character strings in quotes and of the numbers converted to strings by the `num2str()` function. Commas separate the pieces that are to be glued together to make the overall string. Note the use of ... to indicate to Matlab that we want to continue the line of code started in line 59 on the next line. This makes the code more readable and in particular makes it nicer for printing. Note also the use of `.^` in line 58. This operator returns a vector whose elements are the squares of the elements of `v`. The `mean()` function then averages these v^2 values.

When we run this program, we consistently get values of $|\langle v \rangle|$ that are less than 1 m s^{-1}. Given that the root-mean-squared velocity at this temperature is about 87 m s^{-1}, this is reasonable. The values of $\langle v^2 \rangle$ are consistently high (~7500 m s^{-1}, compared to the theoretical value of 7122 m s^{-1}). This is due to the step size. We get better agreement if we reduce Δt. This results in more values being stored, which might create memory issues, but it is possible to write a program that only stores every nth point instead of storing every single point generated as ours does. Hint: you need two counters to make this work. Figure 12.6 shows the velocities computed

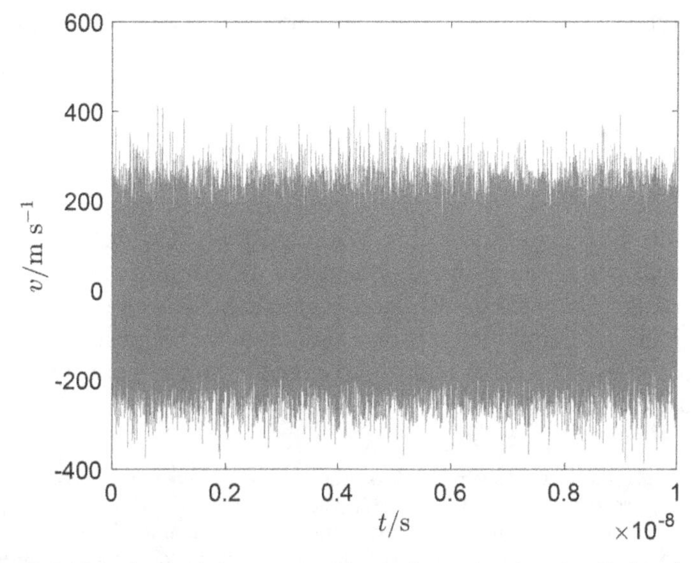

Figure 12.6. The velocity vs time generated by the Langevin solver described in the text.

[4] There are also backward Euler methods, in which the right-hand side of the differential equation is evaluated at the destination of the integration step rather than at its origin. In general, backward Euler methods are more complicated to implement, although they provide greater stability and therefore allow for larger step sizes.

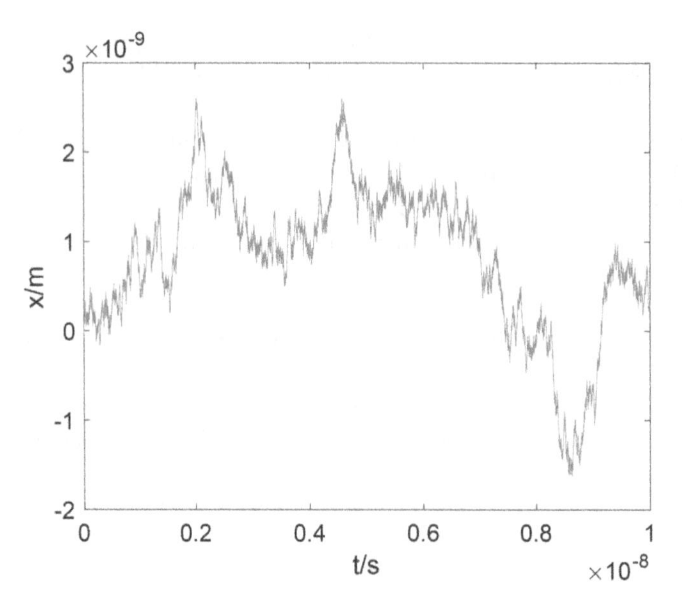

Figure 12.7. The position vs time generated in the same Langevin simulation as in figure 12.6.

during one run of the program. The random force causes the velocity to fluctuate wildly, while the frictional force tends to restore the velocity towards zero. The position, which is an integral of the velocity, is much smoother (figure 12.7). Note that the random dynamics means that these figures will never look quite the same way twice[5].

12.4.3 The Kramers equation

The Langevin equation describes individual diffusive trajectories of molecules. In section 12.4.1, we carried out some ensemble averaging to obtain the statistical properties of the Langevin trajectories. We could push this idea further and try to develop an equation for the probability densities of position x and velocity v. Such an equation exists. It is called the **Kramers equation**. Specifically, let $\rho(x, v, t)\,\mathrm{d}x\,\mathrm{d}v$ be the probability that simultaneous measurements of the position and velocity at time t return values between x and $x + \mathrm{d}x$, and v and $v + \mathrm{d}v$, respectively. The Kramers equation tells us how $\rho(x, v, t)$ evolves in time.

[5] Computers don't generate true random numbers. Rather, they generate sequences of pseudo-random numbers. These sequences can be started in different places so that each simulation is independent of the previous one. However, depending on the system used, running a program immediately after restarting the simulation environment sometimes results in the same sequence of pseudo-random numbers being used. Matlab has this property. You therefore have to be a bit careful about combining results from simulations run in different instances of Matlab. Depending on the exact sequence of events before you run the simulations, you may generate exactly the same sequence of pseudo-random numbers each time, so your results won't be independent of each other, which is a necessary property for any statistical analysis. To avoid this behavior, you can insert `rng shuffle` in your program before you generate any random numbers. Note that you only use this command once in your program. `rng shuffle` uses the system's clock to initialize the pseudo-random number generator, which guarantees that you will get a different sequence each time you run the program.

The derivation of the Kramers equation is time-consuming, so I just present it here without proof:

$$\frac{\partial \rho}{\partial t} + v\frac{\partial \rho}{\partial x} - \frac{1}{m}\frac{\partial V}{\partial x}\frac{\partial \rho}{\partial v} = \frac{f}{m}\left[\frac{\partial}{\partial v}(v\rho) + \frac{k_B T}{m}\frac{\partial^2 \rho}{\partial v^2}\right].$$

The variables appearing in this equation have the same meanings as in the Langevin equation treatment.

We can apply the Kramers equation to the intramolecular motions of molecules provided we interpret the symbols a little bit. In particular, we need to replace the mass m by a reduced mass μ associated with a motion. The Kramers equation would then describe the evolution of a probability density associated with an intra-molecular motion while the molecule is interacting with a solvent that is generating drag forces opposing motion as well as random forces from collisions. The potential energy $V(x)$ could be interpreted as the potential energy associated with motion along a particular internal coordinate of the molecule.

For the sake of fixing a picture in our minds, think about an isomerization reaction $A \rightleftharpoons B$. The system has a potential energy surface that depends on the $3N - 6$ internal coordinates of the molecule. One of these internal coordinates, say x, will convert A into B. This is the reaction coordinate. Along this reaction coordinate, the PES reduces to a potential energy curve, probably a double well, one well of which corresponds to A and the other to its isomer B (figure 12.8). As a rule, when the atoms of A rearrange into B, the required intramolecular motions require solvent molecules in the immediate neighborhood to move, causing drag. This suggests that we can treat motion along a reaction coordinate as the motion of an imaginary particle experiencing drag as it moves through a solvent with the added force due to the potential energy profile along the reaction coordinate.

If it's tricky to derive the Kramers equation itself, it's just as tricky to obtain solutions. In fact, there isn't a general solution that we can use. However, a useful

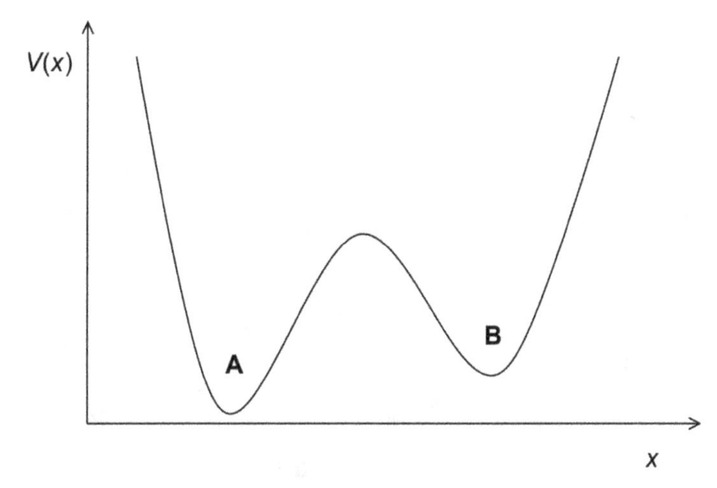

Figure 12.8. A double-well potential as a simplified picture of the potential energy surface for an isomerization reaction $A \rightleftharpoons B$.

special case can be obtained. When we solve the Kramers equation for a double-well potential and calculate the rate of crossings from reactants to products, a correction (i.e. a transmission coefficient) to the TST rate constant is obtained in the case of medium to high friction. This is the regime that is typically appropriate for reactions in solution. The Kramers transmission coefficient is

$$\kappa_K = \left[1 + \left(\frac{f}{2\mu^{\ddagger}\omega^{\ddagger}}\right)^2\right]^{1/2} - \frac{f}{2\mu^{\ddagger}\omega^{\ddagger}}. \tag{12.15}$$

In this equation, $\omega^{\ddagger}$ is the frequency associated with the reactive mode and $\mu^{\ddagger}$ is the corresponding reduced mass. This expression for κ_K is plotted in figure 12.9. Note that κ_K decreases as f increases. At low velocities, the frictional coefficient is always proportional to the solvent viscosity, η. We saw an example of this in the Stokes–Einstein equation, $f = 6\pi r\eta$. The Kramers equation thus predicts that the transmission coefficient should decrease as solvent viscosity increases, an observation that has been confirmed experimentally. Roughly speaking, this is because high friction attenuates any momentum along the reaction coordinate as the system approaches the transition state, which has the effect of making the random force relatively more important. As the frictional coefficient increases, crossing of the transition state becomes a more and more random event, which decreases the rate of crossings in the reactant-to-product direction.

There is a significant problem in applying equation (12.15): we don't know how to compute f for motion along a reaction coordinate. We will have to improvise. This will require looking at the motion along the reaction coordinate and asking what moves? And how large a profile do the atoms involved in the motion present to the solvent?

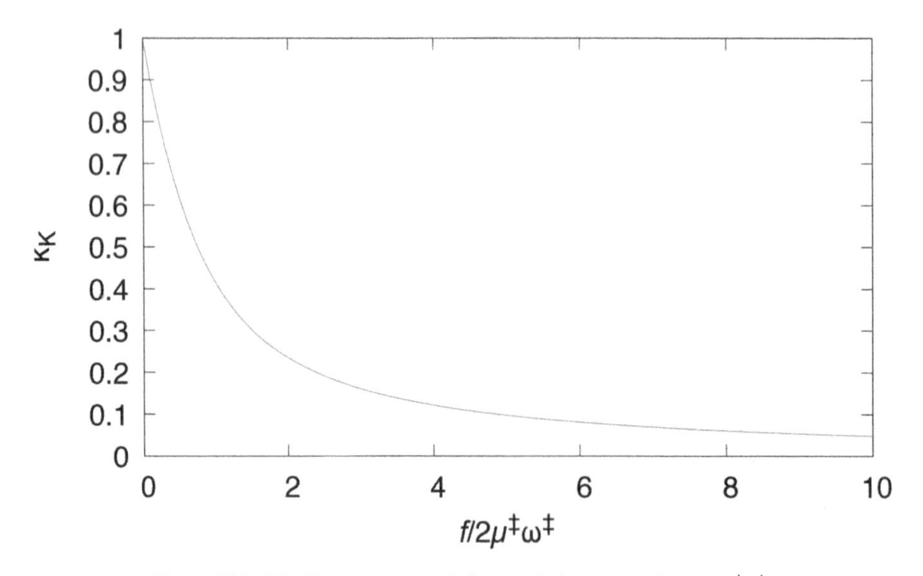

Figure 12.9. The Kramers transmission coefficient plotted vs $f/2\mu^{\ddagger}\omega^{\ddagger}$.

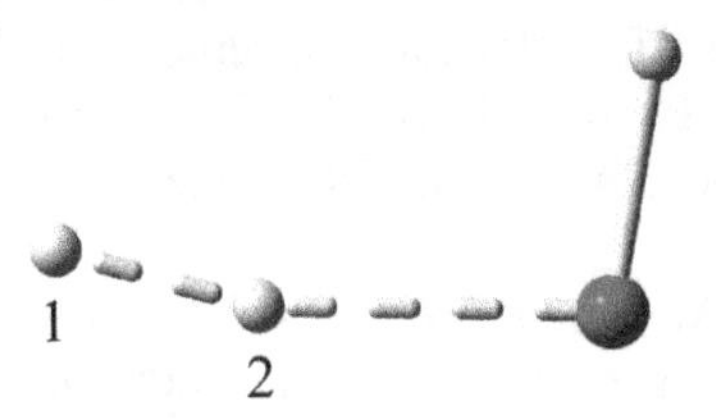

Figure 12.10. *The transition state of the reaction of* H_2 *with* $\cdot OH$.

Example 12.4. *We will consider a simple reaction in an aqueous solution,* $H_2 + \cdot OH \rightarrow \cdot H + H_2O$. *If you find the transition state (figure* 12.10*) and then animate the reaction coordinate in GaussView, you will see that the reaction coordinate is essentially an antisymmetric stretch in which hydrogen atom 2 (using the numbering in the figure) has, by far, the largest amplitude of motion. We can guess that the frictional coefficient will essentially be that of a hydrogen atom. The van der Waals radius, which gives a rough idea of how close two atoms can approach each other before they repel, is likely the appropriate measure of radius to use here. For a hydrogen atom, the van der Waals radius is approximately 1.10 Å. At* 25 °C, $\eta(H_2O) = 8.91 \times 10^{-4}$ Pa s. *Therefore*

$$f = 6\pi r\eta = 1.85 \times 10^{-12}\ \mathrm{N\ s\ m^{-1}}.$$

Looking through the Gaussian output of the transition-state calculation, we find $\tilde{\nu}^\ddagger = 912\ \mathrm{cm^{-1}} \equiv$ *and* $\mu^\ddagger = 1.1155\ \mathrm{amu} \equiv 1.8523 \times 10^{-27}$ kg. *We can convert the wavenumber to an angular frequency by* $\omega^\ddagger = 2\pi c\tilde{\nu}^\ddagger = 1.72 \times 10^{14}\ \mathrm{s^{-1}}$.

We now have everything we need to calculate the Kramers transmission coefficient:

$$\frac{f}{2\mu^\ddagger\omega^\ddagger} = 2.90.$$

$$\begin{aligned}\kappa_K &= \left[1 + \left(\frac{f}{2\mu^\ddagger\omega^\ddagger}\right)^2\right]^{1/2} - \frac{f}{2\mu^\ddagger\omega^\ddagger}\\ &= 0.17.\end{aligned}$$

We see from this calculation that the effect of drag on the motion across a transition state can be significant.

Further reading

Many details of the reaction conditions can have an effect on the kinetics of a reaction in solution. In particular, the ionic strength can have a large effect. The theory for dilute solutions is covered in

- Roussel M R 2012 *A Life Scientist's Guide to Physical Chemistry* (Cambridge: Cambridge University Press) section 17.4 https://doi.org/10.1017/CBO9781139017480w

There are many reviews of important concepts in TST and related theories for calculating rate constants. I particularly like the following, which covers not only TST in the gas phase and in solution but also master equation methods, among other topics:

- Fernández-Ramos A, Miller J A, Klippenstein S J and Truhlar D G 2006 Modeling the kinetics of bimolecular reactions *Chem. Rev.* **106 4518–84** https://doi.org/10.1021/cr050205w

If you're interested in the application of TST in enzyme kinetics, I recommend

- Gao J, Ma S, Major D T, Nam K, Pu J and Truhlar D G 2006 Mechanisms and free energies of enzymatic reactions *Chem. Rev.* **106 3188–209** https://doi.org/10.1021/cr050293k
- Truhlar D G 2015 Transition state theory for enzyme kinetics *Arch. Biochem. Biophys.* **582 10–7** https://doi.org/10.1016/j.abb.2015.05.004

Exercise

12.1 In example 12.1, we analyzed the data of Branton and coworkers [1] for the reaction

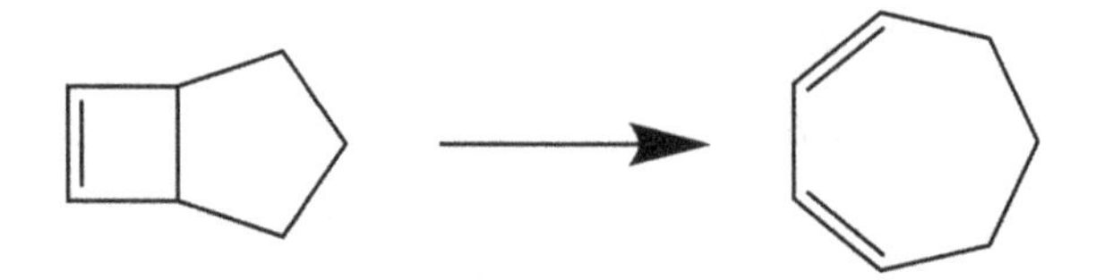

(a) Using the calculated data from example 12.1, calculate $\Delta^{\ddagger}\Delta_{solv}G^{\circ}$ for this reaction. Explain the meaning of the sign of the result. If possible, put the size of the result in perspective.

(b) Experiments in solution gave the following rate constants:

T/°C	256.2	259.9	265.2	268.5	273.8	280.4	283.5	288.4
$k/10^{-4}\ s^{-1}$	0.513	0.705	1.11	1.43	2.18	3.37	4.39	6.40

Determine the enthalpy and entropy of activation for this reaction.

(c) TST suggests that, instead of a potential energy barrier to reaction, we should think in terms of a free-energy barrier. Based on your calculations, for the reaction in solution, which is the more important contributor to the free-energy barrier, the enthalpic contribution or the entropic contribution? Can you generate a plausible hypothesis to explain why this might be the case?

12.2 The hydrogen malonate anion $HOOCCH_2COO^-$ has the following as one of its equilibrium structures:

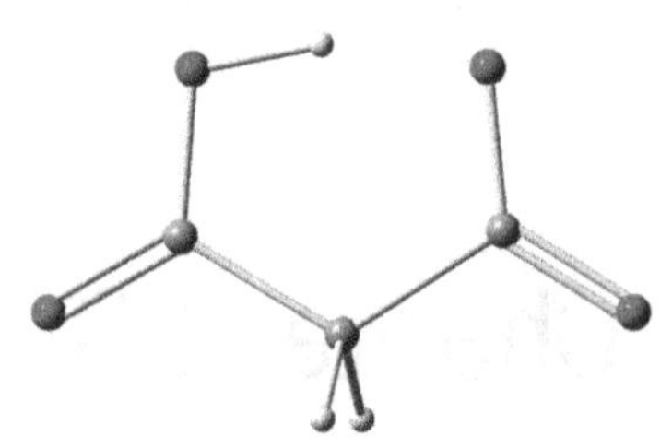

(a) Using Gaussian, reproduce this equilibrium geometry.
(b) Calculate the rate constant for the transfer of the acidic proton from one oxygen to the other in the gas phase at 25 °C and 1 bar using TST. Make sure to discuss your choice of computational method.
(c) I specified the pressure in part (b). Does it matter? Why or why not?
(d) Do you expect tunneling to be important in this reaction? Why or why not?
(e) If you wanted to calculate the rate constant for this reaction in aqueous solution, what would you need to do and why?

References

[1] Branton G R, Frey H M, Montague D C and Stevens I D R 1966 Thermal unimolecular isomerization of cyclobutenes. Part 7. Bicyclo [3,2,0]-hept-6-ene *Trans. Faraday Soc.* **62** 659–63

[2] Wassermann A 1938 Kinetics of bimolecular association in the gaseous and condensed phases *Trans. Faraday Soc.* **34** 128–37

[3] Baliga B T, Withey R J, Poulton D and Whalley E 1965 Effect of pressure on the hydrolysis of methyl acetate and ethylene oxide in acetone+water mixtures: quantities of activation at constant volume, and the isokinetic pressure *Trans. Faraday Soc.* **61** 517–30

IOP Publishing

Chapter 13

Marcus electron-transfer theory

In this chapter, we treat simple electron-transfer reactions. These are reactions in which the main process, and possibly the only observable process, is the transfer of an electron. Many simple examples of electron-transfer reactions come to us from inorganic chemistry. For example,

$$[Fe(OH_2)_6]^{2+} + [Ru(bpy)_3]^{3+} \rightleftharpoons [Ru(bpy)_3]^{2+} + [Fe(OH_2)_6]^{3+}$$

But electron-transfer processes show up in every area of chemistry. Understanding these reactions is therefore of central importance to the chemical sciences. Like many of the other topics we have studied in kinetics, the theory we study in this chapter, which was developed by Rudy Marcus (the M in RRKM theory), is valuable not only for its ability to predict rates of reaction or to correlate experimental data but perhaps more importantly for the insights it grants us into the underlying physicochemical processes.

Before we proceed, it is useful to know that electron-transfer (ET) reactions can roughly be divided into two groups:

Inner-sphere ET: a ligand that bridges between the two complexes conveys the electron(s) from one metal center to the other. The ligands in the coordination spheres of the metal ions are reorganized in the process.

Outer-sphere ET: a solvent molecule often acts as a transfer agent for the electron(s). There is no reorganization of the ligands. The reaction shown above is of this type.

The theory of outer-sphere ET is simpler to present, so we focus our attention there.

13.1 A mechanistic decomposition of electron-transfer reactions

By now, decomposing a complex process into a series of steps should be a familiar trick. The two reactants in an electron-transfer reaction are the electron donor, D,

doi:10.1088/978-0-7503-5321-2ch13

and the electron acceptor, A. For an outer-sphere process where the net effect of the reaction is only to transfer an electron from D to A, we can write

$$D + A \underset{k_{-d}}{\overset{k_d}{\rightleftharpoons}} \{DA\} \xrightarrow{k_{ET}} \{D^+A^-\} \xrightarrow{k_u} D^+ + A^-.$$

The first step in this mechanistic decomposition is the formation of the encounter pair, {DA}. Electron transfer follows, and finally the two products separate. The rate constants k_d and k_{-d} are familiar quantities from the theory of diffusion-influenced reactions. So is k_u, which is the same kind of rate constant as k_{-d}, although it will turn out that we do not need the value of this rate constant.

We want to determine how the rate of reaction depends on the concentrations of the reactants and therefore how the experimental rate constant depends on the rate constants in our description of the reaction. As you might guess, we apply the steady-state approximation to the intermediates. From the mechanism, we have $v = k_u[\{D^+A^-\}]$. The intermediates are {DA} and $\{D^+A^-\}$. Applying the steady-state approximation to the latter, we get $k_u[\{D^+A^-\}] = k_{ET}[\{DA\}]$. The rate of reaction can therefore be rewritten

$$v = k_{ET}[\{DA\}].$$

Applying the steady-state approximation to {DA}, we now get

$$[\{DA\}] = \frac{k_d[D][A]}{k_{-d} + k_{ET}}.$$

The rate of reaction is therefore

$$v = k_{obs}[D][A],$$

where the observed rate constant is

$$k_{obs} = \frac{k_{ET}k_d}{k_{-d} + k_{ET}}.$$

The rate constant for an outer-sphere electron-transfer reaction thus depends on two rate constants we know how to estimate, k_d and k_{-d}, and on the first-order rate constant for electron transfer within the encounter pair, k_{ET}. We clearly need to focus our attention on the latter.

13.2 Harmonic model

The model we will study here is not the one originally considered by Marcus, but it contains the essential elements of the Marcus theory, and it will help us understand the key quantities that enter into the theory.

Since chemical potential is, in essence, the molar free energy of a substance in a system, and since free energy is a measure of the ability of a system to do work, we can imagine a single-molecule chemical potential that measures the ability of one molecule, again in the context of a particular system, to do work. The chemical potential so conceived depends on the distortion of the molecule from its equilibrium configuration by random motions, both intrinsic vibrational motions and motions induced by collisions with the solvent. This chemical potential is minimized at the equilibrium geometry. We can typically describe the shape of a curve near a minimum as a parabola. In this case, we can write

$$\mu_i \approx \mu_i^\circ + \frac{1}{2} f (q - q_i)^2,$$

where q is a reaction coordinate and q_i is the value of q at the equilibrium geometry of complex i, which can be either the reactant $R \equiv \{DA\}$, or the product $P \equiv \{D^+A^-\}$ of the electron-transfer step. The constant f characterizes the shapes of the parabolas. It should be different for each of the complexes, but for mathematical simplicity, we take it as being the same for both here.

We can plot the parabolic chemical potentials together, which we do in figure 13.1. The minimum requirement for reaction is that the system must overcome the free-energy barrier to the reaction, denoted $\Delta^{\ddagger}G$. However, just reaching the free energy of activation is not quite enough. Because electron transfer is generally a very fast process, the nuclei do not have time to change their positions while the electrons are being transferred. Moreover, since the charge distribution changes, there must be changes involving the solvent molecules near the encounter pair following the electron transfer. The solvent effects are of two kinds: there are fast changes in the charge distributions in nearby solvent molecules and slower changes involving the reorientation of entire molecules when the solvent is polar. The reaction is much more efficient if these changes occur prior to electron transfer. Thus, we really want to consider the **reorganization energy** λ, which is the work required to reconfigure $\{DA\}$ and the surrounding solvent to a $\{D^+A^-\}$-like state. Because of the thermal energy in the system and the constant collisions of molecules with one

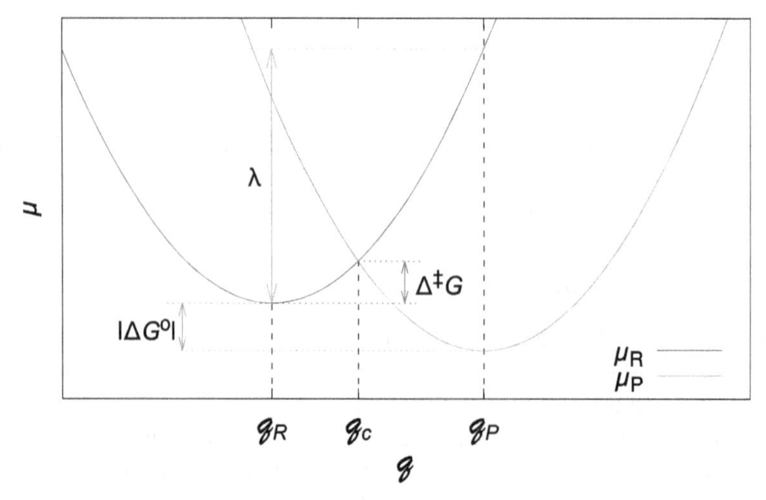

Figure 13.1. The parabolic dependence of the chemical potential near the equilibrium geometries of the reactant, $R \equiv \{DA\}$, and the product, $P \equiv \{D^+A^-\}$, of an electron-transfer reaction. Note that ΔG° is negative, as illustrated here.

another, such a configuration can occasionally arise by random chance. Once this happens, electron transfer should be rapid.

The figure also shows the intersection between the two chemical-potential parabolas at a coordinate q_c (c for crossing). This intersection, which is the transition state of the system, can be determined by solving the equation $\mu_R = \mu_P$. Note that $\Delta G^\circ = \mu_P^\circ - \mu_R^\circ$. The calculation of the transition-state coordinate proceeds as follows:

$$\mu_R^\circ + \frac{1}{2}f(q_c - q_R)^2 = \mu_P^\circ + \frac{1}{2}f(q_c - q_P)^2.$$

$$\therefore \frac{1}{2}f(q_c - q_R)^2 = \Delta G^\circ + \frac{1}{2}f(q_c - q_P)^2.$$

$$\therefore -fq_cq_R + \frac{1}{2}fq_R^2 = \Delta G^\circ - fq_cq_P + \frac{1}{2}fq_P^2.$$

$$\therefore fq_c(q_P - q_R) = \Delta G^\circ + \frac{1}{2}f\left(q_P^2 - q_R^2\right).$$

$$\therefore q_c = \frac{\Delta G^\circ}{f(q_P - q_R)} + \frac{q_P + q_R}{2}.$$

The reorganization energy is easy to compute:

$$\lambda = \mu_R(q_P) - \mu_R^\circ.$$

$$\therefore \lambda = \frac{1}{2}f(q_P - q_R)^2.$$

We now calculate the free energy of activation:

$$\Delta^{\ddagger}G = \mu_R(q_c) - \mu_R(q_R) = \frac{1}{2}f(q_c - q_R)^2$$

$$= \frac{1}{2}f\left(\frac{\Delta G^\circ}{f(q_P - q_R)} + \frac{q_P + q_R}{2} - q_R\right)^2$$

$$= \frac{1}{2}f\left(\frac{\Delta G^\circ}{f(q_P - q_R)} + \frac{q_P - q_R}{2}\right)^2$$

$$= \frac{1}{2f(q_P - q_R)^2}\left[\Delta G^\circ + \frac{1}{2}f(q_P - q_R)^2\right]^2,$$

$$\therefore \Delta^{\ddagger}G = \frac{1}{4\lambda}(\Delta G^\circ + \lambda)^2. \tag{13.1}$$

Equation (13.1) is the central equation of the theory. It already contains a significant surprise: you have probably been told over and over again that kinetics and thermodynamics are independent. The free-energy change ΔG°, which provides the thermodynamic driving force for a reaction, can be very negative without the reaction necessarily being fast. This is because for most classes of reaction, the difference in free energy between the reactants and the products is independent of the height of the barrier separating them. This is not so for electron-transfer reactions. We can see that there is a surprisingly simple relationship between $\Delta^{\ddagger}G$ and ΔG° for these reactions.

We now borrow an idea from transition-state theory to obtain the electron-transfer rate constant:

$$\begin{aligned} k_{\mathrm{ET}} &= k^{\ddagger}K^{\ddagger} \\ &= k^{\ddagger}\exp\left(\frac{-\Delta^{\ddagger}G}{k_{\mathrm{B}}T}\right). \\ \therefore\ k_{\mathrm{ET}} &= k^{\ddagger}\exp\left(\frac{-(\Delta G^{\circ}+\lambda)^2}{4\lambda k_{\mathrm{B}}T}\right) \end{aligned}$$

or

$$\ln k_{\mathrm{ET}} = \ln k^{\ddagger} - \frac{(\Delta G^{\circ}+\lambda)^2}{4\lambda k_{\mathrm{B}}T}. \tag{13.2}$$

This equation makes a remarkable prediction: because of the quadratic dependence of $\ln k_{\mathrm{ET}}$ on ΔG°, the rate constant should initially increase but then decrease as ΔG° becomes more negative (figure 13.2). So now we see that not only does $\Delta^{\ddagger}G$ depend on ΔG° for electron-transfer reactions, but there is a regime in which increasing the thermodynamic driving force is paradoxically responsible for a *decrease* in the rate of reaction. We call the region where increasing the thermodynamic driving force causes an increase in k_{ET} the **normal** region, and the region where increasing the driving force causes a decrease in the rate constant the **inverted** region.

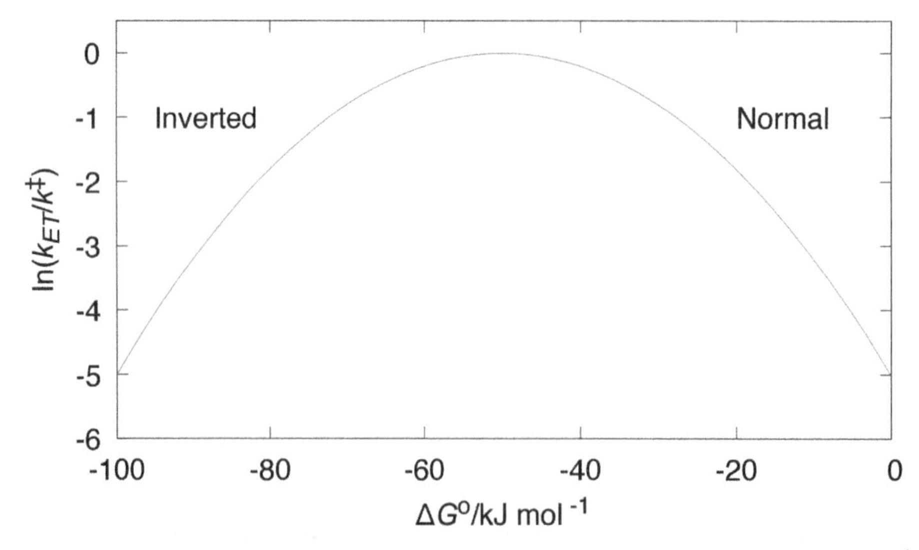

Figure 13.2. A plot of equation (13.2), showing the dependence of k_{ET} on ΔG° for $\lambda = 50\ \mathrm{kJ\ mol^{-1}}$ at $T = 300$ K. In the 'normal' region, increasing the thermodynamic driving force (decreasing ΔG°) increases the value of k_{ET}. In the 'inverted' region, further increases in the thermodynamic driving force cause a decrease in the rate constant.

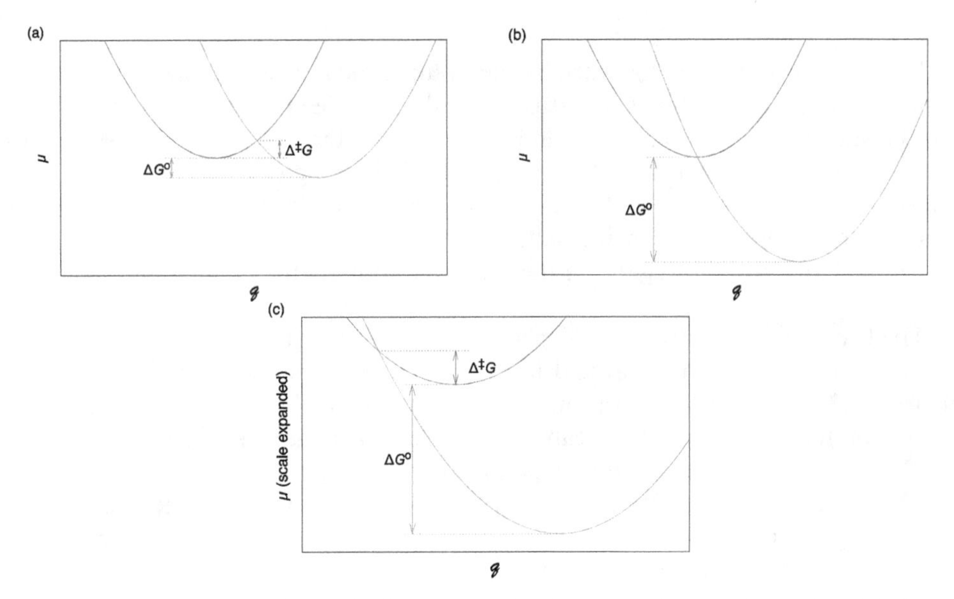

Figure 13.3. Parabolic intersections (a) in the normal region, (b) at the minimum of $\mu_R(q)$, and (c) in the inverted region. Note that the scale of the ordinate in panel (c) is expanded relative to those of panels (a) and (b).

The existence of both normal and inverted regions in electron transfer can be understood as a simple consequence of geometry. First consider figure 13.3(a), which shows the intersections of the reactant and product chemical potential curves in the normal region. Decreasing ΔG° corresponds to moving the product parabola downwards. This decreases $\Delta^\ddagger G$ and therefore increases the rate of reaction. As we keep decreasing ΔG°, we eventually reach the situation in figure 13.3(b), where the two parabolas intersect at the minimum of $\mu_R(q)$, i.e. at q_R. In this very particular case, $\Delta^\ddagger G = 0$, and the reaction proceeds at the maximum possible rate. If we decrease ΔG° further, we cross over into the inverted regime pictured in figure 13.3(c), where continuing to decrease ΔG° necessarily increases $\Delta^\ddagger G$ and therefore slows the reaction down.

Although Marcus predicted the existence of the inverted region in the mid-1950s, the first unambiguous experimental demonstration of its existence had to wait until 1988 [1]. The Closs and Miller experiments simplified the problem of obtaining k_{ET} by measuring the rates of intramolecular charge transfer in a series of compounds with the following structure:

Donor

A

In this structure, A is the variable acceptor. Studying intramolecular electron transfer eliminates the effects of diffusion, so that the electron-transfer process itself can be studied in isolation. Closs and Miller varied the acceptors tethered to this structure, which allowed them to vary ΔG° while keeping other factors reasonably constant. In particular, the reorganization energy should be about the same for suitably chosen acceptors. Their results are shown in figure 13.4. We clearly see both the normal and inverted regions in their data, confirming the predictions of Marcus theory.

The Closs and Miller data do display one discrepancy from the theory presented here, which is that the normal and inverted regions are not symmetric. (Compare figures 13.2 and 13.4.) This turns out to be due to the involvement of the vibrational modes of the acceptor, which can be included in a more sophisticated model, yielding excellent agreement with experiment.

There is another issue with the simple Marcus theory as presented here, and it is a familiar one. Equation (13.2) predicts that $k_{\rm ET} \rightarrow 0$ as $T \rightarrow 0$. However, this is not the case, as seen in figure 13.5. There is some scatter in the data, but we get a reasonable fit to Arrhenius behavior down to about $T = 125$ K. The rate constant then becomes roughly independent of temperature. For the last point shown at $T \approx 86$ K, $k_{\rm ET} \approx 307$ s^{-1}. A further point at a much lower temperature not shown in the graph[1] corresponds to $k_{\rm ET} \approx 300$ s^{-1}. Clearly, the rate constant is reaching a limiting value.

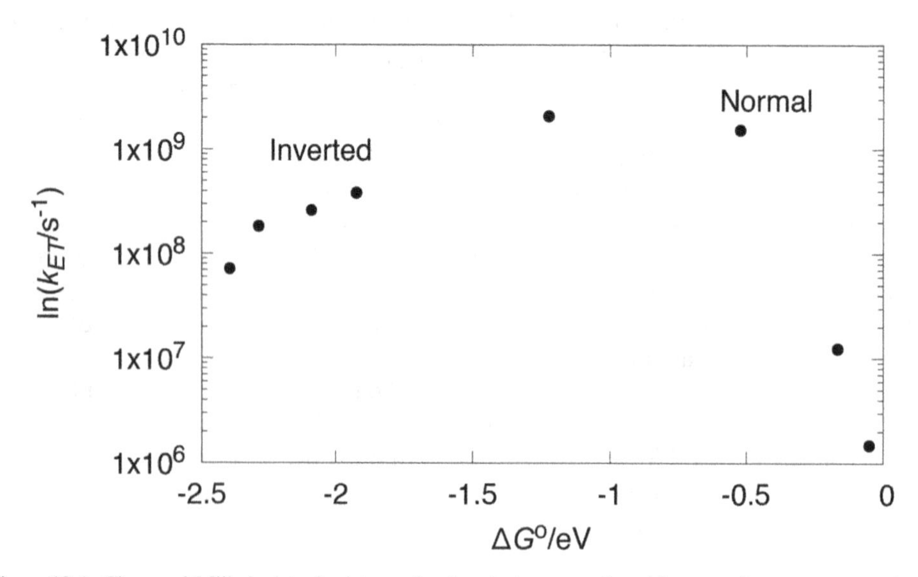

Figure 13.4. Closs and Miller's data for intramolecular electron transfer with a set of acceptors spanning a range of ΔG° values [1].

[1] The presentation in the paper by Barbara *et al* (1996), from which the data were extracted, made it impossible to determine the exact temperature of the low-temperature point.

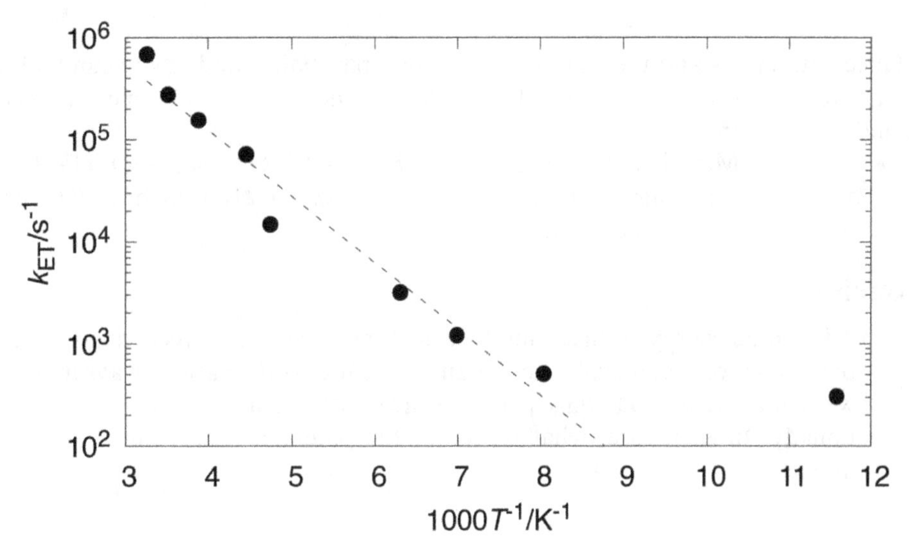

Figure 13.5. An Arrhenius plot of the electron-transfer rate constant for the oxidation of cytochrome in *Chromatium vinosum*. Data from D DeVault, *Quantum Mechanical Tunneling in Biological Systems*, Cambridge University Press, Cambridge, 1984; reproduced in a review paper by Barbara *et al* (1996) [2].

Perhaps you will have guessed by now that tunneling allows electron transfer to proceed at low temperatures, bypassing the need to overcome an activation barrier. We should have expected this, given that electrons are extremely light particles which, all other things being equal, tends to favor larger tunneling probabilities.

Marcus electron-transfer theory and its extensions now provide the basis for our understanding of the kinetics of inorganic redox reactions, of processes at electrode surfaces, and of the functioning of redox enzymes. Marcus' 1992 Nobel Prize in Chemistry was awarded in recognition of the practical and theoretical importance of this theory.

Further reading

There are many derivations of the Marcus electron-transfer equation in the literature, and the quantity λ is often represented graphically in ways that are at variance with its definition. I found the following expository article particularly clear:

- Silverstein T P 2012 Marcus theory: thermodynamics CAN control the kinetics of electron transfer reactions *J. Chem. Ed.* **89 1159–67** https://doi.org/10.1021/ed1007712

If you're interested in digging deeper into electron-transfer processes, including experimental approaches to their study, I recommend

- Barbara P F, Meyer T J and Ratner M A 1996 Contemporary issues in electron transfer research *J. Phys. Chem.* **100 13 148–68** https://doi.org/10.1021/jp9605663

Once you understand the Marcus intersecting-parabolas model, you can look at other applications of the same idea. For example, tunneling at low temperatures can be understood using the same model:

- Schleif T, Merini M P, Henkel S, and Sander W 2022 Solvation effects on quantum tunneling reactions *Acc. Chem. Res.* **55 2180–90** https://doi.org/10.1021/acs.accounts.2c00151

Exercise

13.1 Potential energy surfaces and their features have been a recurring theme in our studies of theoretical reaction kinetics. Discuss the various ways in which we made use of potential energy surfaces either conceptually or computationally. In each case, what part(s) of the potential energy surface is (are) important?

References

[1] Closs G L and Miller J R 1988 Intramolecular long-distance electron transfer in organic molecules *Science* **240** 440–7

[2] Barbara P F, Meyer T J and Ratner M A 1996 Contemporary issues in electron transfer research *J. Phys. Chem.* **100** 13148–68

IOP Publishing

Foundations of Chemical Kinetics
A hands-on approach
Marc R Roussel

Appendix A

Matlab programming

The intention of this chapter is not to teach Matlab programming in any reasonable sense of the word 'teach.' There are modules within the book where some instruction takes place, and if you really get serious about Matlab programming, you will need more than I can provide here. Rather, this chapter provides a summary of some key Matlab language features and useful functions.

Except where noted, every command described here works in both Matlab and Octave.

A.1 Code and comments

As a rule, each line of Matlab code contains a complete statement, which you can think of as a sentence. Some statements open or close a block of code that is only to be executed under some conditions or that repeats; see section A.5. Most lines of code, however, can be viewed as standalone sentences.

On occasion, a line of code is too wide for the editor window or would be too wide for printing. In these cases, we can use ... to indicate that a line of code, i.e. a statement, is continued on the next physical line.

Matlab ignores anything on a line after a % character. As suggested in the text, use comments liberally to explain your code.

A.2 Output and its suppression

Many Matlab commands produce output that is typically displayed on the screen. Ending a line with a semicolon prevents that command's output from being printed. Screen output is a particularly slow operation, so suppressing output can speed up a program considerably, depending on how often output is generated and how complex the output is.

In addition to Matlab's normal screen output, it is possible to display specific material on the screen using the `disp()` function. The argument of `disp()` can be a character string (enclosed in single quotes) or a variable.

doi:10.1088/978-0-7503-5321-2ch14

A.3 Variables

Variables are containers attached to a name. Variables in Matlab can contain many different kinds of things. As the name might suggest to you, Matlab was originally designed to be a language for handling matrices, so matrices, vectors, and scalars (single numbers) are the most commonly stored objects. As for variable names, they can be almost anything. The general rule is that the first character of a variable name should be a letter, and subsequent characters can be chosen from the set of letters, digits, and the underscore character. Both upper- and lower-case letters are allowed. Note that names are case-sensitive, so x and X are different variables. However, it's a bad idea to write a program in which you use variables whose names only differ in case, as this can lead to bugs that are difficult to track down. The following are legal variables names in Matlab:

```
x, X1, x_a
```

For (perhaps) obvious reasons, arithmetic operators and other characters with special meanings cannot be included in variable names. The following are **not** legal variable names:

```
_x, 1x, x+y, xy
```

As a rule, try to use meaningful variable names. They make your code easier to understand.

To store something in a variable, just use the equal sign. For example,

```
x = 1
```

stores the scalar 1 in the variable x. If you then type

```
x = 2
```

the original value of x is replaced by 2. You can also perform arithmetic on a number and store it back into the same variable. For example,

```
x = x^2
```

squares the current value of x (^ raises a number to a power) and stores the result back in x. So, if x initially had a value of 2, it would have the value 4 after running this line of code.

The notation m(i,j) accesses the element of a matrix m at row i, column j. For vectors ($n \times 1$ or $1 \times n$ matrices), v(i) accesses the ith element.

Matlab arrays are extensible. This means that if you try to place a value in an element that does not exist, Matlab increases the size of the matrix to make this possible. Otherwise undefined elements are set to zero in the process. Consider

```
>> x=1
x =

     1

>> x(4) = 1
x =

     1     0     0     1
```

The first command creates a simple scalar variable. However, the second command is taken to mean that x should be thought of as a vector and increased in size so that a fourth element can be stored. Extensible arrays are both a great convenience, since we don't necessarily need to know the size of an array ahead of time, and a source of subtle bugs. Note also that the memory reallocation necessary to make this work can make a program inefficient.

The clear keyword deletes all variables from memory. It is often a good idea to start a new program with clear in order to avoid side-effects due to (e.g.) arrays left over from a previous run of the program.

A.4 The range operator (:)

There are two variations on the range operator. The simplest version consists of two numbers separated by a colon, e.g. 1:4, which expands to the list 1, 2, 3, 4. The rule is that the range starts at the first value and that the values are incremented by one until we reach the value following the colon. Note that the second value in the range must be at least as large as the first, or the range is considered to be empty.

In the second variation, we also specify a step size. Thus, 1:0.2:4 would count from one to four in steps of 0.2. If you need to count backwards, you must use this version of the range operator, e.g. 4:-1:1 expands to 4, 3, 2, 1.

The range operator can be used to operate on parts of a vector or matrix. For example, if wup is a scalar, you could type

```
a(5:8) = wup*N(2:5);
```

wup*N(2:5) is a four-element vector calculated using elements two to five of N. The result is stored in the four elements of the vector a numbered five through eight.

The only constraint is that the quantity on the right-hand side has to have the same dimensions as the sub-matrix that the result is stored in.

The keyword `end` can be used to specify the last position of a vector. Thus, `a(end)` means the last element of `a`. It is possible to add or subtract values from `end`. For example, `a(end-1)` refers to the penultimate element of `a`. The `end` keyword can also be used in ranges such as `a(2:end)` or `a(1:end-1)`.

A.5 Control structures

In computer science, a control structure is a set of instructions that control which lines of code are executed. All Matlab control structures are blocks of code that start with a keyword (`if`, `while`, `for`) and end with the keyboard `end`[1].

The ability to run code only if certain conditions are met is central to a lot of programming[2]. The most fundamental form of conditional execution is provided by `if` statements. The general syntax is the following:

```
if condition
    code to execute
elseif another condition
    alternative code
else
    default code
end
```

The `elseif` and `else` blocks are optional. Moreover, you can have several `elseif` statements if necessary. When Matlab enters an `if` structure, it proceeds as follows:

- It checks whether the first condition is true. If so, it executes the code between the `if` line and the next block of the `if` structure (`elseif`, `else`, or `end`), and then it exits the structure.
- If the first condition is false, it checks the condition given by the first `elseif` statement (if any). Again, if this condition is true, it executes the code following the `elseif` line and the next block of the structure, then it exits the structure.
- Subsequent `elseif` conditions are verified in order.
- If none of the `if` or `elseif` conditions are true, then the code following `else` (if any) is executed.

Note that **no more than one block of code in an `if` control structure is executed on encountering an `if` block.**

Matlab also provides looping structures, which are also a form of conditional execution:

---

[1] You may note that end has two quite distinct meanings in Matlab: It can be used in array (vector or matrix) indexing, as in the previous section. It also terminates control structures.

[2] In fact, a computer language with conditional execution is Turing complete, i.e. it can in principle be used to calculate any computable quantity.

`while` *condition* The block of code following `while` and up to the matching end statement is repeated as long as the *condition* remains true.

`for var=values` The block of code following `for` and up to the matching end statement is repeated for each value in `values`, setting `var` to each value in succession. The `values` are most often a range created using the colon (range) operator.

A.6 Functions

Functions are listed alphabetically in each of the sections below.

A.6.1 Mathematical functions

`ceil(val)` The ceiling of a value (`val`), i.e. the result of rounding to the nearest integer towards $+\infty$.

`cumsum(vect)` This calculates the partial sums of the values stored in the vector. Thus, `cumsum(a)` returns the vector $(a_1, a_1 + a_2, \ldots, \sum_\ell a_\ell)$.

`floor(val)` The floor of a value (`val`), i.e. the result of rounding to the nearest integer towards $-\infty$.

`log(val)` The natural logarithm of `val`. Note that there is no `ln()` function in Matlab.

`log10(val)` The base-10 logarithm of `val`. Note that `log()` returns the natural logarithm in Matlab.

`sqrt(val)` The square root of `val`.

A.6.2 Random numbers

`rand(a,b)` Returns an `a`×`b` matrix of random numbers, each of which is a random floating-point number between zero and one.

`randi([n,m])` Returns a random integer between `n` and `m`, inclusive.

`randn(a,b)` Returns an `a`×`b` matrix of normally distributed random numbers with unit variance.

A.6.3 Statistical functions

`mean(vals)` Returns the mean (average) of the values in the vector `vals`.

`coeffs = polyfit(x,y,degree)` fits a polynomial of the selected degree to (x, y) data.

`coeffs` is a vector of coefficients stored in descending order of exponent.

`std(vals)` Returns the standard deviation of the values in the vector `vals`.

A.6.4 Searching

`find()` is an extremely powerful tool for locating quantities that match a particular condition. In the simplest case, `find(`*condition*`)` returns a list of vector indices that match the given condition. For example, if `v` is a vector, `find(v > 4)` returns the indices of all of the elements of `v` greater than four.

An optional second argument of `find()` limits the number of indices returned. This is often used in combination with a third argument that sets the direction of search. This argument has two possible values, `'first'` and `'last.'` Thus, `find(v > 4,2,'last')` locates the last two values (or fewer if there aren't two) in `v` that are greater than four.

A.6.5 Special matrices

`zeros(m,n)` Creates an `m×n` matrix of zeros.

A.6.6 Plotting

- `axis equal` Makes the value of a unit the same along each axis. In other words, the distance between zero and one will be the same on each axis. This is particularly useful when the two axes represent quantities in the same units (e.g. two length axes).
- `colorbar()` Adds a color scale to a plot where this makes sense, e.g. with `contour()`.
- `contour(x,y,z)` Generates a contour plot of the matrix of z values with axes defined by `x` and `y`. The length of `x` must match the number of columns of `z`, and the length of `y` must match the number of rows of `z`.
  - `contour()` has several optional arguments, of which the following may be particularly useful:
    - A number as the fourth argument tells Matlab how many contours to draw.
    - The pair of arguments `'Showtext','On'` adds values of z directly onto the contours.
- `errorbar(x,y,erry)` Plots the (x, y) points with y error bars stored in `erry`.
- `figure` Starts a new figure. This function has many options, but not all combinations are allowed.

  `figure(n)` If figure n doesn't exist, a new figure with this number is created.

  If figure n already exists, the figure is cleared for a new plot.

  `figure('DefaultAxesFontSize',i)` Creates a figure with default font size i.
- `hist(vals,nbins)` Creates a histogram of the values in `vals` using `nbins` bins.
- `histogram(vals)` Creates a histogram of the values in `vals` (Matlab only).
- `hold` tells Matlab to keep adding to a plot rather than erasing and starting over.

  `hold off` tells Matlab that the plot is complete.

- `plot(x,y)` plots the points whose x and y coordinates are contained in the two input vectors. Additional arguments control the appearance of the plot. A few useful options are listed below, but many others are available:
  `plot(x,y,'o')` plots a circle at each point.
  `plot(x,y,'.')` plots a dot at each point.
  `plot(x,y,'s')` plots a square at each point.
  `plot(x,y,'d')` plots a diamond at each point.
  `plot(x,y,'-')` connects the points with lines.
  `plot(x,y,'--')` connects the points with dashed lines.
  `plot(x,y,':')` connects the points with dotted lines.
  `plot(x,y,'r')` plots points and/or lines in red. Other colors are available: (g)reen, (b)lue, (c)yan, (m)agenta, (y)ellow, blac(k), (w)hite.

Note that options can be combined. For example, `'o:k'` plots both circles and connecting dotted lines in black.

- `xlabel(xlabl),ylabel(ylabl)` These functions create, respectively, an x or a y label for a plot.
- `xlim([xmin xmax]),ylim([ymin ymax])` Sets the limits of the x or y axes.

Commands that put text on a plot can take optional arguments to control various attributes of the text. One of the most useful attributes is the `interpreter`. This specifies a language for interpreting the labels. One of the most useful interpreters is the `latex` interpreter, which allows you to use a subset of the LaTeX language to generate high-quality labels (Matlab only). To use this, add `'interpreter', 'latex'` to the arguments of a text labeling command. An alternative is the TeX interpreter, obtained by `'interpreter','tex'`. This is a little less powerful but does allow for subscripts and superscripts, among other features.

A.7 Sending formatted text to the screen

Again, we only touch on some rudiments here.

The `disp(string)` function displays its argument, typically a character string. A simple character string can be made enclosing text in single quotes, e.g.

```
string = 'This is a string.'
```

In Matlab, a character string is a vector of characters, so, for example, `string(3)` would be the character `i` in this example. This means that strings can be composed by creating vectors out of shorter strings, e.g.

```
string2 = [string,' Why not?']
```

would contain the string `This is a string. Why not?`.

The `num2str()` function converts a numeric argument to a character string. This allows you to compose messages that contain numeric information, as in lines 59–60 of the code in section 12.4.2.

Index

Printed in the USA
CPSIA information can be obtained
at www.ICGtesting.com
JSHW060712031123
51216JS00004B/95

9 780750 353199